AF598413

Bath Workshop on

Power Transmission and Motion Control

(PTMC 2000)

Edited by

Professor C R Burrows
Director
Centre for Power Transmission and Motion Control
University of Bath, UK

and

Professor K A Edge
Deputy Director
Centre for Power Transmission and Motion Control
University of Bath, UK

Professional Engineering Publishing Limited
London and Bury St Edmunds, UK.

First Published 2000

ISBN 1 86058 264 8

A CIP catalogue record for this book is available from the British Library.

Printed by The Cromwell Press, Trowbridge, Wiltshire, UK

Front cover picture:
Axial piston pump with ring valve.
Back cover picture:
Pneumatic servo system.

Contents

Transmission Lines

Cylinders

Components and Systems

Servos/Servovalves

Preface

The Centre for Power Transmission and Motion Control at the University of Bath has been organizing annual Workshops since 1988. This book comprises the papers presented at PTMC 2000 which was held at the University on 13–15 September 2000. The 21 papers presented here have a principal focus on hydraulic and pneumatic components and systems, and their control. There is a strong emphasis on the latest international research and development in the field. All papers have been subjected to a formal review and we are especially grateful to the reviewers for their efforts and thoroughness.

We are indebted to Mrs J L Phippen and Mrs J R Harvey for their considerable help and assistance in the organization of the event and compilation of materials comprising this book. Staff at Professional Engineering Publishing Limited are also gratefully acknowledged for their support and efficiency in preparing these proceedings.

Professor C R Burrows, Director
Professor K A Edge, Deputy Director
Centre for Power Transmission and Motion Control
Bath, September 2000

Related Titles of Interest

Title	Editor/Author	ISBN
IMechE Engineers' Data Book – Second Edition	C Matthews	1 86058 248 6
Application of Multi-Variable System Techniques (AMST 98)	R Whalley and M Ebrahimi	1 86058 128 5
Combined Power and Process – An Exergy Approach	F Barclay	1 86058 129 3
Regenerator and the Stirling Engine	A J Organ	1 86058 010 6
Power Transmission and Motion Control (PTMC 98)	C R Burrows and K A Edge	1 86058 134 X
Power Transmission and Motion Control (PTMC 99)	C R Burrows and K A Edge	1 86058 205 2
Pumping Sludge and Slurry	IMechE Seminar	1 86058 155 2
Steam Plant – Optimization and Development by Modelling	IMechE Seminar	1 86058 234 6
Turbochargers and Air Management Systems	IMechE Conference	1 86058 150 1

For the full range of titles published by Professional Engineering Publishing contact:

Sales Department
Professional Engineering Publishing Limited
Northgate Avenue
Bury St Edmunds
Suffolk
IP32 6BW
UK

Tel: +44 (0)1284 724384
Fax: +44 (0)1284 718692

Modelling and Control

Modelling and control of vibration in a smart structure

A R PLUMMER
Instron Limited, UK
W J MANNING and **M C LEVESLEY**
University of Leeds, UK

Abstract
The term smart structure implies a structure which has integrated sensors and/or actuators. For vibration control, sensors are required to measure motion, and actuators are required to influence the motion; a microprocessor or other form of 'intelligence' is required to interpret the sensor signals and provide appropriate actuator demands.

In this paper a simple smart structure with a pair of piezoceramic actuators is studied. Strain gauges are used for feedback. Two vibration control strategies are compared:

- a simple derivative feedback scheme
- a model based control scheme, with model parameters estimated from experimental data using least squares, and pole-placement controller design.

The latter method is shown to give a reduced settling time.

One important feature of piezoceramic actuators is their limited extension. This manifests itself as actuator saturation in the control scheme, and means that for a typical vibration suppression application several cycles are required before enough energy can be removed to bring the vibrating structure to rest. As the actuator saturation has such a significant effect on overall response, its impact on stability is included in the control system analysis.

1 INTRODUCTION

Structures in dynamic mechanical systems are becoming lighter to reduce weight and inertia, improve compactness, and reduce cost. However, this is accompanied by a reduction in stiffness of these structures, and consequently a reduction in the frequencies at which structural resonances occur. To counteract this trend, an ability to actively control vibration is attractive.

The smart structure approach to vibration control consists of embedding actuators and sensors within the structure [1]. A microprocessor provides the intelligence to process sensor signals and drive the actuators such that the structural dynamics are altered in a desired way. To be of practical use for complex structures, the controller should have the ability to tune itself, and not be dependent on the provision of a dynamic model of the structure derived from physical analysis. However, many controller design studies in the past for this sort of system have relied on physical modelling.

The approaches which have been used to find physical models of the system behaviour include dynamic analysis of the modal response [2,3], finite element analysis [4] and modelling the response as flexural waves [5]. In comparison, the estimation of models from experimental data has not been widely reported. Artificial Neural Networks have been used to 'learn' a non-linear model of the system with either state variables [6], or previous system inputs and outputs [7] as the input vector. Though these neural networks can produce good models, the number of parameters that need to be calculated in each learning pass is large, resulting in a slow identification procedure and a high computational burden, and there is no algorithmic design approach to determine their structure.

In the present work the plant is modelled using a linear discrete-time transfer function expressed in the backward shift operator z^{-1}. Estimating such a model from experimental data, using least squares for example, is well understood.

In this paper, the vibration control of a simple cantilevered beam is studied. Piezoceramic strips are bonded to the beam for actuation, and strain gauges are used for feedback. Section 2 describes the experimental rig, and Section 3 presents a mathematical model of the plant. Results and analysis of two types of controller are contained in Section 4 (derivative feedback control) and Section 5 (pole-placement control).

2. EXPERIMENTAL SMART FLEXIBLE STRUCTURE

Figure 1 shows a flexible beam, clamped at one end. When perturbed it exhibits significant vibration in the horizontal plane only. Two piezoceramic actuators are bonded near the clamped end of the beam and electrically connected in parallel. The full bridge strain gauge arrangement is attached to the beam in close proximity to the piezo actuators. An analogue interface card is used to monitor the amplified strain signal and transmit the control signal to a voltage amplifier, which drives the piezoactuators in the range ± 90V. A saturation block is implemented in software to prevent the drive signal exceeding this value and damaging the piezoceramics. The digital control schemes are implemented on a PC and developed under the Matlab/Simulink/Real Time Workshop environment. Specific details of the experimental equipment are listed in Table 1 .

Figure 2 shows the frequency response of the beam at the strain gauge sensor, and at the tip where a laser displacement sensor is used to monitor the displacement. Note that the strain gauge sensor happens to be near a point of zero bending for the second mode shape (45Hz). Figure 3 shows the free response of the beam.

Table 1 Characteristics of the smart structure

Beam	Dimensions: Material:	350x24x1 mm Aluminium
Actuator (piezoceramic)	Dimensions: Material: Piezoelectric Transverse Strain Coefficient: Youngs Modulus: Density: Maximum safe voltage: Amplifier gain:	72x24x0.5 mm PZT, Navy type II -190×10^{-12} m/V 4.9×10^{10} N/m^2 7750 Kg/m^2 90V 20
Sensor (Full Strain Gauge Bridge)	Gauge Factor: Resistance: Location on Beam: Amplification:	2.1 120Ω 100 mm from clamp 1200
Processor (Pentium PC)	Sample rate:	100Hz

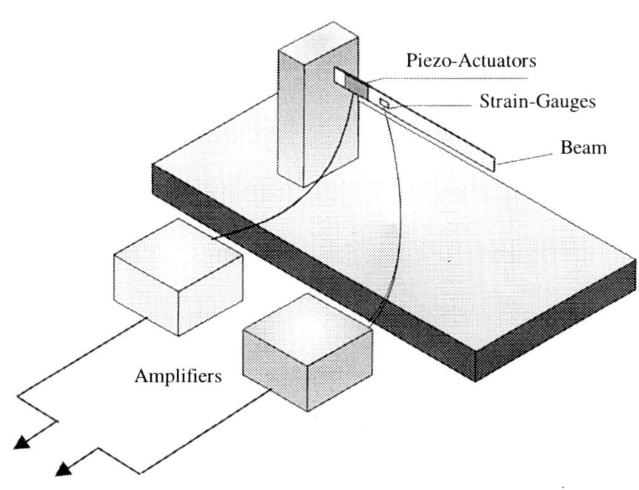

Fig 1. Experimental smart structure

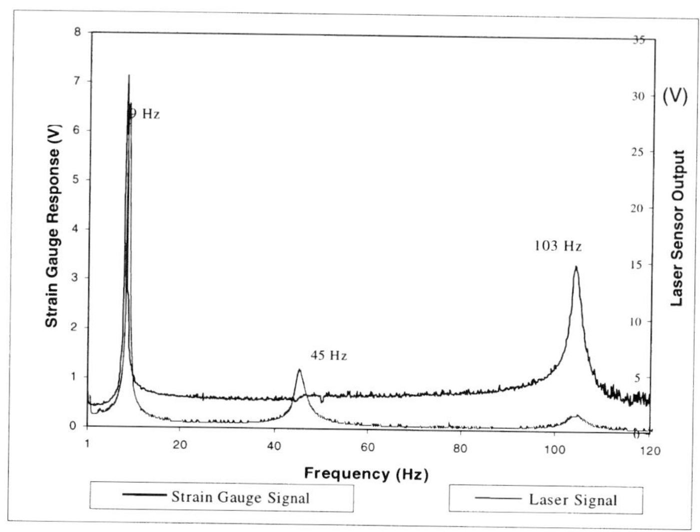

Fig. 2 Frequency response of smart structure

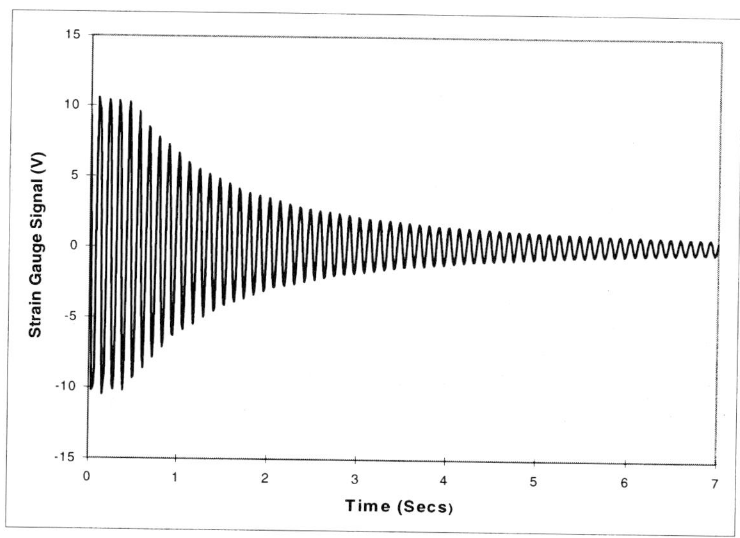

Fig. 3 Free response of smart structure

3 MODELLING

To help understand the behaviour of the smart structure, a model of the system is required. If y is a displacement related measurement on a structure (in this case strain), and f is an applied force, these variables can be related by a sum of second order transfer functions, each one representing one mode of vibration:

$$y = \left(\frac{\beta_1}{s^2 + \alpha_{11}s + \alpha_{10}} + \frac{\beta_2}{s^2 + \alpha_{21}s + \alpha_{20}} + \cdots + \frac{\beta_n}{s^2 + \alpha_{n1}s + \alpha_{n0}} \right) f \tag{1}$$

In this case the first mode of vibration is at a substantially lower frequency than the other modes, and so the model can be approximated by replacing the higher modes with their steady state gains:

$$y = \left(\frac{\beta_1}{s^2 + \alpha_{11}s + \alpha_{10}} + C \right) f \tag{2}$$

where

$$C = \sum_{i=2}^{n} \frac{\beta_i}{\alpha_{i0}} \tag{3}$$

As the strain gauge is not co-located with the actuator, a constant force gives zero strain, so the steady state gain is zero. Thus:

$$\frac{\beta_1}{\alpha_{10}} + C = 0 \tag{4}$$

Combining (2) and (4) gives:

$$y = \frac{Cs(s + \alpha_{11})}{s^2 + \alpha_{11}s + \alpha_{10}} f \tag{5}$$

In the frequency range of interest, the relationship between actuator control signal (u) and output force (f) can be approximated by a gain and a short delay:

$$f = Pe^{-ds}u \tag{6}$$

So (5) can be written in terms of u:

$$y = \frac{C_1 s(s + \alpha_{11})e^{-ds}}{s^2 + \alpha_{11}s + \alpha_{10}} u \tag{7}$$

By inspection of the free and forced responses of the system, it is possible to estimate the parameters in this transfer function:

$\alpha_{10} = 2980$	(54.6 rad/s natural frequency)
$\alpha_{11} = 1.125$	(0.010 damping ratio)
$d = 0.01$	
$C_1 = 0.12$	

Figure 4 shows that the free response of the model is very similar to that of the real system (Figure 3).

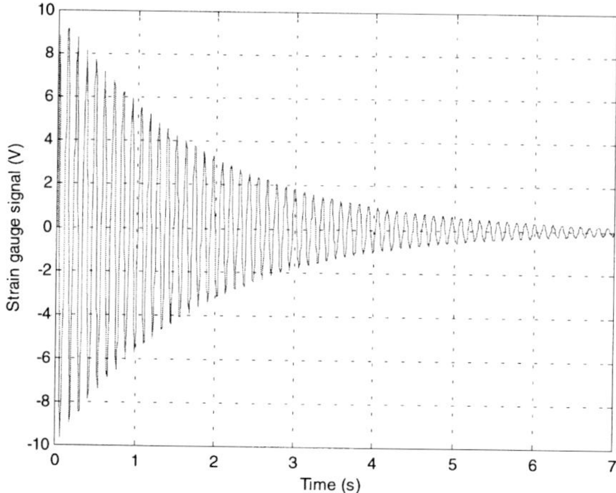

Fig 4. Free response of plant model

4 DERIVATIVE FEEDBACK CONTROL

4.1 Controller design and results

The derivative of the strain gauge signal is an indication of the velocity close to the piezoceramic actuators, and to a first approximation the force output of the actuators is proportional to the drive voltage. Thus assuming the objective of the control system is to increase damping, using negative derivative feedback is a reasonable choice. The control scheme is shown in Figure 5. A 40Hz low pass filter is employed to attenuate noise. The saturation block is implemented in software and is set to ±4.5V, limiting the output of the actuator voltage amplifier to ± 90V.

With a feedback gain of k = 0.02, the settling time is significantly reduced compared to the free response, as seen in Figure 6a. Increasing the feedback gain causes limit cycling. Figure 6b shows the response with k = 0.06. Figure 7 shows that simulating the system with the plant model of Section 3 gives similar responses.

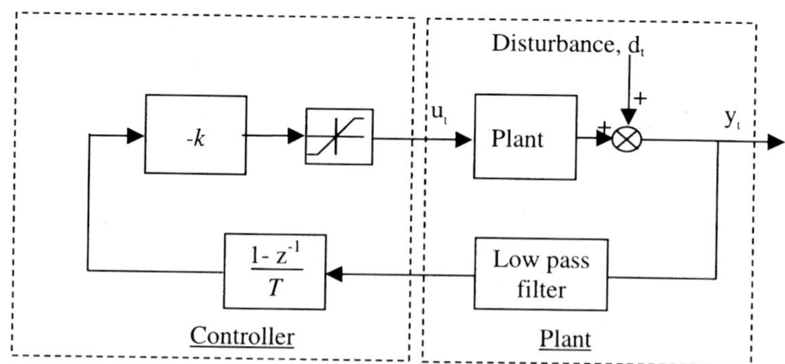

Fig. 5 Derivative feedback controller (sample interval *T*)

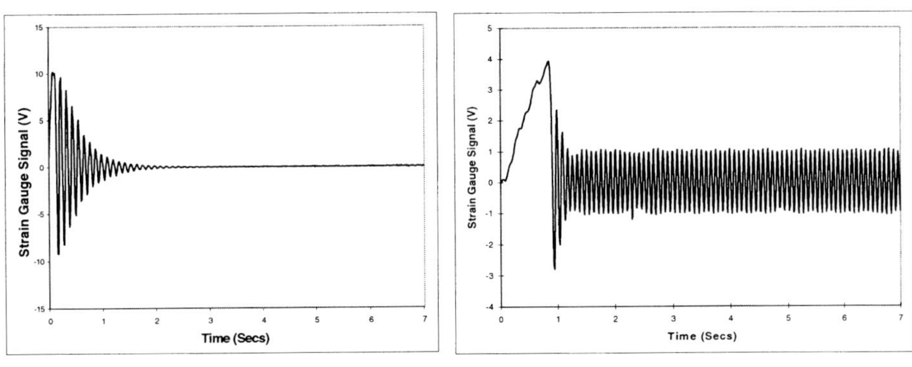

(a) $k = 0.02$ (b) $k = 0.06$

Fig. 6 Response with derivative feedback controller

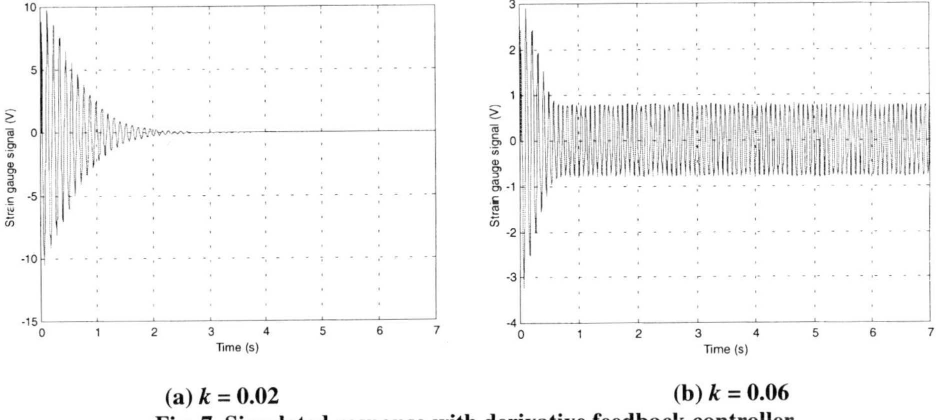

(a) $k = 0.02$ (b) $k = 0.06$

Fig. 7 Simulated response with derivative feedback controller

3.2 Analysis of derivative feedback control

The control scheme of Figure 5 can be analysed by making use of the plant model (equation (7)). Figure 8 is a root locus plot showing the movement of the closed loop pole positions with increasing k. The plot predicts that the closed-loop system will be stable if gain k is less than 0.59. However, this is a misleading result. The actuators are driven to saturation for all but very small amplitudes of vibration – see for example Figure 9 which compares the control signal before and after the saturation element. So the effect of saturation on closed-loop stability must be considered. A sufficient condition for stability can be derived from the off-axis circle criterion for a system with a single static non-linearity [8]. To ensure asymptotic stability, whatever the extent of the saturation, this criterion can be expressed as a forbidden zone on a Nyquist diagram [9,10]. Figure 10 shows that rather than the usual Nyquist stability criterion concerning encirclement of the –1+0j point, the stability test is as follows:

A sufficient condition for global asymptotic stability is that there exists a line of any slope m through point –1+0j which lies wholly to the left (negative real) side of the locus of the open-loop transfer function, M(e$^{-j\omega T}$), on the Nyquist plot.

Figure 11 shows three Nyquist plots for this system for different values of gain *k*. It can be seen that there is no guarantee of stability for values of *k* greater than 0.04.

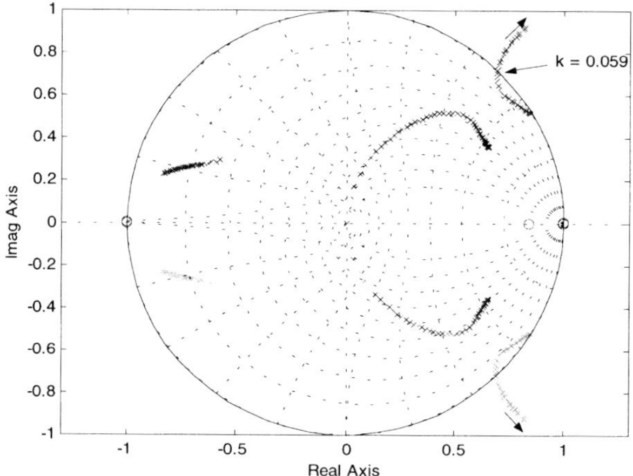

Fig. 8 Root locus for derivative feedback controller

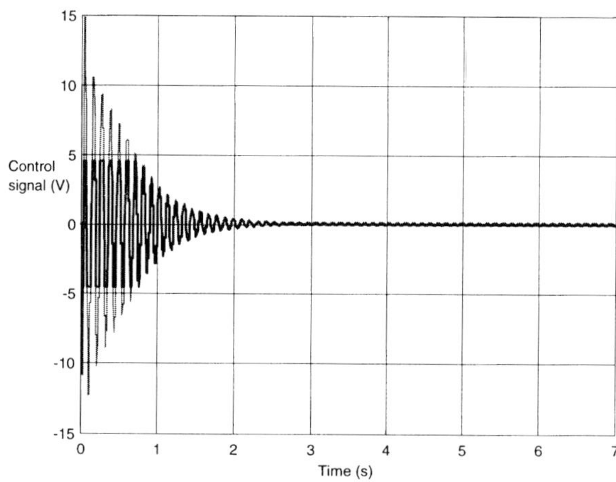

Fig. 9 Saturated and unsaturated control signal (simulation, *k* = 0.02)

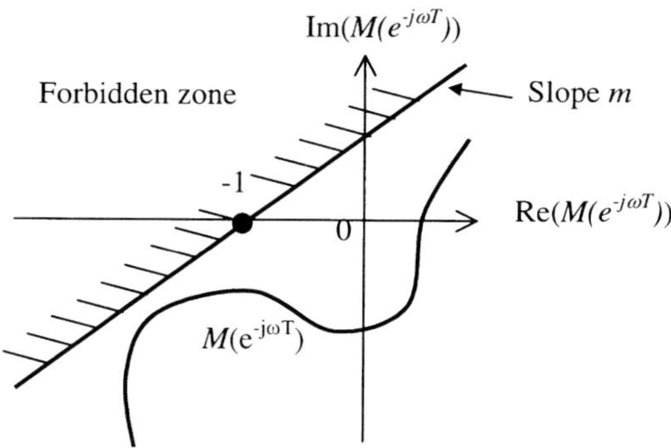

Fig. 10 Nyquist plot: sufficient condition for stability

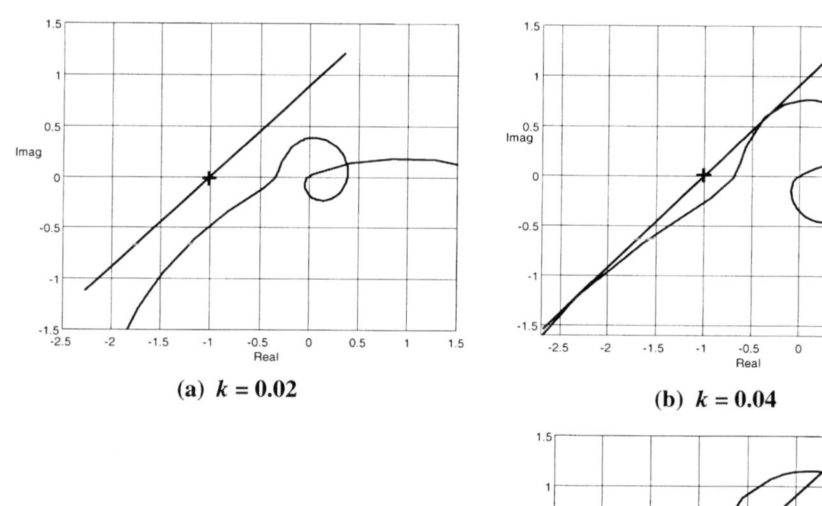

(a) $k = 0.02$

(b) $k = 0.04$

Fig. 11 Nyquist plots for different values of gain k

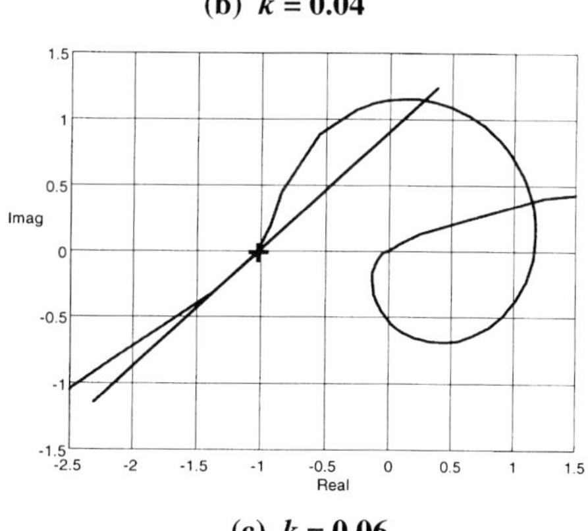

(c) $k = 0.06$

5. POLE-PLACEMENT CONTROL

5.1 Controller design and results

A model-based control scheme may give a better response than the derivative feedback controller. Using the controller structure of Figure 12, neglecting the saturation element and the dynamics of the low pass filter, the closed-loop transfer function is:

$$y_t = \frac{G(z^{-1})B(z^{-1})}{A_m(z^{-1})} d_t \tag{9}$$

where $$A_m(z^{-1}) = F(z^{-1})A(z^{-1}) + G(z^{-1})B(z^{-1}) \tag{10}$$

and where $B(z^{-1})/A(z^{-1})$ is a discrete-time plant model. Equation (10), the Diophantine equation, can be solved for $F(z^{-1})$ and $G(z^{-1})$ given any desired closed-loop characteristic polynomial $A_m(z^{-1})$. To determine a discrete-time plant model, the model of equation (7) could be discretized. However, for general applicability of the approach, it is desirable to have an automatic way of determining model parameters. This has been achieved by collecting experimental input-output data in open-loop, and estimating coefficient values from this data using least squares parameter estimation. Based on the structure of the continuous-time model, a discrete-time model with numerator of degree 3 and denominator of degree 2 is chosen. This assumes that the low pass filter has a sufficiently high bandwidth so that it's dynamics does not materially effect the response. The model with estimated coefficients is shown below:

$$y_t = \frac{-0.0845z^{-1} + 0.1555z^{-2} - 0.0731z^{-3}}{1 - 1.6814z^{-1} + 0.9915z^{-2}} u_t \tag{11}$$

where y_t and u_t are sampled versions of strain y and control signal u.

Two different characteristic polynomials have been tried in this case (Table 2), corresponding to the two pairs of closed-loop pole positions of Figure 13. The closed-loop pole positions have been chosen principally to increase the damping ratio compared to the open-loop response. The experimental responses are shown in Figure 14. Controller I gives a good response, but the additional increase in damping expected from Controller II is not seen due to saturation, and in fact this controller develops a limit cycle.

Table 2. Desired closed-loop pole locations and controller polynomials

Controller	Desired Pole Location on z-plane	Characteristic polynomial A_m	Controller Polynomials $F(z^{-1})$	$G(z^{-1})$
I	0.757 ± 0.480j	$1 - 1.513z^{-1} + 0.803z^{-2}$	$1 + 0.353z^{-1} - 0.445z^{-2}$	$2.189 - 6.030z^{-1}$
II	0.630 ± 0.400j	$1 - 1.261z^{-1} + 0.558z^{-2}$	$1 + 1.067z^{-1} - 1.166z^{-2}$	$7.654 - 15.81z^{-1}$

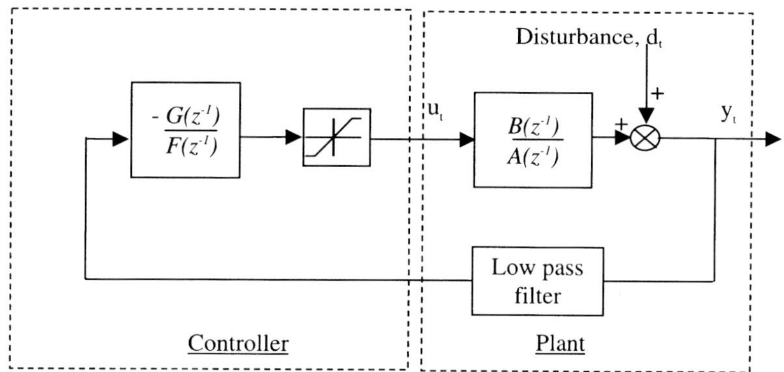

Fig. 12 Pole-placement controller

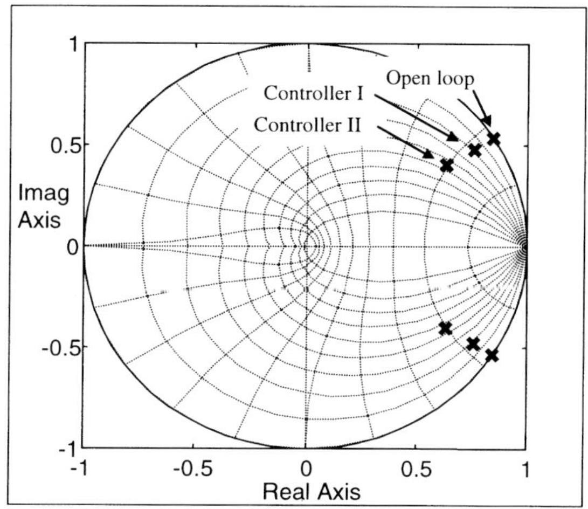

Fig. 13 Pole-placement control: z-plane pole positions

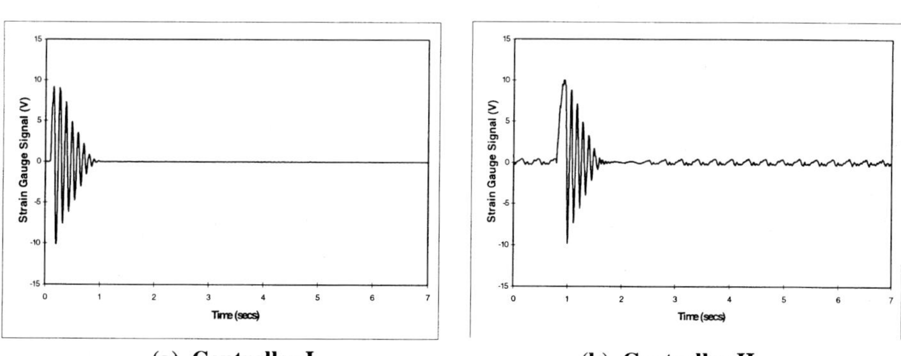

(a) Controller I **(b) Controller II**

Fig. 14 Response with pole placement controller

5.2 Analysis of pole-placement control

Figure 15 shows the simulated responses for controllers I and II. The simulations are similar to the experimental responses. For controller II, the onset of a limit cycle is predicted in simulation, although the difference in frequency of the limit cycle indicates some error in the simulation model.

Figure 16 shows Nyquist plots for the two controllers. The plot for Controller II predicts the instability. The fact that the specified pole positions are stable, but the Nyquist plot indicates instability, is because the plot is calculated from the plant model of equation (7), and includes the low pass filter. Thus the instability can be attributed to the modelling error in the discrete-time model used for controller design, and the choice of pole positions has meant that Controller II has a lower stability robustness than Controller I.

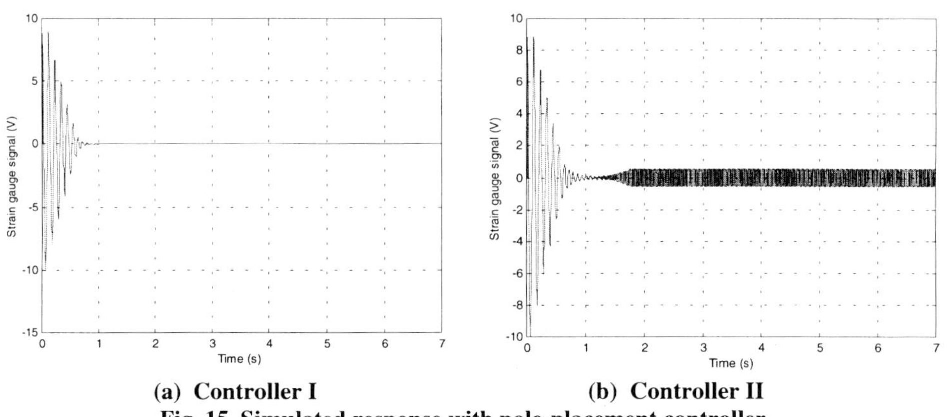

(a) Controller I **(b) Controller II**

Fig. 15 Simulated response with pole-placement controller

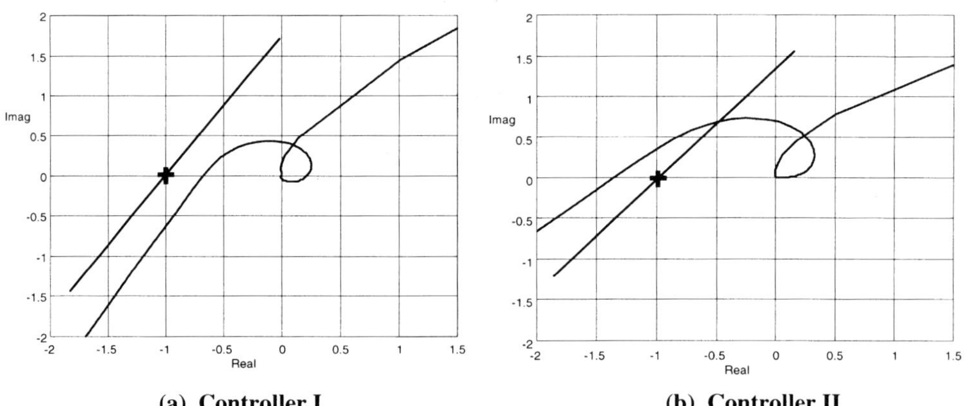

(a) Controller I **(b) Controller II**

Fig. 16 Nyquist plots for pole-placement controllers

6 CONCLUSIONS

Control studies of a simple smart structure have been presented. The objective has been to reduce the settling time seen in the free vibration response by exerting forces on the structure from a pair of surface mounted piezoceramic actuators. From the two types of controller investigated, it can be concluded that:

- a derivative feedback controller can reduce the settling time by about a factor of five.
- a pole-placement controller can reduce the settling time by about a factor of ten.

The pole-placement controller is a model based controller, but it is shown that the co-efficients in a linear discrete-time model can be estimated from experimental data using least squares parameter estimation. This is particularly important as it enables vibration control without detailed *a priori* knowledge of the vibration characteristics of the structure.

Linear analysis of the derivative feedback controller allows limits to be placed on the feedback gain based on the requirement for closed loop stability. However, as the actuators are routinely driven into saturation, it is shown that this linear analysis is deficient. A more conservative bound can be placed on the feedback gain by taking into account the saturation non-linearity.

A key limit on the performance of the pole-placement controller is the accuracy of the model used for controller design. Modelling error, particularly unmodelled dynamics (in this case in the form of a low pass filter) can cause instability despite specifying stable closed-loop poles. Ensuring a choice of pole positions which gives good stability robustness is one way to minimise this risk.

ACKNOWLEDGEMENTS

The authors wish to thank Mr R Scott for providing the continuous-time model parameters, and Dr M D Brown for advice during the conduct of this work.

REFERENCES

1. B Culshaw, Smart Structures and Materials. Artech House, Boston MA, 1996.
2. Chen C-Q, Shen Y-P, Optimal control of active structures with piezoelectric modal sensors and actuators, *Smart Materials and Structures,* 1997, 6(4), 403-409.
3. R Butler, V Rao, A state space modelling and control method for multivariable smart structural systems', *Smart Materials and Structures*, 1996, 5(4) 386-399.
4. S K Ha, C Kielers, F-K Chang, Finite element analysis of composite structures containing distributed piezoceramic sensors and actuators, *AIAA Journal,* 1992, 30(3), 772-780.
5. CR Fuller, SJ Elliot, PR Nelson, Active Control of Vibration. Academic Press, London, 1996.
6. V Rao, R Damle, C Tebbe, F Kern, The adaptive control of smart structures using neural networks, *Smart Materials and Structures,* 1994, 3(3), 354-366
7. C-J Li and AG Ulsoy, High precision measurement of tool-tip displacement using strain gauges in precision flexible line boring, *Mechanical Systems and Signal Processing*, 1999, 13(4), 531-546

8. Cho, Y, Narendra, K An off-axis circle criterion for the stability of feedback systems with a monotonic non-linearity *IEEE Trans on Automatic Control*, 1968, 13(4), 413-416.
9. Plummer, A R. Servosystem design with control signal saturation. *Bath Workshop on Power Transmission and Motion Control (PTMC 99),* University of Bath, Sept 99, 31-44.
10. Plummer, A R, Ling, C S. Stability and robustness for discrete-time systems with control signal saturation. *Proc Instn Mech Engrs, Part I*, 2000, 214, 65-76.

A tool for tracking cause–effect propagation in hydraulic systems

J LARSSON, P KRUS, and **J-O PALMBERG**
Fluid and Mechanical Engineering Systems, University of Linköping, Sweden

Abstract
Due to increased complexity in technical systems, there is a need for distilling the important results out of the huge amount of results produced from computational simulation of large models. That way, the engineers can get a better overall view of the system, not needing to look at irrelevant details. The method used in the paper visualises a simulation run in a way that resembles the executing of statecharts. Statecharts are used by system engineers as a means for modelling, simulating and evaluating primarily software against given specifications. In short, the continuous results produced from numerical tools for simulation of systems of ordinary differential-algebraic equations are made discrete.

1 INTRODUCTION

Technical systems are becoming increasingly integrated, e.g. with the intensive use of software due to demands on energy efficiency, performance and customisability. This means that the complexity of their dynamical behaviour rises. Questions like "What caused that pressure peak at the pump outlet 10 s from the start of simulation?" and "How can the pump regulator and the pressure compensators be better synchronised?" are more and more difficult to answer due to the large number of degrees of freedom in the overall system. Notice that these questions have nothing to do with stability and therefore are not solved using frequency analysis but rather demand qualitative answers originating from the knowledge of how all components interact. Automated FMEA can answer some of the questions above in the case of static behaviour (see (1)). Is there some way of using an existing simulation model, adding some functionality and then getting information that helps in answering the questions above

for dynamic systems? System engineers use statecharts as a helping tool in grasping complex systems. They might look like the example in Figure 1.

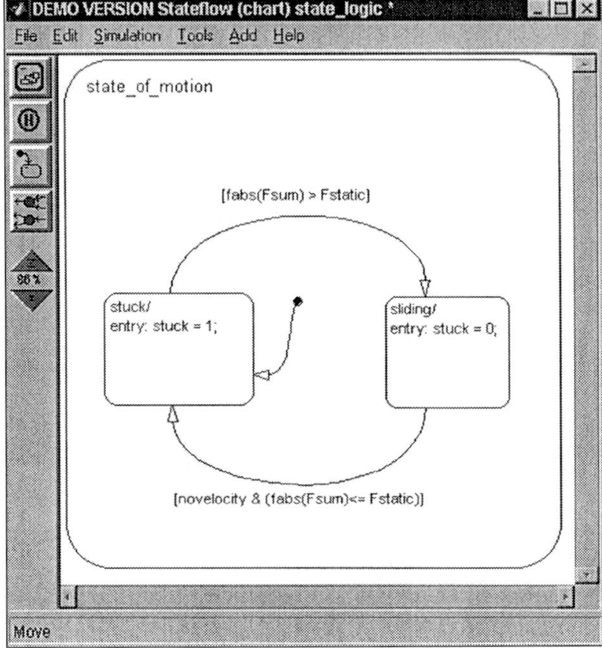

Figure 1: An arbitrary statechart

In the statecharts there are a number of *objects* and each has a *state* that it currently is in. So called *events* can be sent between the objects causing changes to their states. The objects could in a hydraulic system be orifices, volumes and so on. The state of an object is everything the object needs to know to give a new state due to incoming events. For more details see (2).

The algorithms presented in the paper help in visualising the results from simulations of hydraulic systems in a way that resembles of executing statecharts. The paper does not however provide means for automatically generating statecharts from the hydraulic models. The algorithms are to be used in simulation tools that make use of numerical integration routines for ordinary differential equations

2 THE GENERAL IDEA

The systems in the paper consist of volumes and orifices as shown in Figure 2. The pressures in the volumes are calculated as in Eq. (1). It is assumed that the density of the fluid is constant as well as the size of the volumes. A changing pressure is therefore caused by a flow sum differing from zero.

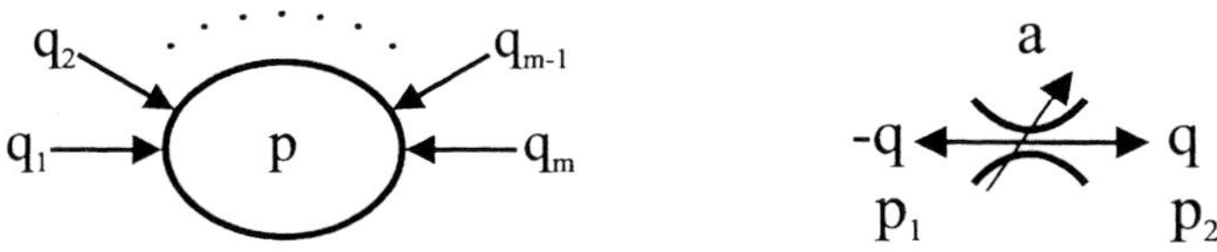

Figure 2: The input- and state variables of a volume and an orifice. Inputs to the volume are the flows q_i that generates the state variable p standing for pressure. The state variable q for the orifice depends on the opening area a and the surrounding pressure p_1 and p_2

$$\frac{dp}{dt} = const \cdot \sum_{i=1}^{m} q_i \qquad (1)$$

For the orifices it is assumed that the flow is turbulent and Eq. (2) can then be used.

$$q = const \cdot a \cdot sign(p_1 - p_2) \cdot \sqrt{|p_1 - p_2|} \qquad (2)$$

Let us look at a simulation example where a number of volumes and orifices are connected in line. The opening area of the orifice in the middle is increased at 1.0 s. What then happens is that the pressure downstream rises and the pressure upstream is reduced. The pressures first change in the volumes surrounding the controlled orifice and then the changes propagate to the outer volumes as shown in the lower part of Figure 3. This is at least the case if each volume contains a time delay of one time step which is the case in the transmission line method (see (5)) that is used here. With a centralised numerical solver, the lower part of Figure 3 would have been produced directly.

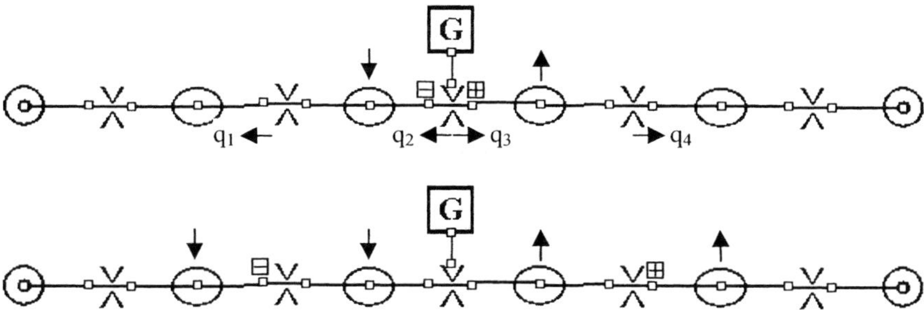

Figure 3: Pressure changes shown as arrows after an increase in the opening area of the middle orifice.

2.1 Introducing the concept of events in simulation of dynamic systems

The plus and minus signs in Figure 3 can be looked upon as events. They indicate that something new has happened in a neighbour component and that something well defined will happen to the state in the current component. For a specific volume i, the events $e_{volume,\ i}$ generated by surrounding orifices are detected during simulation using an algorithm E_{volume}

acting on the flows $\boldsymbol{q}_i$ flowing in and out of the volume i. Bold letters indicate vectors. Events can also be sensed from the surrounding of the orifices in the same manner. So, all events can be written as

$$\boldsymbol{e} = \begin{pmatrix} \boldsymbol{e}_{volume} \\ \boldsymbol{e}_{orifice} \end{pmatrix} = \begin{pmatrix} E_{volume}(\boldsymbol{q}) \\ E_{orifice}(\boldsymbol{p}) \end{pmatrix} \qquad (3)$$

The operation in Eq. (3) could be performed after the simulation is done. That however has some drawbacks such as the huge amounts of data that need to be stored. Also, the analysis tool needs to keep track of which component the results came from, how all components are connected and what the results really mean. A more intuitive way of producing simplified results is to let the components themselves sense what is going on and log important events on a common log file that can be interpreted by any tool that has knowledge about the system structure. This means that the algorithm E_{volume} is used in every part of the system model that deals with calculations as Eq. (1). The inputs to each instance of the algorithm are then the neighbouring flows that in the example always are two.

$$\boldsymbol{e} = \begin{pmatrix} \boldsymbol{e}_{volume} \\ \boldsymbol{e}_{orifice} \end{pmatrix} = \begin{pmatrix} \boldsymbol{e}_{volume,1} \\ \vdots \\ \boldsymbol{e}_{volume,n} \\ \boldsymbol{e}_{orifice,1} \\ \vdots \\ \boldsymbol{e}_{orifice,m} \end{pmatrix} = \begin{pmatrix} E_{volume}\begin{pmatrix} \boldsymbol{q}_{volume,1} \\ \vdots \\ \boldsymbol{q}_{volume,n} \end{pmatrix} \\ E_{orifice}\begin{pmatrix} \boldsymbol{p}_{orifice,1} \\ \vdots \\ \boldsymbol{p}_{orifice,m} \end{pmatrix} \end{pmatrix} \qquad (4)$$

In Eq. (4), n and m are the number of volumes and orifices respectively. $\boldsymbol{q}_{volume,\,i}$ is the set of flows flowing in and out of the volume component i.

The four events are stored in a log file during simulation as shown below. An analysis tool reads the log file after simulation and the events and states are displayed as in Figure 3. How could the log file be created from a simulation in a numerical simulation tool and how should an event be specified?

```
q2- => pressure decrease -714599.03 from 1.0000 converging at 1.0001
q3+ => pressure increase  723980.49 from 1.0000 converging at 1.0001
q1- => pressure decrease -714599.03 from 1.0001 converging at 1.0020
q4+ => pressure increase  723980.49 from 1.0001 converging at 1.0020
```

2.2 Definition of events and how to detect them from the viewpoint of volumes

In the example above, the pressures in the volumes are of interest. In the figure, it is indicated by minus- and plus signs from where the flows came that changed the pressure in a certain way. We would like to know what node the flow came from that was the last one to change the pressure. So, let us use the volume components as event sensors for the log file.

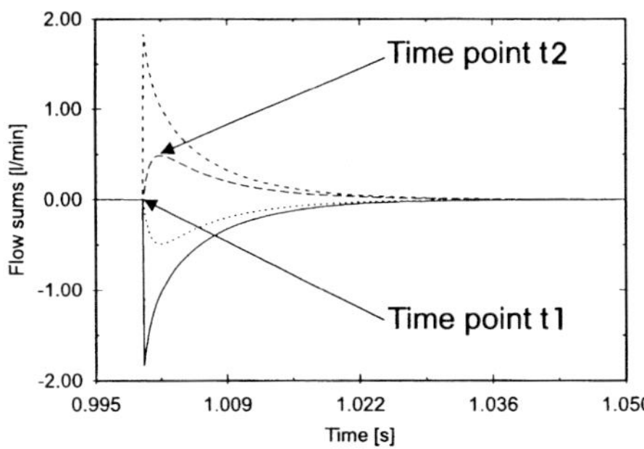

Figure 4: The flow sums in all volumes in Figure 1

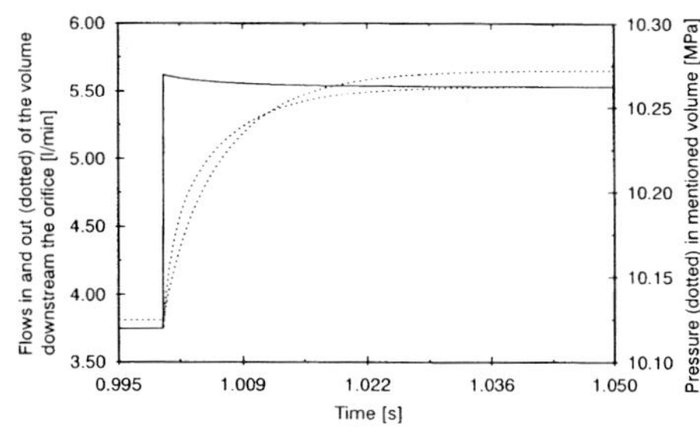

Figure 5: The in- and outgoing flow for the volume closest to the right of the controlled orifice in Figure 1

The flow sum is shown for all volumes in Figure 4. As the orifice opens, a step in the flows to the two surrounding volumes occurs as shown in Figure 5. A softer flow change will occur to the outermost volumes since the preceding pressures don't change instantly. What is common for all volumes is that the flow sums diverge from zero during some time and then begin to converge to equilibrium again where the flow sum is zero. During diversion, the flow causing the diversion is called *active* in the paper. In this case the flows from the controlled orifice are active for an infinitely short time period whereas the flows from the inner volumes to the outer ones are active for 0.002 s. The latter time is t_2-t_1 in Figure 4. Exactly when a diversion occurs, the volume component should check what node the causing flow came from and log the event to a file. The flow with the largest derivative must be the most active flow in the case of a total of two flows, since the volume acts as a low-pass filter (see Figure 5). The statements above can be summarised in the following rule:

> A certain flow is *active* if it has the largest absolute value of its derivative compared to all flows and if the absolute value of the flow sum is increasing. It could be said that the active flow is pushing the pressure in the volume from its static value

In the general case where the number of nodes is larger than two, several flows can be equally active. This hasn't been considered in this paper, but can easily be implemented by stating to the log file that some flows with derivatives close enough to one another were all active. Now, however it is not enough just to compare the absolute values of the flow derivatives in choosing the active one. Lets look at the system model in Figure 6 with the flow derivatives and the pressures of the right volume shown in Figure 7.

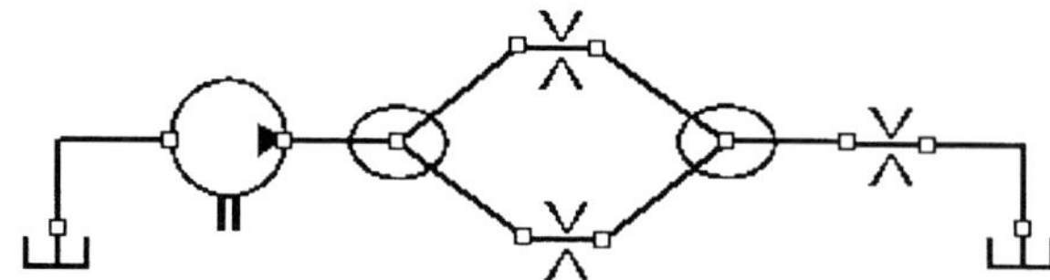

Figure 6: System with two orifices in parallel causing problems with using just the largest derivative of the flow as the choice for active flow

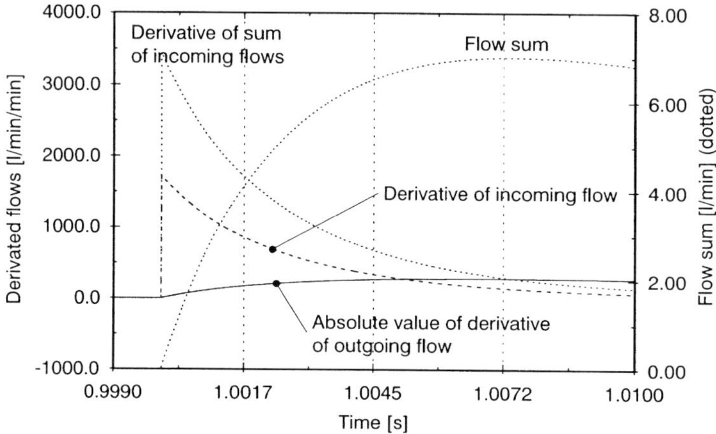

Figure 7: The flows and pressure of the right volume in Figure 6

The incoming flows to the right volume increase and causes an increase in the outgoing flow due to the increased pressure in the volume. The incoming flows should both be regarded as active until 1.0072 s, where the flow sums begins to decrease to equilibrium. After that no flow is active. Since the paper does not consider the case of several active flows, the algorithm should at least pick one of the incoming ones. As can be seen the absolute value of the derivative of the outgoing flow becomes larger than the corresponding values of every other flow *before* the flow sum begins to decrease. This is due to the fact that the derivative of the outgoing flow is generated from the sum of the derivatives of the incoming flows. However, the increase of the outgoing flow couldn't have caused the increase of the flow sum and should therefore be left out in the comparison. A new rule can therefore be added to the former one.

> Only flows with the same sign of their derivatives as that of the flow sum are to be compared.

Since a flow can be active for many time steps in sequence in a simulation, it is only noted the first time. If however after that the flow sum has decreased in magnitude, another flow has been active or the change of the mentioned flow has another sign, it can be noted again if the mentioned flow becomes active. The rules described above are summed up in Table 1 showing the algorithm used in the simulations in the paper.

```
if n = 1 do to_message = false                       % n is the time index
q_sum_n = Σq_i,n                                      % i=1..m where m is number of flows
if | q_sum_n | < lowest_q_sum do q_sum_n = 0.0        % entering the volume

if | q_sum_n | > | q_sum_n-1 | do {                   % Flow sum is diverging from zero
        to_message = true

        % Find the node of the flow change that is larger than the flow changes of the
        % other nodes and that has the same derivative as the flow sum. This node causes the
        % change of the other flows

        max_q_diff = 0.0
        for i = 1..m do {
            q_diff = q_i,n - q_i,n-1                  % Derivative of flow in node i
            if | q_diff | > max_q_diff and sign(q_diff)=sign(q_sum_n - q_sum_n-1) do
                max_q_diff = q_diff                   % Largest flow derivative so far is stored
                causing_node_n = i                    % Node number of flow mentioned above
        }

        % Find out if the current node was responsible and also had the same sign on the
        % flow change the last simulation run n-1. Then don't continue, otherwise create a
        % new event

        direction_n = sign(q_causing_node,n - q_causing_node,n-1)
        if not ( direction_n = direction_n-1 and causing_node_n = causing_node_n-1 ) do {
            if n > 1 do {
                Write message to file telling that the flow in node causing_node_n caused a
                pressure change of p_n - last_pressure starting at last_time
            }
            last_pressure = p_n
            last_time = time
        }
}

if | q_sum_n | ≤ | q_sum_n-1 | do {                   % Flow sum is converging to zero
        if ( | q_sum_n | > 0.0 and to_message = true ) do {
            to_message = false
            Write message to file telling that the flow from the node that was responsible the
            last time for increasing the flow sum now is reducing its difference from the other
            flows
        }
}
```

Table 1: Algorithm used for finding the flows causing changes to the pressure in a volume

The algorithm works equally well for all components where the only state variable is equal to an integral of a sum. In the case of a hydraulic motor with inertia the corresponding equation would be as Eq. (5) below where the variables can be found in Figure 8.

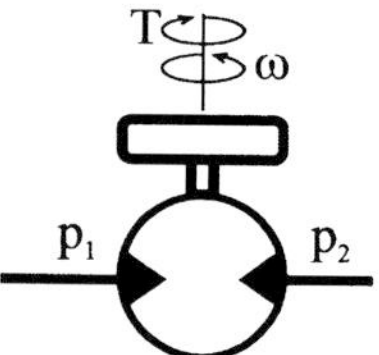

Figure 8: A hydraulic motor where T is applied torque [Nm], ω is rotational speed [rad/s] and p_1 and p_2 are inlet and outlet pressures respectively

$$\frac{d\omega}{dt} = const \cdot (p_1 - p_2 - T\frac{2\pi}{D_m}) \text{ where } D_m \text{ is the displacement } [m^3/\text{revolution}] \tag{5}$$

In this case the derivative of the torque term would be compared with the derivatives of the surrounding pressures in finding the active variable causing the latest change to ω. In the case of static components such as orifices, linearisation is used to sort out the contributing terms to a specific variable such as the flow in the case of an orifice. This is shown in the next chapter. The orifices will then be able to tell whether it was the pressure difference or the opening area that caused a specific change to the flow.

3 A MORE COMPLEX EXAMPLE

As shown to the left in Figure 9, the example is a pressure compensator. Its function is to keep the pressure difference p_{red}-p_L constant. That in turn gives a flow that is constant with respect to system pressure p_s and load pressure p_L. The load case is such that a step is given to p_s from 20 to 21 MPa.

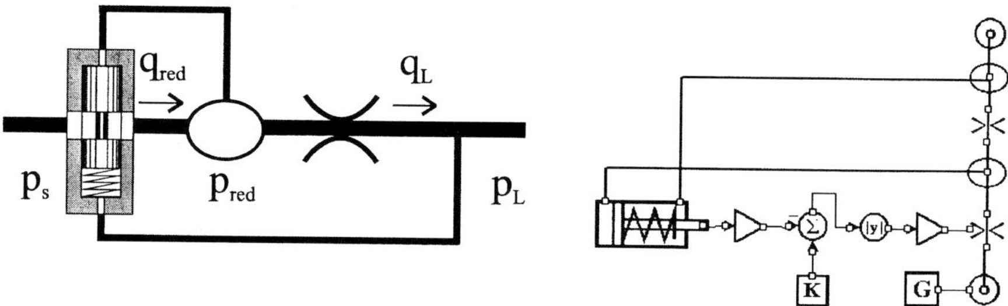

Figure 9: The real system

Figure 10: The corresponding simulation model made in Hopsan

The simulation model is shown in Figure 10. It has been done in the simulation tool Hopsan (see (3) and (4)). The lower orifice and the volume above it are used as sensing components since they are of most interest. The piston doesn't use any flow from the volumes since that would make the system too complex to demonstrate. The inertia of the spool is not modelled, neither the spring characteristics nor flow forces. Before looking at the results, let's look at how the algorithm used in the orifice component works.

The flow q_{red} is derived from Eq. (2) to find out if it is the opening area or the pressure difference that causes the change to the flow.

$$\frac{dq_{red}}{dt \cdot const} = \frac{da}{dt} sign(p_s - p_{red})\sqrt{|p_s - p_{red}|} + sign(p_s - p_{red}) \cdot a \cdot \frac{\frac{d}{dt}(p_s - p_{red})}{2\sqrt{|p_s - p_{red}|}} \qquad (6)$$

The change in flow from the current working point 0 would then be after linearisation

$$\frac{\Delta q_{red}}{const} = \underbrace{\Delta a \cdot sign(p_{s0} - p_{red0})\sqrt{|p_{s0} - p_{red0}|}}_{\Delta q_a} + \underbrace{a_0 \cdot sign(p_{s0} - p_{red0})\frac{\Delta p_s - \Delta p_{red}}{2\sqrt{p_{s0} - p_{red0}}}}_{\Delta q_p} \qquad (7)$$

The change Δa is calculated as below. The same procedure is used for Δp_s and Δp_{red}.

$$\Delta a_n = (1-\alpha)(a_n - a_{n-1}) + \alpha \Delta a_{n-1} \text{ where } 0 < \alpha < 1 \qquad (8)$$

The purpose of the constant α is to make the flow changes smoother. If not used, the changes get exaggerated and an unnecessary large number of events are created when Δq_a and Δq_p alternate in being the largest. A value of 0.1 has been used for α in all simulations. Here follows the whole algorithm used in the orifice simulation component. Some extra mechanisms are needed for the algorithm to function for the last time step of the simulation. As can be seen, if the derivatives of the flow terms are below the constant value lower_limit_on_change they are set to zero. This is because of the numerical inaccuracy that always exists. The noise level is however very low in the simulations performed by the author so this constant is set once and for all. Below is the complete algorithm that is used.

$$\Delta a_n = (1-\alpha)(a_n - a_{n-1}) + \alpha \Delta a_{n-1}$$
$$\Delta p_{s,n} = (1-\alpha)(p_{s,n} - p_{s,n-1}) + \alpha \Delta p_{s,n-1}$$
$$\Delta p_{red,n} = (1-\alpha)(p_{red,n} - p_{red,n-1}) + \alpha \Delta p_{red,n-1}$$

$$\Delta q_{a,n} = \Delta a_n \sqrt{p_{s,n-1} - p_{red,n-1}} \cdot sign(p_{s,n-1} - p_{red,n-1})$$
$$\Delta q_{p,n} = a_{n-1} \frac{\Delta p_{s,n} - \Delta p_{red,n}}{2\sqrt{p_{s,n-1} - p_{red,n-1}}} \cdot sign(p_{s,n-1} - p_{red,n-1})$$

if $|\Delta q_{a,n}| <$ lower_limit_on_change **do** $\Delta q_{a,n} = 0.0$
if $|\Delta q_{p,n}| <$ lower_limit_on_change **do** $\Delta q_{p,n} = 0.0$

```
if | Δq_a,n | > | Δq_p,n | and ( last_change = "Δq_p" or n = 1 ) do {
        if n > 1 do {
            Write message to file telling that the pressure last_pressure_change has
            caused a flow change of q_n-1-last_flow starting from time point last_time
        }
        last_time = time
        last_flow = q_n-1
        last_change = "Δq_a"
}

if | Δq_p,n | > | Δq_a,n | and ( last_change = "Δq_a" or n = 1 ) do {
        if n > 1 do {
            Write message to file telling that the opening area has
            caused a flow change of q_n-1-last_flow starting from time point last_time
        }
        last_time = time
        last_flow = q_n-1
        last_change = "Δq_p"

        if | Δp_s,n | > | Δp_red,n | do
            last_pressure_change = "p_s"
        else
            last_pressure_change = "p_red"
}
```

Table 2: The algorithm used for finding the cause of a change to the flow in an orifice

A simulation is performed and the following messages are given by the orifice component

```
Ps =>              Flow increase  7.9044E-5 from 1.0001E-2 to 1.0080E-2
Opening area =>    Flow decrease -7.7912E-5 from 1.0080E-2 to 2.0000E-2
```

The two timepoints of 1.0001E-2 and 1.0080E-2 correspond to t_1 and t_3 in Figure 11 and Figure 12. What can be seen is that the spool gets ahead of the pressure difference increase at t_2 and changes the flow back to its original value.

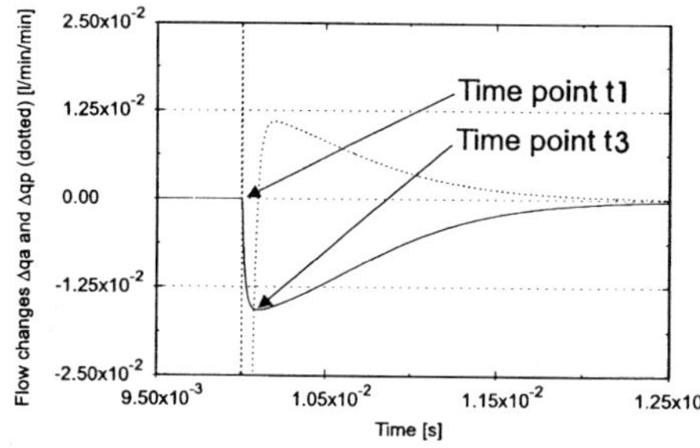

Figure 11: Flow change contributions

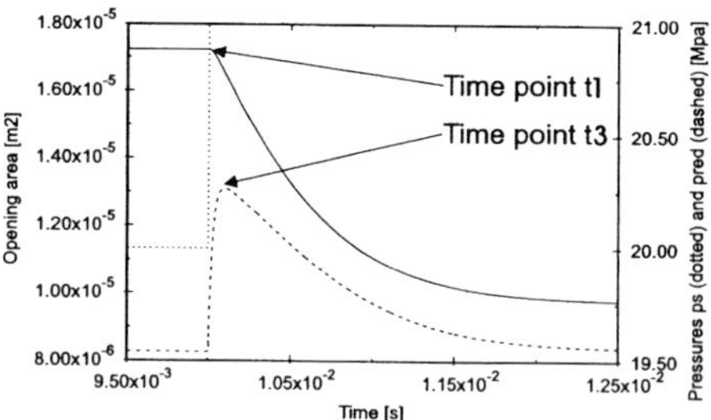

Figure 12: Opening area and pressures

The following messages are given by the volume component

```
qred => pressure incr.  723980.49 from 1.0002E-2 converging at 1.0003E-2
qred => pressure decr. -714599.03 from 1.0101E-2 converging at 1.0294E-2
```

The two points of time 1.0002E-2 and 1.0101E-2 correspond to t_2 and t_4 in Figure 13 and Figure 14. The flow q_{red} immediately reduces its gained difference from q_L as can be seen in Figure 14 and on the time of convergence in the upper of the log file rows above. The flow q_{red} however is not only reduced to q_L, but steadily continues below the mentioned flow. This should not have happened if the flow wasn't controlled in one way or another. A new event is therefore created and q_{red} is held responsible since it had the largest derivative at that time.

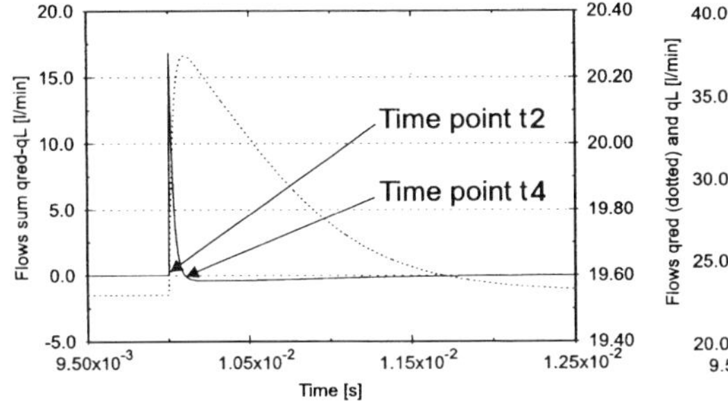

Figure 13: Reduced pressure pulse due to the step in supply pressure

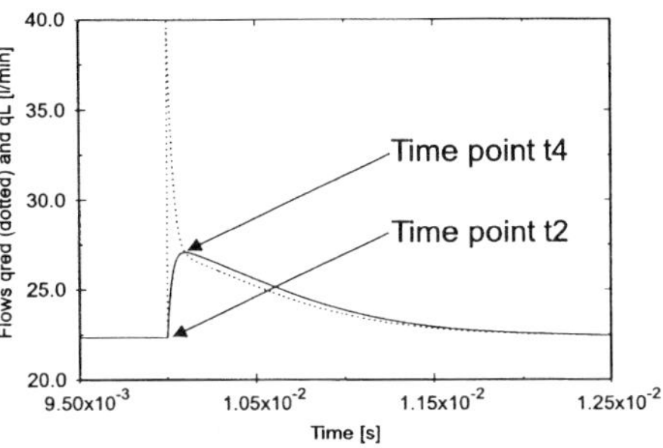

Figure 14: The flows in and out of the reduced volume

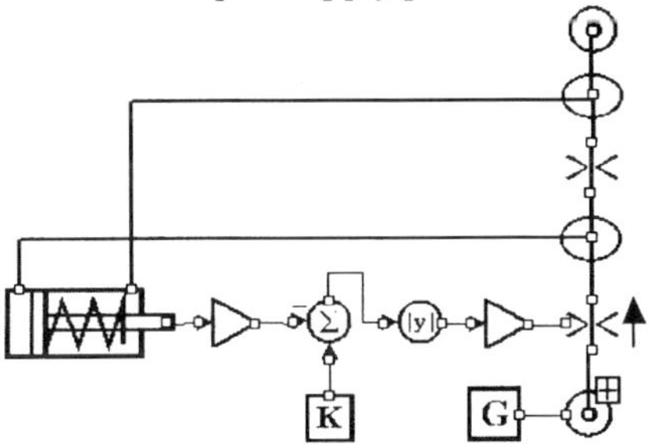

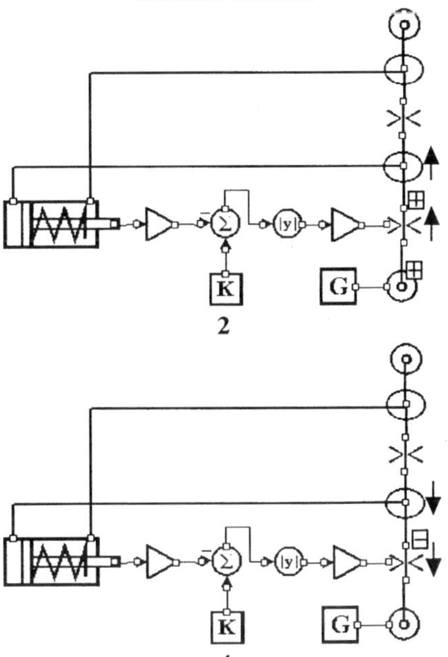

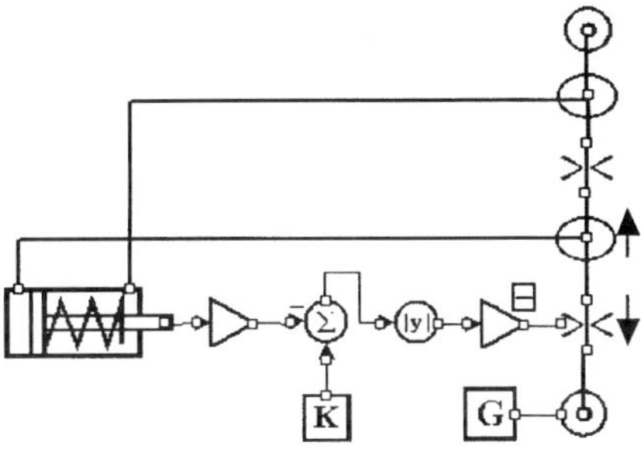

```
1. Ps    => Flow increase     7.9044E-5 from 1.0001E-2 to        1.0080E-2
2. qred  => Pressure incr.   723980.49 from 1.0002E-2 conv. at 1.0003E-2
3. area  => Flow decrease    -7.7912E-5 from 1.0080E-2 to        2.0000E-2
4. qred  => Pressure decr.  -714599.03 from 1.0101E-2 conv. at 1.0294E-2
```

Figure 15: The output of the analysis tool after having read the log file from the pressure compensator model

The whole scenario is shown graphically above together with the logs sorted in time. The pictures are similar to the presentation given when using Hopsan as an analysis tool on the log file.

The actual number of events created is often much larger than the ones presented. This is due to numerical errors but could also be of natural causes and can be filtered out. The filter in this paper, functions so that it takes away all events where the change to a variable is below a certain value that needs to be set. In the volume e.g. all events are used where the pressure is changed more than 1000 Pa. A more robust way would be for each volume component to sort out those events where the pressure was changed a certain magnitude below the largest changes in that volume throughout the simulation. The same principle could be used for the orifices.

4 CONCLUSIONS

A simple approach for converting continuous simulation results into a finite number of events has been shown. The main difficulty has been to find a suitable definition for events in dynamical systems and in eliminating events generated due to numerical noise in flow and pressure variables. The algorithms provide information useful for visualisation of the dynamic behaviour that in an intuitive way shows which components currently are controlling the situation, how excitations propagate through the system. The events can be sorted after importance by evaluating the size of the change they caused. In short, algorithms have been presented that have potential of providing very useful results for further analysis in the case of large simulation models for both better understanding as well as fault diagnosis.

5 REFERENCES

(1) D.R. BULL, J.S. STECKI, K.A. EDGE AND C.R. BURROWS, "Failure Mode and Effect Analysis of a Valve-Controlled Hydrostatic Drive", Fluid Power Engineering: Challenges and Solutions, 1998, pp. 131-144

(2) RUMBAUGH, J., BLAHA, M., PREMERLANI, W., EDDY, F. and LORENSEN, W. (1991) *Object-Oriented Modeling and Design*, (Prentice-Hall International, Inc.)

(3) "HOPSAN a Simulation Package User's Guide", http://hydra.ikp.liu.se/hopsan.html, 1998.

(4) LARSSON J., "GDynmoc User's guide", IKP-R-1088, Dept. of Mech. Eng., Div. of Fluid and Mech. Eng. Systems, Linköping University, Linköping, Sweden, 1999.

(5) JOHNS P. B. AND O'BRIEN M., "Use of transmission line modelling (t.l.m) method to solve nonlinear lumped networks," The Radio Electron and Engineer, vol. 50, pp. 59-70, 1980.

Experimental validation of mathematical models of hydraulic pipelines with high-frequency excitation

B MANHARTSGRUBER
Institute of Mechanics and Machine Design, University of Linz, Austria

Abstract. Laminar pipe flow of compressible fluids has been studied by numerous authors in the past. While the one-dimensional wave equation was sufficient for the early work on wave-propagation and waterhammer, the second order effects of viscosity, heat-transfer, etc. were introduced in the early 1930s. This two-dimensional theory leads to the concept of frequency-dependent friction which is crucial for the simulation of fluid power transmission lines.
The aim of this paper is to provide reliable experimental results of a hydraulic pipeline using a test rig with simple boundary conditions and a rigid mounting of the pipe.

NOTATION

p	$N/m^2, bar$	hydrostatic pressure	ρ	kg/m^3	mass density
$\vec{\mathbf{v}}$	m/s	velocity vector	ν	m^2/s	kinematic viscosity
$\mathbf{x}$	m	coordinates in $\mathbb{R}^3$	η	$kg\,m^{-1}s^{-1}$	dynamic viscosity
r	m	radial coordinate	d	m	inner diameter
x	m	axial coordinate	D	m	outer diameter
t	s	time	L	m	pipe length
r^*		normalised r	E_{fl}	N/m^2	bulk modulus
Q	$m^3/s, l/\min$	volumetric flow rate	p_0	$N/m^2, bar$	pressure at the inlet
c_0	m/s	wave speed	p_1	$N/m^2, bar$	pressure at the end
Z	$kg\,m^{-4}s^{-1}$	char. impedance			
Γ		propagation factor			

INTRODUCTION

While the input-output characterisitics of transmission lines with frequency-dependent friction can easily be described in the frequency domain, the arising transcendental transfer functions cannot be exactly represented in the time domain. Numerous authors have studied the problem of approximation of the frequency domain behaviour in the time domain. The first solution to this problem uses a discretization in time and space, resulting in various modifications of the method of characteristics (Zielke 1968, Suzuki, Taketomi & Sato 1991). The second approach uses modal approximations in order to approximate the transcendental functions by rational functions which can be represented by ordinary differential equations in the time domain (Hsue & Hullender 1983, Hullender, Woods & Hsue 1983, Piché & Ellman 1995).
Most papers concentrate on the approximation of a given frequency domain model. The experimental validation of the models is carried out via transient experiments in the time domain. Some papers give experimental results in the frequency domain for the low frequency region. In order to verify the frequency-dependent friction theory, experimental results over a large frequency range are needed. This paper gives experimental results up to the eighth eigenmode of a pipe with a length of 5600 mm and an inner diameter of 16 mm.

FREQUENCY DOMAIN THEORY

The following derivation of the two-dimensional theory is standard and can be found in the literature (Goodson & Leonard 1972). It is carried out in detail in order to show the essential assumptions and simplifications of this theory.

Derivation of momentum and continuity equations

The following derivation of the equations of momentum and continuity is restricted to the case of a straight pipe with a circular and constant cross section. The pipe is treated as a rigid body in this derivation. The effect of the radial deflection is usually accounted for by reducing the bulk modulus of viscosity resulting in a decrease of the wave speed. The bulk modulus E_{fl} is replaced by an effective bulk modulus (Goodson & Leonard 1972).

Using a standard cartesian coordinate system

$$\mathbf{x} = \begin{bmatrix} x_1 & x_2 & x_3 \end{bmatrix}^T \quad \in \quad \mathbb{R}^3,$$

the equation of momentum is

$$\frac{D}{Dt}(\rho\,\vec{\mathbf{v}}) = -\nabla p\,(\mathbf{x},t) + \eta\left(\nabla^2\vec{\mathbf{v}} + \frac{1}{3}\frac{\partial}{\partial \mathbf{x}}(\nabla\vec{\mathbf{v}})\right) \tag{1}$$

in which ρ denotes the mass density, $\vec{\mathbf{v}}\,(\mathbf{x},t)$ the velocity vector, $p\,(\mathbf{x},t)$ the hydrostatic pressure and η the dynamic viscosity. The equation of continuity reads

$$\frac{D}{Dt}\rho = -\rho\nabla\vec{\mathbf{v}}\,. \tag{2}$$

Equation (2) can be used to simplify equation (1) yielding the equation of momentum in the form

$$\rho\frac{\partial}{\partial t}\vec{\mathbf{v}} = -\frac{\partial}{\partial \mathbf{x}}p + \eta\left(\nabla^2\vec{\mathbf{v}}+\frac{1}{3}\frac{\partial}{\partial \mathbf{x}}\left(\nabla\vec{\mathbf{v}}\right)\right). \tag{3}$$

The linear state equation

$$\rho = \rho_0\left(1+\frac{p}{E_{fl}}\right)$$

gives a relationship between pressure p and density ρ. The time derivative $\frac{\partial\rho}{\partial t}$ and the gradient $\nabla\rho$ can be expressed as

$$\begin{aligned}\frac{\partial}{\partial t}\rho &= \frac{\partial\rho}{\partial p}\frac{\partial p}{\partial t} = \frac{1}{E_{fl}}\frac{\partial p}{\partial t},\\ \nabla\rho &= \frac{\partial\rho}{\partial \mathbf{x}} = \frac{\partial\rho}{\partial p}\frac{\partial p}{\partial \mathbf{x}} = \frac{\rho_0}{E_{fl}}\frac{\partial p}{\partial \mathbf{x}}.\end{aligned}$$

This gives the equation of momentum

$$\rho\frac{\partial}{\partial t}\vec{\mathbf{v}} = -\frac{\partial}{\partial \mathbf{x}}p + \eta\left(\nabla^2\vec{\mathbf{v}}+\frac{1}{3}\frac{\partial}{\partial \mathbf{x}}\left(\nabla\vec{\mathbf{v}}\right)\right) \tag{4}$$

and the equation of continuity

$$\frac{\partial p}{\partial t} + \vec{\mathbf{v}}\cdot\frac{\partial p}{\partial \mathbf{x}} = -E_{fl}\frac{\boldsymbol{\rho}}{\rho_0}\frac{\partial\vec{\mathbf{v}}}{\partial \mathbf{x}}. \tag{5}$$

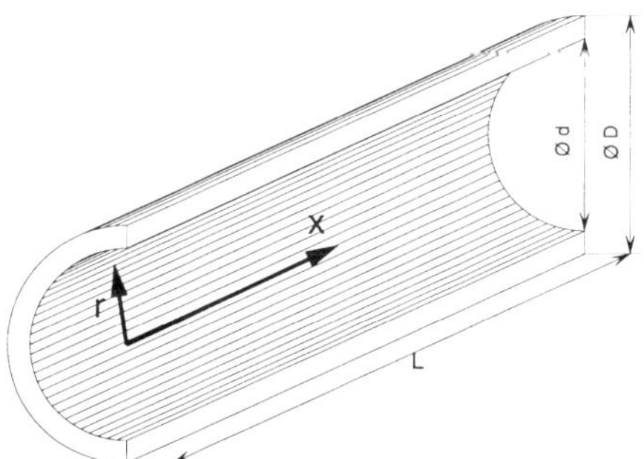

Fig. 1: Coordinate system in the pipe.

While this is a general model for three dimensions in space we expect the flow in the pipe to be symmetric with respect to the pipe axis. Therefore, we will now develop a two-dimensional model. Figure 1 shows a straight pipe of length L with constant inner diameter d and outer diameter D, respectively. The transversal coordinate is x and the radial coordinate is r. The state of the system at time t is described by the pressure distribution $p(x,r,t)$ and velocity distribution $v(x,r,t)$.

In cylindrical coordinates, the equation of momentum is

$$\rho\frac{\partial v}{\partial t} = -\left(\frac{\partial p}{\partial x} + \frac{1}{r}\frac{\partial p}{\partial r}\right) + \eta\left(\frac{1}{r}\frac{\partial v}{\partial r} + \frac{\partial^2 v}{\partial r^2} + \frac{\partial^2 v}{\partial x^2} + \frac{1}{3}\frac{\partial^2 v}{\partial x^2}\right) . \tag{6}$$

The wave-length of the interesting oscillations is assumed to be much larger than the pipe diameter. Hence, the velocity gradient in axial direction can be neglected since it is very small compared to the velocity gradient in radial direction. Furthermore, the pressure gradient in radial direction is neglected, resulting in the simplified equation of momentum

$$r^2\frac{\partial^2 v}{\partial r^2} + r\frac{\partial v}{\partial r} - \frac{r^2}{\nu}\left(\frac{\partial v}{\partial t} + \frac{1}{\rho}\frac{\partial p}{\partial x}\right) = 0 , \tag{7}$$

where the kinematic viscosity ν is defined as

$$\nu = \frac{\eta}{\rho} .$$

If the velocity is small compared to the sonic speed

$$\vec{\mathbf{v}} \cdot \frac{\partial p}{\partial \mathbf{x}} \ll \frac{\partial p}{\partial t} ,$$

the convective term in the equation of continuity can be neglected. Furthermore, if the changes in density are small

$$\frac{\rho}{\rho_0} \approx 1 ,$$

the equation of continuity takes the simple form

$$\frac{\partial p}{\partial t} = -E_{fl}\frac{\partial v}{\partial x} . \tag{8}$$

Laplace Transform

To get rid of the time derivative, the Laplace transform is applied to equation (7) yielding

$$r^2\frac{\partial^2 \hat{v}}{\partial r^2} + r\frac{\partial \hat{v}}{\partial r} - \frac{s\,r^2}{\nu}\left(\hat{v} + \frac{1}{s\,\rho}\frac{\partial \hat{p}}{\partial x}\right) = 0 .$$

Since $\hat{p}$ is independent of r, the number of variables can be reduced by setting

$$\hat{u} = \hat{v} + \frac{1}{s\,\rho}\frac{\partial \hat{p}}{\partial x} .$$

This results in the equation

$$r^{*2}\frac{\partial^2 \hat{u}}{\partial r^{*2}} + r^*\frac{\partial \hat{u}}{\partial r^*} + r^{*2}\hat{u} = 0 . \tag{9}$$

where the radial coordinate has been scaled according to

$$r^* = \sqrt{-\frac{s}{\nu}}r = j\sqrt{\frac{s}{\nu}}r .$$

Together with the boundary conditions $\hat{v}(R) = 0$ (zero velocity at the wall) and $\hat{v}(0) < \infty$

(limited velocity in the pipe axis) the Bessel differential equation (9) yields the solution

$$\hat{u}(r^*) = \frac{1}{s\,\rho}\frac{\partial \hat{p}}{\partial x}\frac{J_0(r^*)}{J_0(R^*)} \qquad \Rightarrow \qquad \hat{v}(r^*) = \frac{1}{s\,\rho}\frac{\partial \hat{p}}{\partial x}\left(\frac{J_0(r^*)}{J_0(R^*)} - 1\right)$$

where the pipe radius R is given by

$$R = \frac{d}{2}, \qquad R^* = j\sqrt{\frac{s}{\nu}}R$$

In fluid power pipelines the interesting variables are pressure p and volumetric flow rate Q. Thus, the velocity is integrated

$$\hat{Q} = \int_0^R 2\pi\, r\, \hat{v}(r)\; dr = \frac{R^2\pi}{s\,\rho}\,\frac{\partial \hat{p}}{\partial x}\,\frac{J_2(R^*)}{J_0(R^*)} \tag{10}$$

to get the Laplace transform $\hat{Q}(x)$ of the flow rate at the axial position x. The frequency domain representation of equation (8) is

$$s\,\hat{p} = -\frac{E_{fl}}{R^2\pi}\frac{\partial \hat{Q}}{\partial x}\,. \tag{11}$$

Differentiation of equation (11) with respect to x and substituton into equation (10) gives

$$\frac{\partial^2 \hat{Q}}{\partial x^2} + \frac{J_0(R^*)\,s^2}{J_2(R^*)\,c_0^2}\hat{Q} = 0 \qquad \text{or equally} \qquad \frac{\partial^2 \hat{p}}{\partial x^2} + \frac{J_0(R^*)\,s^2}{J_2(R^*)\,c_0^2}\hat{p} = 0 \tag{12}$$

where the sonic speed c_0 is given by

$$c_0 = \sqrt{\frac{E_{fl}}{\rho}}\,.$$

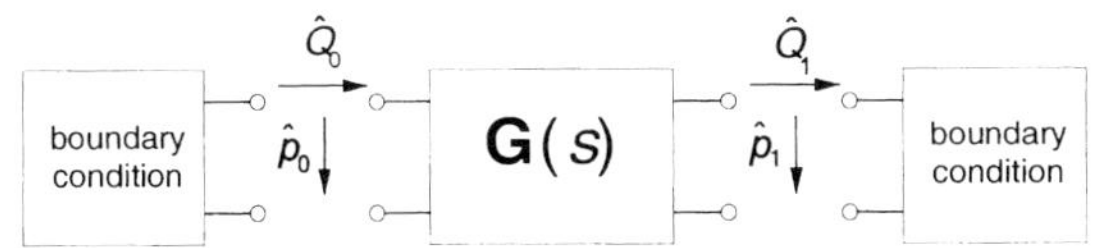

Fig. 2: Two-port representation of the input-output behaviour.

Input-Output Behaviour

From a control system point of view, the input-output behaviour of the pipe is important. The input-output model relates the two variables pressure p and flow rate Q at the two ends of the line. The Laplace transforms of these four variables are denoted by

$$\hat{Q}_0(s) = \hat{Q}(x,s)\Big|_{x=0} \qquad \hat{p}_0(s) = \hat{p}(x,s)|_{x=0}$$
$$\hat{Q}_1(s) = \hat{Q}(x,s)\Big|_{x=L} \qquad \hat{p}_1(s) = \hat{p}(x,s)|_{x=L}$$

Figure 2 shows the two-port representation of the hydraulic transmission line. The general solution of the ordinary differential equation (12) gives the transfer matrix **G** in the form

$$\begin{bmatrix} \hat{p}_1(s) \\ \hat{Q}_1(s) \end{bmatrix} = \underbrace{\begin{bmatrix} \cosh(\Gamma) & -Z\sinh(\Gamma) \\ -\frac{1}{Z}\sinh(\Gamma) & \cosh(\Gamma) \end{bmatrix}}_{\mathbf{G}} \begin{bmatrix} \hat{p}_0(s) \\ \hat{Q}_0(s) \end{bmatrix}. \tag{13}$$

where the characteristic impedance Z is given by

$$Z = Z_0\sqrt{-\frac{J_0(R^*)}{J_2(R^*)}}, \qquad Z_0 = \frac{\sqrt{E'\rho}}{R^2\pi}$$

and the propagation operator Γ reads as

$$\Gamma = \frac{s\,L}{c_0}\sqrt{-\frac{J_0(R^*)}{J_2(R^*)}}.$$

EXPERIMENTAL SETUP

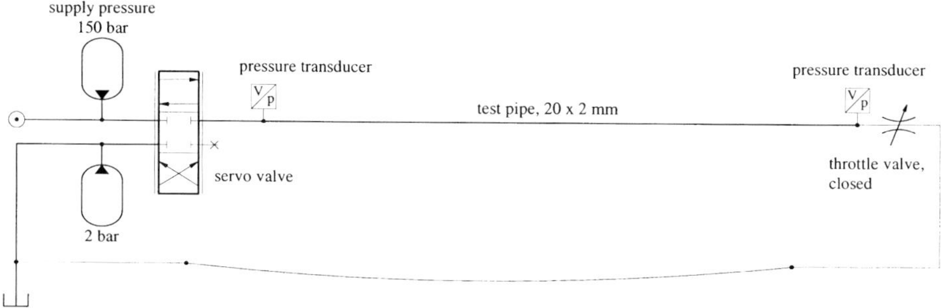

Fig. 3: Schematic diagram of the test rig for the closed pipe experiments.

Experimental work on fluid power systems requires the measurement of pressure and/or flow. The measurement of volumetric flow is limited to mean flow and slowly varying flow rates. Furthermore, most flow sensors use fluid-structure interaction disturbing the flow to be measured. Pressure sensors feature a higher bandwidth at negligible disturbance of the system under test. Since wave propagation in pipelines of fluid power systems is a high frequency phenomenon, only pressure sensors can be applied. Therefore, a closed pipe arrangement according to figure 3 is used. The flow rate at the closed end is zero resulting in the two equations

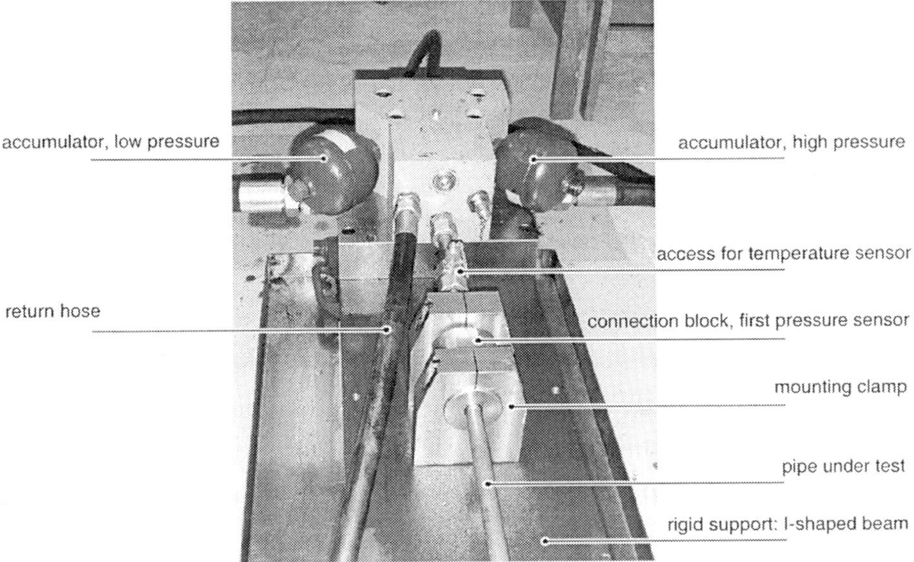

Fig. 4: The valve block (valve mounted to the rear) with the two accumulators and the first pressure measurement bock.

$$\begin{bmatrix} \hat{p}_1(s) \\ 0 \end{bmatrix} = \begin{bmatrix} \cosh(\Gamma) & -Z\sinh(\Gamma) \\ -\frac{1}{Z}\sinh(\Gamma) & \cosh(\Gamma) \end{bmatrix} \begin{bmatrix} \hat{p}_0(s) \\ \hat{Q}_0(s) \end{bmatrix}. \tag{14}$$

This gives the well known transfer function

$$\frac{\hat{p}_1(s)}{\hat{p}_0(s)} = \frac{1}{\cosh(\Gamma)}.$$

The experimental setup features a steel pipe of outer diameter $D = 20$ mm and inner diameter $d = 16$ mm. The length between the two pressure sensors is $L = 5600$ mm. The distance between the valve and the first pressure transducer is 500 mm. The inner diameter of this connection is 16 mm, too. While this means that the actuation is not close to the input of the pipe under test, the separation of the excitation and the first measurement point should help to smooth out disturbances of the axisymmetric flow in the pipe.

FREQUENCY DOMAIN EXPERIMENTS

The most effective way of measuring frequency responses is to use Fast Fourier Transforms (FFT). However, there are also some drawbacks of the FFT method:

- One problem is the proper choice of the excitation signal. Especially for systems with sharp resonances the excitation signal has to be designed carefully to catch rapid changes in the frequency response and to avoid the over-excitation of resonances.

- Every discrete Fourier transform is subject to error if the excitation signal contains frequencies which do not exactly fit into the frequency grid defined by the sampling process.
- The collected data has to be filtered. While high frequency disturbances, which cause aliasing problems, can effectively be removed, low frequency noise will inevitably cause errors in the computed frequency response. There are numerous methods to address this problem. See, for example (Welch 1967), (Cooley, Lewis & Welch 1970) or (Ljung 1987). The basic principle of these methods is the smoothing of the frequency response via weighted averaging of a number of FFTs. The problem with all of the variance reduction methods is the tradeoff between the resolution of the computed frequency response and the sensitivity to noise. Furthermore, the choice of the parameters for the smoothing process (number of FFTs, window length, etc.) is crucial for the quality of the results.

For these reasons the FFT method was not used for the experiment. The most reliable and easy to use way of measuring frequency responses is to test one harmonic frequency at a time and plot the results over a specified frequency range. This is the method which is used when frequency responses are tested by means of a signal generator and an oscilloscope. Since this is ineffective and subject to operator error, the measurement process has to be automated.

Frequency response measurement - mathematical background

Let

$$G(s) = \frac{\hat{p}_1(s)}{\hat{p}_0(s)}$$

be the transfer function from pressure p_0 to pressure p_1. The harmonic excitation

$$p_0(t) = a_0 \cos(\omega t) + b_0 \sin(\omega t)$$

results in the harmonic response

$$p_1(t) = a_1 \cos(\omega t) + b_1 \sin(\omega t)$$

with

$$G(s)|_{s=j\omega} = \frac{a_1 + j\, b_1}{a_0 + j\, b_0}\,.$$

The coefficients a_0, a_1, b_0, b_1 are given by

$$\begin{aligned} a_0 &= \frac{\omega}{\pi} \int_{-\frac{\pi}{\omega}}^{\frac{\pi}{\omega}} p_0(t) \cos(\omega t)\, dt, & b_0 &= \frac{\omega}{\pi} \int_{-\frac{\pi}{\omega}}^{\frac{\pi}{\omega}} p_0(t) \sin(\omega t)\, dt, \\ a_1 &= \frac{\omega}{\pi} \int_{-\frac{\pi}{\omega}}^{\frac{\pi}{\omega}} p_1(t) \cos(\omega t)\, dt, & b_1 &= \frac{\omega}{\pi} \int_{-\frac{\pi}{\omega}}^{\frac{\pi}{\omega}} p_1(t) \sin(\omega t)\, dt. \end{aligned}$$

The measured signals

$$\tilde{p}_0(t) = p_0(t) + d_0(t) \qquad \text{and} \qquad \tilde{p}_1(t) = p_1(t) + d_1(t)$$

contain disturbances $d_0(t)$ and $d_1(t)$. To improve the noise rejection of the method, the integral is carried out over M periods, resulting in the estmation

$$\tilde{a}_0 = \frac{\omega}{\pi} \int_0^{M\frac{2\pi}{\omega}} \tilde{p}_0(t) \cos(\omega t)\, dt, \qquad \tilde{b}_0 = \frac{\omega}{\pi} \int_0^{M\frac{2\pi}{\omega}} \tilde{p}_0(t) \sin(\omega t)\, dt, \tag{15a}$$

$$\tilde{a}_1 = \frac{\omega}{\pi} \int_0^{M\frac{2\pi}{\omega}} \tilde{p}_1(t) \cos(\omega t)\, dt, \qquad \tilde{b}_1 = \frac{\omega}{\pi} \int_0^{M\frac{2\pi}{\omega}} \tilde{p}_1(t) \sin(\omega t)\, dt \tag{15b}$$

for the four coefficients a_0, a_1, b_0, b_1. If the noise signals d_0 and d_1 are uncorrelated with the excitation signal, the error of the estimation approaches zero as M increases.
The internal electronics of most pressure sensors contains a low-pass filter. If this filter is similar for both pressure sensors, the filter does not influence the measured transfer function because it is cancelled out. The pressure transducers used for the experiment (HYDAC type HDA-3844-B-250) showed identical behaviour (phase error $\pm$ 2°, magnitude error 0.1 dB) up to 1200 Hz. This was tested by mounting two pressure sensors at the same point in the pipeline.

Frequency response measurement - implementation

The system has to be excited with the frequency

$$f_{exc} = \frac{\omega}{2\pi}.$$

The signals are measured with a digital signal processing system. With the sampling frequency f_S, the continuous time signals are replaced by

$$\tilde{p}_{0,k} = \tilde{p}_0(t)\big|_{t=\frac{k}{f_S}}, \qquad \tilde{p}_{1,k} = \tilde{p}_1(t)\big|_{t=\frac{k}{f_S}}, \qquad k = 0 \ldots M\frac{f_S}{f_{exc}}.$$

Thus, the ratio of the sampling frequency f_S and the excitation frequency f_{exc} must be an integer. Furthermore, the excitation signal must be exactly periodic and in phase with the sampling process. In order to fulfil this requirements, the excitation signal is generated within the signal processing system. A sampling frequency of 65536 Hz was chosen for the experiments. This value (2^{16}) is a natural choice since it results in a sampling time which can be realized without any roundoff error.
The approximation of the continuous time integrals (15a) and (15b) on the grid defined by the sampling process is carried out using the Gill-Miller method (Gill & Miller 1972). The resulting coefficients are used to compute magnitude and phase of the frequency response at $\omega = 2\pi\, f_{exc}$. This procedure can be repeated for a number of different excitation frequencies.
The test rig uses a MOOG D760 super high response servovalve with a 90° frequency of 290 Hz. The excitation signal is a sinusoidal current into the torque-motor of the servo valve driven by an operational amplifier. The offset value of this current is used to maintain a constant mean pressure in the test pipe. This is crucial for the whole experiment because of the influence of the pressure on the bulk modulus of elasticity.
The amplitude of the excitation signal has to be chosen big enough to ensure a proper signal to noise ratio but not too big since this would cause over-excitation of resonance frequencies resulting in cavitation.
In order to guarantee a constant mean pressure and a proper amplitude of the excitation signal, both parameters are controlled during the experiment.

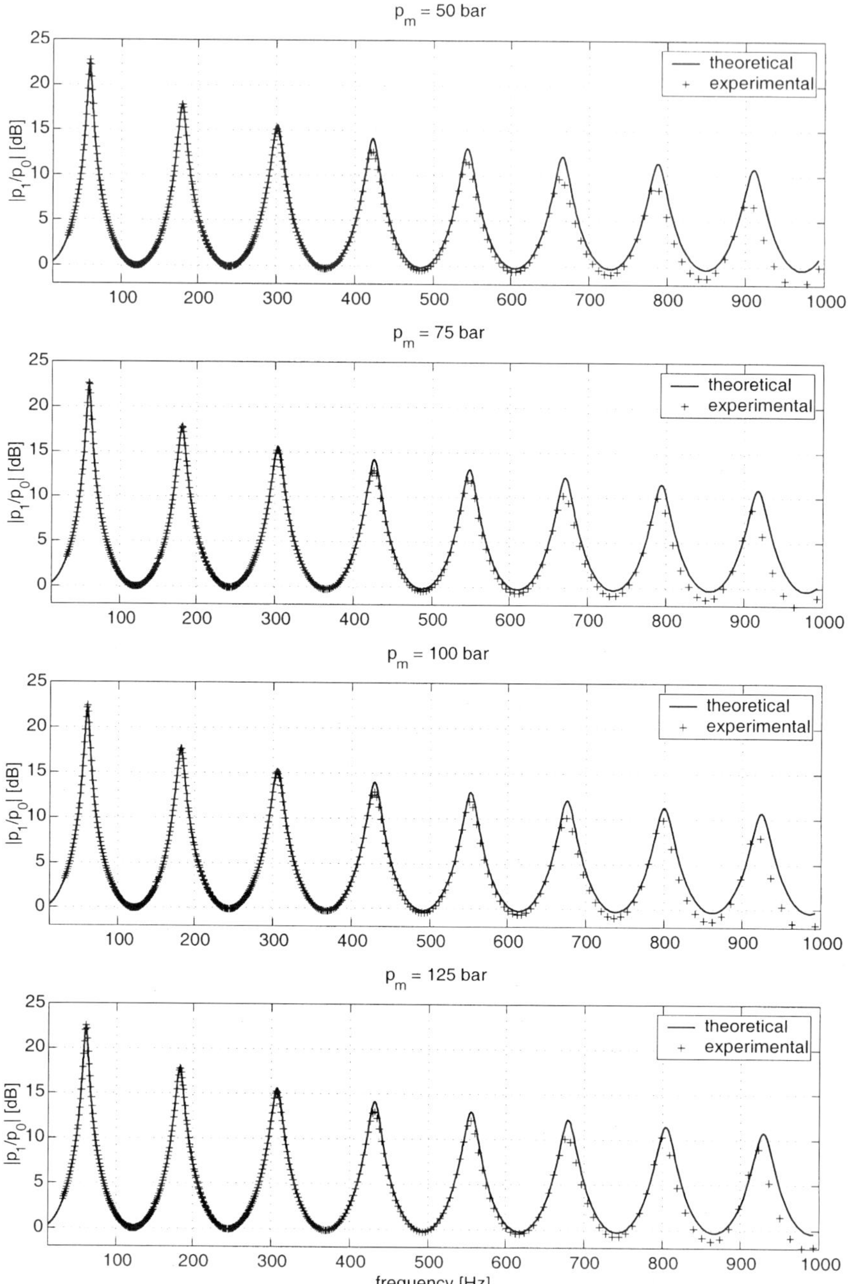

Fig. 5: Magnitude of $G(s) = \hat{p}_1/\hat{p}_0$ for 50, 75, 100, and 125 bar of mean pressure.

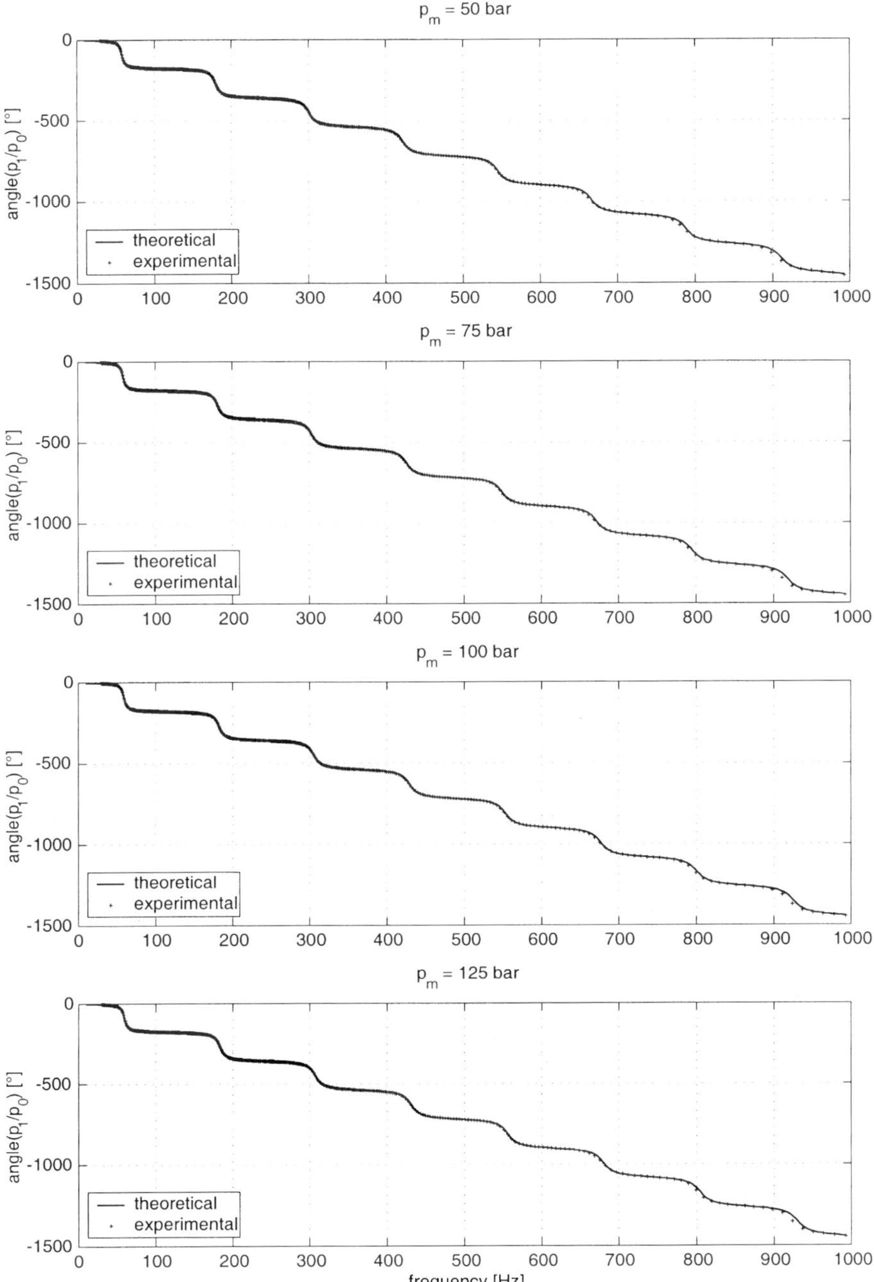

Fig. 6: Phase of $G(s) = \hat{p}_1/\hat{p}_0$ for 50, 75, 100, and 125 bar of mean pressure.

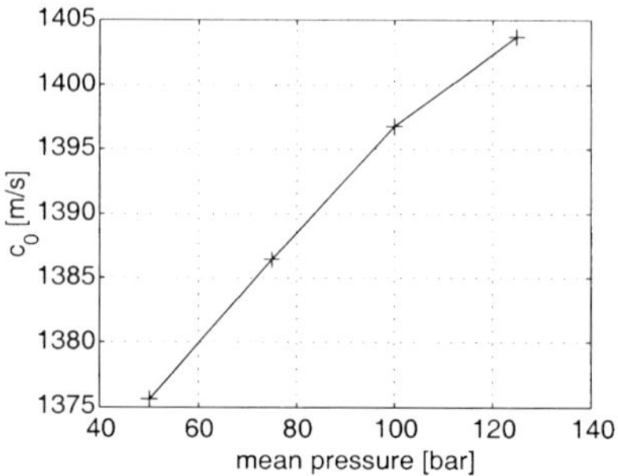

Fig. 7: Influence of the mean pressure on the speed of wave propagation.

Results

The magnitude and phase plot of the transfer function from the pressure at the inlet of the test pipe to the pressure at the closed end are shown in figure 5 and 6, respectively. The crosses give the data from an experiment using $M = 200$ periods for cancelling out the noise signal. While the geometric dimensions of the test pipe are known, the speed of wave propagation c_0 and the kinematic viscosity ν have to be computed from the experimental results. The two free parameters were used to minimize the quadratic error between the experimental results and the theoretical transfer function. This optimization was carried out over the frequency range up to 1000 Hz.

Independent of the mean pressure, the high frequency damping is even stronger than due to the two-dimensional frequency domain theory. The experiment was carried out with a pressure amplitude (absolute maximum of the deviation of p_0 or p_1 from the mean pressure value) of 5 bar. Results are shown for mean pressure values of 50, 75, 100 and 125 bar. The results do not change if the pressure amplitude is decreased. Amplitude values between 0.5 and 5 bar have been tested.

The influence of the mean pressure on the computed wave speed is shown in figure 7.

CONCLUSIONS & OUTLOOK

The proposed method yields very accurate results in the frequency domain. The increase of the damping with increasing frequency is stronger than due to the simple two-dimensional theory. This deviation is a linear effect since it does not vary with the excitation amplitude.

Further experimental work will cover the following topics:

- variation of the pipe diameter. Experiments with an inner diameter of 6 and 28 mm are scheduled in addition to the current results for an inner diameter of 16 mm.
- superposition of mean flow.
- investigation of the transition into turbulence. This experiments will be carried out using the small inner diameter of 6 mm. The current test rig does not reach the turbulent region due to the restriction of the pump flow.
- improvement of the pressure measurement by use of faster transducers in order to measure even higher eigenmodes

REFERENCES

Cooley, J., Lewis, P. & Welch, P. (1970). The application of the fast fourier transform algorithm to the estimation of spectra and cross-spectra., *Journal of Sound and Vibration* **12**(3): 339–52.

Gill, P. & Miller, G. (1972). An algorithm for the integration of unequally spaced data., *Computer Journal* **15**: 80–83.

Goodson, R. & Leonard, R. (1972). A survey of modeling techniques for fluid line transients, *ASME Journal of Basic Engineering* **94**(2): 474–482.

Hsue, C.-Y. & Hullender, D. (1983). Modal approximations for the fluid dynamics of hydraulic and pneumatic transmission lines, *in* M. Franke & T. Drzewiecki (eds), *Fluid Transmission Line Dynamics 1983*, ASME, pp. 51–77.

Hullender, D., Woods, R. & Hsue, C.-Y. (1983). Time domain simulation of fluid transmission lines using minimum order state variable models, *in* M. Franke & T. Drzewiecki (eds), *Fluid Transmission Line Dynamics 1983*, ASME, pp. 78–97.

Ljung, L. (1987). *System Identification - Theory For the User*, Prentice Hall, Englewood Cliffs, New Jersey.

Piché, R. & Ellman, A. (1995). A fluid transmission line model for use with ODE simulators, *in* C. Burrows & K. Edge (eds), *Eighth Bath International Fluid Power Workshop*.

Suzuki, K., Taketomi, T. & Sato, S. (1991). Improving zielke's method of simulating frequency-dependent friction in laminar liquid pipe flow, *ASME Journal of Fluids Engineering* **113**: 569–573.

Welch, P. (1967). The use of fast fourier transform for the estimation of power spectra: A method based on time averaging over short, modified periodograms., *IEEE Trans. Audio Electroacoustics* **15**: 70–73.

Zielke, W. (1968). Frequency-dependent friction in transient pipe flow, *ASME Journal of Basic Engineering* **90**(1): 109–115.

Control

The 'Smalve' – integrated digital control of a hydraulic servo drive

M GARSTENAUER and **M KURZ**
Department of Foundations of Machine Design, Johannes Kepler University of Linz, Austria

ABSTRACT

In this article we describe an integrated, fully digital control concept for a linear drive nicknamed "Smalve" (="SMArt vaLVE"). Some problems with state of the art implementations of digital servo control in hydraulic drives are illuminated and alternative implementations are proposed.

To develop this integrated controller, we will first derive a mathematical model of the proportional solenoid valve and then discuss controller design for the valve and drive. Finally experimental data is shown to compare conventional control strategies to the integrated control concept.

NOTATION

Table 1 explains the symbols that are used in the following sections.

Table 1 Notation

Valve spool velocity	v	Back-EMF coefficient	k_v
Valve spool position	x	Solenoid force	F_S
Solenoid current	i	Force/current coefficient	k_i
Solenoid voltage	u_S	Moving mass (spool+armature)	m
Pulse width of the power amp. input	pw	Damping coefficient	d
Supply voltage	U_{Supply}	Nondim. damping coefficient	ζ
Solenoid resistance	R	Stiffness of the valve spring	c
Magnetic flux within the solenoid	Φ	Coulomb friction force	F_C
Number of turns	N	Flow force	F_F
Solenoid inductance	L	Natural frequency	ω

1 INTRODUCTION

Advances in microelectronics revolutionise the design and operation of modern drives in many applications. (1) gives an overview of the state of the art and current trends concerning the application of electronics in hydraulic drives.

The hydraulic linear servo drive is a well suited candidate for digital control. It is the most popular fluid power application. Digital position sensors become more and more standard due to their significantly better resolution (fewer noise problems compared to analogue transducers). Also, this fact favours the use of digital controllers. The linear drive is very well investigated concerning modelling and control (see e.g. (2)-(5)). (6) discusses many problems that occur with state of the art implementations of servo drives (e.g. as shown in Fig. 1), i.e.

- large variety of components with a large number of different interfaces
- lots of wiring and connectors (important sources of errors)
- high cost of components, commissioning and maintenance

In summary, the introduction of digital electronics has even deteriorated the "user friendliness" of hydraulic drives significantly. Troubleshooting is getting more and more difficult and engineers with a sound knowledge of both hydraulics and electronics are required to "get it all working together".

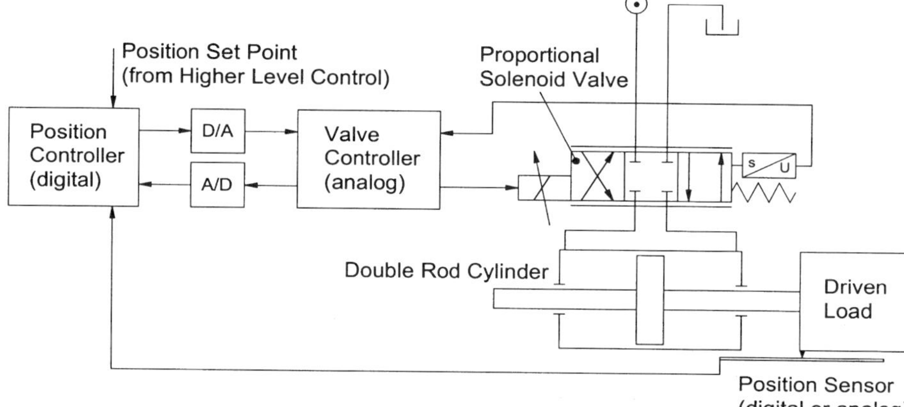

Fig. 1 State of the Art Implementation of a Digitally Controlled Linear Servo Drive

An application in a continuous casting plant featuring these problems motivated our work. For this drive, easy and quick maintainability is of major importance. In this environment, every hour of maintenance causes extreme costs. A method to reduce downtime is the implementation of modular drive units that can be easily replaced and serviced rapidly. It is also advantageous if the drive unit has only a few interfaces with the outside world. This demands that drive cylinder, position sensor, control valve and controller form a functional unit, as shown in Fig. 2.

It is possible to place all control functions even directly on the drive. This results in short wiring that can also be integrated into the mechanical hardware (excellent protection). Furthermore, no sensors have to be connected by the "end user".

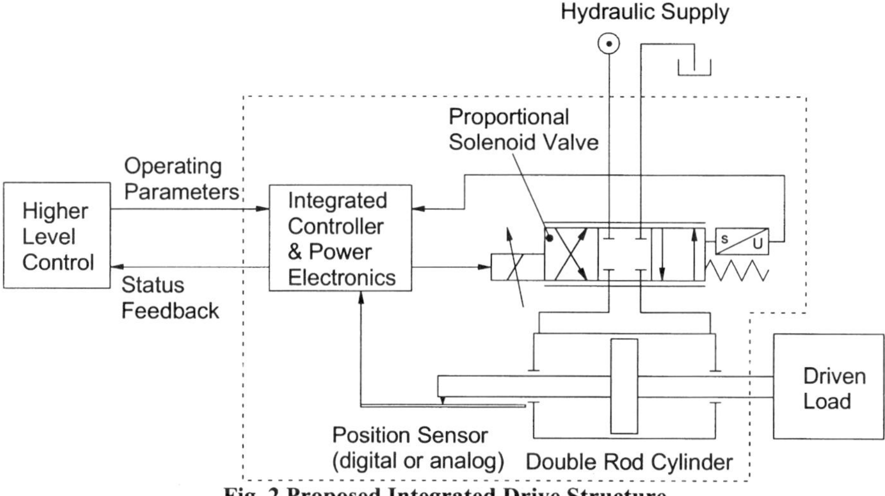

Fig. 2 Proposed Integrated Drive Structure

Another obvious advantage of such a setup is the low number of interfaces to the exterior world, i.e.

- mechanical connection to the driven load
- supply of hydraulic power (hoses or pipes)
- supply of electric power (to run the electronics and the valve solenoid)
- communication to the higher level process controller

The actual implementation of the first three items is simple. However, for communication purposes, a robust, low cost, and standardised interface is required. Field busses, e.g. CAN, InterBus-S, Profibus, etc. fulfil these demands to a great extent. There are several efforts to standardise the communication via field busses in the context of hydraulic drives, see e.g. (6).

In our opinion CAN is especially attractive due to its widespread use in the car industry. This brings component cost down and makes a large variety of CAN-enabled microelectronic components available. The standardisation of CAN for industrial control applications is also developing rapidly with two major types (CANOpen and DeviceNet). Sensors for industrial applications with CAN interface are becoming more widely available, and hydraulic components that can be controlled via CAN are starting to appear (e.g. (7)).

To achieve a widespread acceptance of an integrated drive concept, the following basic requirements were postulated:

- Direct acting proportional solenoid valves without hardware modifications (e.g. no additional sensors) must be used (Due to cost and maintenance reasons, high performance flapper nozzle type servo valve are not an option in many applications).
- The control law must be implemented on a low cost standard microcontroller.

2 PROPORTIONAL SOLENOID VALVE MODELING

As a control valve, we decided to use Bosch's size 6 servo solenoid valve with DC position output (8), which comes with a separate analogue valve controller with position feedback. It is a medium cost valve that is widely used in industrial servo drive applications.

In (9) several control strategies for a similar valve are investigated, and it is quite obvious that simple PID control is not sufficient to achieve the desired dynamic behaviour and thus we propose a model based controller (see section 3). To be able to design an appropriate control algorithm, mathematical models of the valve and the driving electronics are necessary. The model should describe all essential dynamic properties of the valve, but we may tolerate some uncertainty in model parameters or nonlinear effects, since they will be dealt with by the robust control law to be designed.

2.1 Valve hardware

Figure 3 shows a cross section of the valve in use.

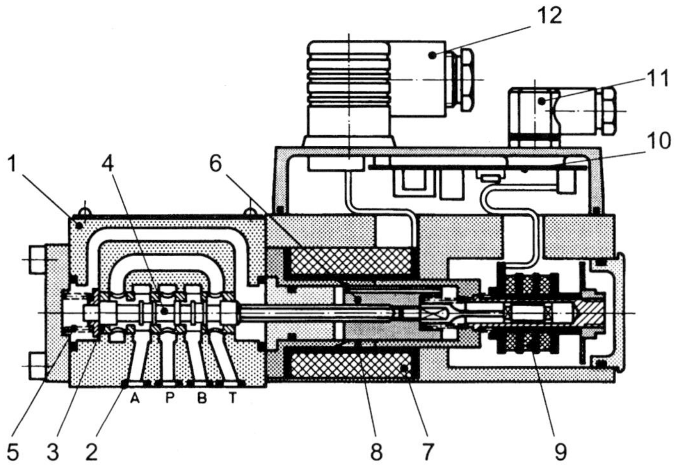

Fig. 3 Cross section of the Bosch size 6 proportional solenoid servo valve

The valve consists of a hydraulic part with valve body 1 and hydraulic ports 2. The valve sleeve 3 and the valve spool 4 form the control orifices. The spring 5 pushes the spool against the solenoid armature 6. This armature is the connecting element between the mechanical and electrical parts of the valve. Current through the magnet winding 7 generates the magnetic field. A magnetic isolator is used to achieve a magnet force that is independent of the armature position. The armature position is measured by a LVDT 9, the necessary electronics 10 to drive the sensor and to generate a voltage that is proportional to the spool position is also contained in the valve housing. There are two connectors to the valve controller, one for the position signal 11 and one for the magnet spool input 12.

From a systems point of view the valve model can be described as shown in Figure 4.

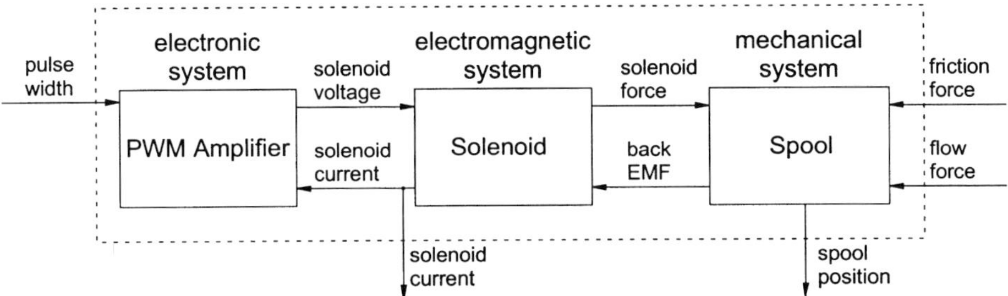

Fig. 4 Valve and power electronics – subsystems and interactions

The system input (pulse width) is translated to an input voltage at the solenoid by the PWM amplifier. Due to this voltage, a current flows through the solenoid. This current generates a magnetic flux through the armature thus creating an attracting force between solenoid housing and armature. This force acts on the valve spool against a spring. Motion of the spool with velocity v induces a voltage in the solenoid (back EMF). Friction and flow forces also act on the spool. Easily measurable quantities are solenoid current and spool position. In the following sections, a mathematical model of the valve is developed.

2.1 Power amplifier

To achieve sufficiently fast control action, a double switch forward amplifier (see block diagram in Fig. 5) was used. It is more complex than a single switch circuit, but allows fast changes in solenoid current (both rising and falling).

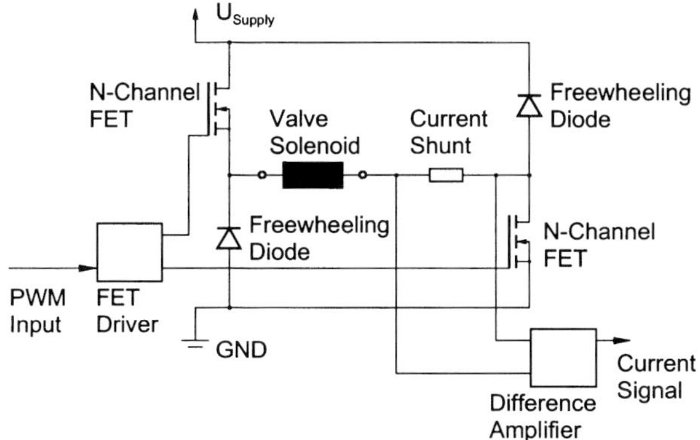

Fig. 5 Two switch forward PWM amplifier with current measurement

As long as the switching frequency of the amplifier is much higher than the significant system dynamics, one can replace the PWM signal by a continuous voltage across the solenoid. Its output voltage can be approximated by

$$u_S = pwU_{Supply} + (1 - pw)(-U_{Supply}) = (2\,pw - 1)U_{Supply} \tag{1}$$

as long as the solenoid current is larger than some minimum value (which is true under all normal operating conditions of the valve). *pw* denominates the pulse width of the PWM input.

2.2 Electromagnetic subsystem

The voltage/current relationship at the solenoid reads as

$$u_S = u_R + u_L = R\,i + N\frac{d\Phi}{dt} \tag{2}$$

The total magnetic flux depends on current and armature position. This means that

$$\frac{d\Phi}{dt} = \frac{\partial\Phi}{\partial x}\frac{dx}{dt} + \frac{\partial\Phi}{\partial i}\frac{di}{dt} = \frac{\partial\Phi}{\partial x}v + \frac{L}{N}\frac{di}{dt} \tag{3}$$

Substituting (3) into (2) and solving for the derivative of solenoid current gives

$$\frac{di}{dt} = \frac{1}{L}\left(-R\,i - N\frac{\partial\Phi}{\partial x}v + u_S\right) = \frac{1}{L}\left(-R\,i - k_v\,v + u_S\right) \tag{4}$$

Measurements have shown that the parameters L and k_v are not constant across the operating range of the valve.

The solenoid is designed to provide a force that is proportional to the spool current and independent of position. Measurements have proven this relation, it can be described by

$$F_S = k_i\,i \tag{5}$$

2.3 Mechanical subsystem – spool motion

The equation of motion for the spool and armature can be written as

$$m\frac{dv}{dt} + d\,v + c\,x = F_S - F_C - F_F \tag{6}$$

with solenoid force F_S, Coulomb friction F_C and fluid forces F_F. In nondimensional form this equation reads as

$$\frac{dv}{dt} + 2\zeta\,\omega\,v + \omega^2\,x = \frac{1}{m}\left(F_S - F_C - F_F\right) \tag{7}$$

The spring stiffness c and the mass of moving parts (spool and armature) m can be measured directly. An approximation for Coulomb's friction F_C can be identified from the open loop behaviour of the valve (with current control).

Determining the damping coefficient d would require significant experimental effort, especially considering the fact that it is influenced by the actual fluid flow in the valve. For controller design we estimate the nondimensional damping coefficient ζ and calculate d as

$$d = 2\zeta\sqrt{c\,m} \tag{8}$$

The flow force F_F is an external force that is determined by pressure difference and flow rate at the metering edges, it acts as an unknown disturbance.

The spring features significant preload, i.e. the working range of spool position is between 2.1 mm and 4.1 mm from the uncompressed spring position. This is also the operating range of the position sensor.

2.4 Mathematical valve model

Linearising the model equations, all three subsystems can now be combined to a linear state space representation

$$\begin{bmatrix} \dot{v} \\ \dot{x} \\ \dot{i} \end{bmatrix} = \begin{bmatrix} -\frac{d}{m} & -\frac{c}{m} & \frac{k_i}{m} \\ 1 & 0 & 0 \\ -\frac{k_v}{L} & 0 & -\frac{R}{L} \end{bmatrix} \begin{bmatrix} v \\ x \\ i \end{bmatrix} + \begin{bmatrix} 0 \\ 0 \\ \frac{1}{L} \end{bmatrix} u_S \tag{9a}$$

$$\begin{bmatrix} x \\ i \end{bmatrix} = \begin{bmatrix} 0 & 1 & 0 \\ 0 & 0 & 1 \end{bmatrix} \begin{bmatrix} v \\ x \\ i \end{bmatrix} \tag{9b}$$

with the input solenoid voltage u_S and the two outputs solenoid current i and spool position x. Table 2 gives the parameters that were used in the design of the controller. The values of inductance L and back EMF coefficient k_V are given for two positions (+80 % and –80 % of the spool position).

Table 2 Valve parameters

Moving mass (spool+armature) m	77 g
Spring stiffness c	15 N/mm
Damping coefficient ζ	0.1
Supply voltage V_{Supply}	24 V
Solenoid resistance R	3.3 Ω
Inductance @ +80% L_{+80}	30 mH
Inductance @ -80% L_{-80}	45 mH
Force/current coefficient k_i	33 N/A
Back-EMF coefficient @ +80% $k_{v,+80}$	70 Vs/m
Back-EMF coefficient @ -80% $k_{v,-80}$	35 Vs/m

3 CONTROLLER DESIGN

3.1 Valve controller

The mathematical model of the valve, derived in the previous section, can now be used to design a position controller for valve spool position. An obvious choice in this case (single input/multi output plant) is state feedback control. As we can only measure spool position x and solenoid current i and we do not want to modify the valve hardware, the third state spool velocity v needs to be determined by a state estimator. Two forward gains Nx and Nu are used to modify the controller structure to accommodate reference signals r (see (11)).

As Coulomb friction is not contained in the linear model, the state feedback controller does not compensate it. But we have already acquired an estimation of the spool velocity, and one can add a friction compensator that contributes

$$u_C = k_F \, sign(\hat{v}) \tag{10}$$

to the control input.

To account for model uncertainties and the disturbance due to flow forces, an error integrator was also added.

To achieve equally good dynamic behaviour at all operating points, gain scheduling was introduced. We used two nominal operating points (+80 % and –80% spool position) and designed two state gain matrices. The actual gain is determined using the sum of both controller gain matrices weighted by the current position set point.

Figure 6 shows the structure of the valve controller. Bold lines show vector valued signals. We give no formal proof of stability here.

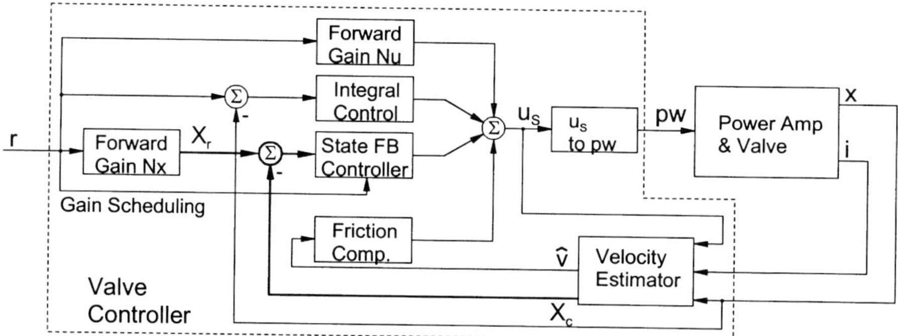

Fig. 6 Structure of the valve controller

3.2 Drive controller

To control the position of the drive, we use a modified P-controller. The position set point is filtered by a rate limiter to achieve a reference trajectory with bounded velocity. This nonlinear filter also includes a desired velocity output which we use for velocity feed forward. The velocity feed forward significantly enhances the dynamic behaviour of the drive. Figure 7 shows this control structure.

Of course, it is possible to implement all other kinds of position control algorithms, although this may be limited by the computing power of the microcontroller used.

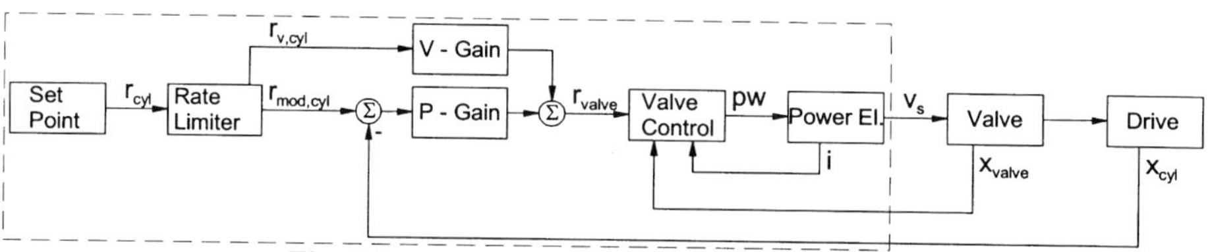

Fig. 7 Structure of the overall drive controller

3.3 Controller Implementation

At this stage, the control algorithm is implemented on a Texas Instruments TMS 320 signal processor. As a hardware platform we use a DS1102 controller board from dSPACE (12). The software was developed using Matlab/Simulink and dSPACE's Real Time Interface. A sampling rate of 2.5 kHz was used.

We plan to transfer the control algorithm to an Infineon C167CR microcontroller (13), which offers most of the peripherals on chip that are favourable for hydraulic drive control (e.g. A/D converter, PWM unit, incremental encoder position counter and synchronous serial interface). There is also a CAN bus interface on chip, which is another attractive feature for communication purposes. The control law has to be transformed to fixed point arithmetic, as there is no hardware support for floating point operations on chip.

4 EXPERIMENTAL RESULTS

4.1 Valve control

We first evaluate the performance of the valve controller that was designed in the previous section. A series of step responses was used to compare the analog and digital valve controller with and without oil flow. For the test with oil, a constant pressure of 70 bar was applied at the pressure port of the valve.

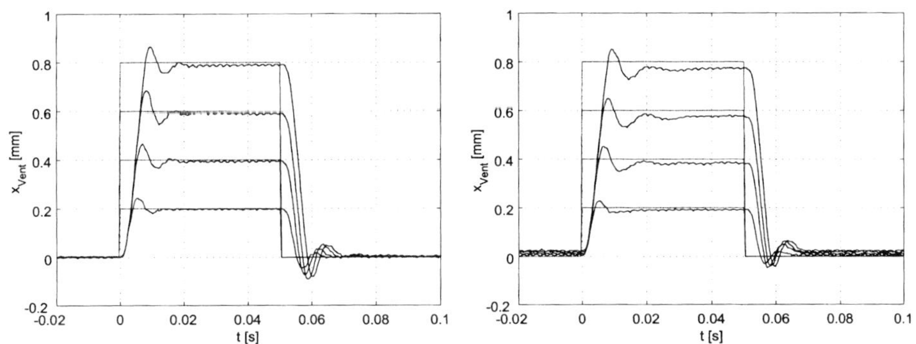

Fig. 8 Step responses with Bosch's analog valve controller:
a – without fluid flow , b – with Δp of 70 bar across the valve input ports

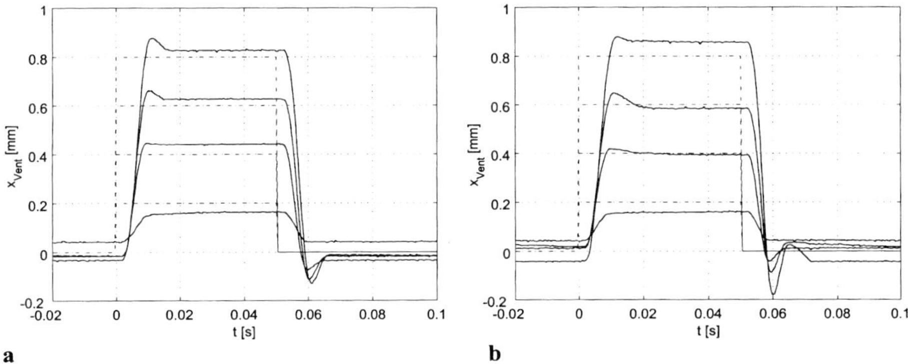

a **b**

Fig. 9 Step responses with digital state feedback control:
a – without fluid flow , b – with Δp of 70 bar across the valve input ports

Figures 8 and 9 show the results of the step response test using the analog Bosch controller and the proposed digital implementation. One can see that the dynamics are similar. Running without fluid flow, the digital controller shows better damping, but with fluid flow, damping deteriorates especially during closing.
A high frequency dither signal was added to the reference input. This reduced steady state error significantly.

4.2 Drive control

4.2.1 Drive setup

To evaluate the performance of the overall system we used a standard hydraulic servo drive setup as shown in the circuit diagram in Fig. 2. Fig. 10 shows a picture of the actual drive.

Fig. 10 Hydraulic servo drive

Table 3 lists its key parameters.

Table 3 Test drive parameters

Moving mass	500 kg
Rod diameter	28 mm
Piston diameter	40 mm
Working stroke	250 mm
Supply pressure	150 bar

The position sensor in use was a Sony Magnescale MD 20 with incremental encoder interface (resolution: 0.5 μm).

4.2.2 Position control experiments

To evaluate the performance of the integrated drive controller, a benchmark experiment was carried out. As a reference experiment, we implemented a conventional digital controller as shown in Fig. 1 (using Bosch's analog valve controller) with the same sampling rate of 2.5 kHz. Fig. 11 a shows the reference input and position signals for a filtered square wave positioning sequence. The velocity was limited to 0.4 m/s.

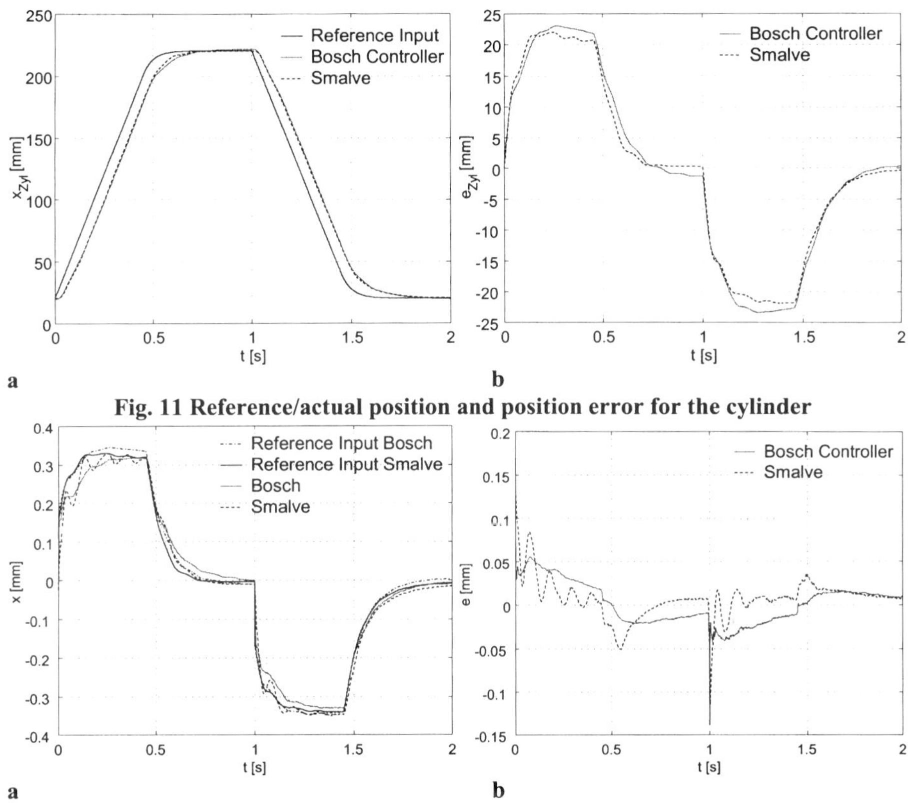

Fig. 11 Reference/actual position and position error for the cylinder

Fig. 12 Reference/actual position and position error for the valve

Fig. 11 b shows position errors for both control strategies, which are very much in the same range.

Fig. 12 a shows the position of the valve during this motion cycle. In Fig. 12 b the position errors of the valve are compared. One sees that the digital implementation exhibits slightly more oscillation than the analog valve controller.

4 CONCLUSIONS AND OUTLOOK

We have demonstrated that full digital control of a hydraulic servo drive with reasonable sampling rates and a moderately complex control structure is feasible. No modification of the hydraulic hardware and/or standard sensors is necessary.

The next steps towards practical applications are the implementation on a low cost microcontroller and more experimental studies with a broader range of loads.

As high performance microcontrollers/DSPs become cheaper and cheaper, it will be feasible to implement advanced control concepts that require more computing power.

Projects to standardise interfaces for such integrated hydraulic drives are currently under way and will make the application of modern digital control in hydraulic drives easier and cheaper. More sophisticated "end users" will also be able to adapt the digital control system to the specific needs of the application with "software only" modifications.

The additional flexibility also presents a chance to reduce the number and variety of necessary hydraulic hardware components which could also lead to larger production quantities and thus lower cost hydraulic systems.

REFERENCES

(1) **Leutner, V; Müller, U.; Feuser, A.; Köckemann, A. (1998)**:
Elektronik in der Fluidtechnik.
Ölhydraulik und Pneumatik, Jahrgang 42, Nr. 6; ISSN 0341-2660

(2) **Merritt, H. E. (1967)**:
Hydraulic Control Systems.
John Wiley & Sons, New York; ISBN 0 471 59617 5

(3) **Anderson, W. (1988):**
Controlling Electrohydraulic Systems.
Marcel Dekker, New York

(4) **Plummer, A. R.; Vaughan, N. D. (1996)**:
Robust adaptive control for hydraulic servosystems.
Transactions of the ASME. Journal of Dynamic Systems, Measurement, and Control 118

(5) **Manhartsgruber, B.; Scheidl, R. (1998)**:
Non-linear control of hydraulic servo-drives based on a singular perturbation approach.
in: Burrows/Edge (eds.): Proceedings of PTMC'98, Bath; ISBN 1-86058-134-X

(6) **Baldy, M. (1998)**:
Geräteprofil für elektrohydraulische Zylinderantriebe am InterBus-S.
Ölhydraulik und Pneumatik, Jahrgang 42, Nr. 4; ISSN 0341-2660

(7) **Claus, M.; Faller, M.; Mertlik, R. (1999)**:
Der CAN-Bus als Partner der Hydraulik.
Ölhydraulik und Pneumatik, Jahrgang 43, Nr. 2; ISSN 0341-2660

(8) **N.N. (1998)**:
Data sheet: Industrial Hydraulics – Servo Solenoid Valves.
Robert Bosch GmbH – Automation Technology, Stuttgart

(9) **Gamble, J. B., Vaughan, N. D. (1996)**:
Comparison of Sliding Mode Control With State Feedback and PID Control Applied to a Proportional Solenoid Valve.
Journal of Dynamic Systems, Measurement, and Control, Vol. 118, September 1996; ISSN 0022-0434

(10) **Vaughan, N. D., Gamble, J. B. (1996)**:
The Modeling and Simulation of a Proportional Solenoid Valve.
Journal of Dynamic Systems, Measurement, and Control, Vol. 118, March 1996; ISSN 0022-0434

(11) **Franklin, G. F., Powell, J. D., Workman, M. (1997)**:
Digital Control of Dynamic Systems.
Addison Wesley, Menlo Park; ISBN 0-201-82054-4

(12) **N.N. (1993):**
dSPACE DS1102 User's Guide.
dSPACE Technologie GmbH, Paderborn

(13) **N.N. (2000):**
C167CR Derivatives – 16-Bit Single-Chip Microcontroller User's Manual.
V 3.0, Infineon AG, Munich

A fixed cascade controller applied to a hydraulic actuator including the servovalve dynamic

M A B CUNHA, R GUENTHER, E R DE PIERI, and **V J NEGRI**
Federal University of Santa Catarina, Florianópolis, Brazil

Abstract
In this paper a fixed cascade controller is proposed based on a 4th order hydraulic actuator nonlinear model including the servovalve dynamics as a first order system. The cascade methodology consists in dividing the model of the hydraulic actuator into two subsystems: a mechanical subsystem and a hydraulic one. In this way, different strategies can be used in each subsystem. Using the Lyapunov's direct method, the exponential stability of the closed loop system is demonstrated under the assumption that the system parameters are known. Experimental results illustrate the main characteristics of the proposed controller.

NOTATION

A	cylinder piston cross sectional area	v	total fluid volume cylinder and lines
B	viscous friction coefficient	V	nonnegative function
d	input disturbance	u	control law
$f(y)$	position dependent function	x_v	spool position
$g(x_v,P_\Delta)$	pressure and spool position dependent function	x_{vd}	desired spool position
K_h	hydraulic constant	y	actuator piston position
K_D	controller gain	z	a measure of the tracking error
K_P	controller gain	y_d	desired trajectory
K_v	controller gain	$\dot{y}_r$	a reference velocity
K_{val}	valve constant	$\tilde{y}$	position tracking error
M	total mass	ρ	a trajectory error vector
P_Δ	pressure difference	β	bulk modulus
P_S	supply fluid pressure	ϕ_1, ϕ_2	positive constants
$P_{\Delta d}$	desired pressure difference	λ	controller gain
$\tilde{P}_\Delta$	pressure difference tracking error	ω_v	valve bandwidth

1 - INTRODUCTION

Due to the ability to provide a power density that is unmatched by any other commercial technology the number of applications of hydraulic actuators has increased in the last few years. Some of the more common application sectors include agriculture, forestry, aviation, construction, leisure, manufacture, materials handling, marine, transport and robotics [1]. Some of these applications require high-performance as do robot manipulators and in these cases the use of a classical control approach is limited due to the undesirable characteristics of hydraulic actuators.

The hydraulic actuator dynamics are highly nonlinear. In the servovalve, the flow/pressure characteristic is nonlinear and there are hard nonlinearities such as saturation, dead-zone, hysteresis. If the servovalve dynamic is analysed using a bode diagram, the dynamic response plot depends on the amplitude of the input signal. The friction on the actuator is position dependent and varies with positive and negative velocities. Additionally, due to hydraulic compressibility, hydraulic actuators have lightly damped dynamics that limit the bandwidth of the system. From the control point of view, the system is subjected to uncertain parameters (for instance, bulk modulus) and unstructured uncertainties (like pipeline dynamics).

In order to overcome these difficulties many kinds of controllers for hydraulic actuators using different strategies have been proposed in the literature. A review of research on the control of fluid power systems up to 1996 is presented in Edge [1]. In Yao et al. [2,3] backstepping methodology controllers are proposed and simulation results are presented. One of the major drawbacks of the backstepping methodology is that the successive derivatives may limit the real time implementation on a physical system [4]. To overcome this problem, Alleyne [4] has developed a systematic approach for controlling hydraulic actuators and applied it to a force control problem. Another way is to use the cascade approach presented by Guenther and De Pieri [5] which is based on the order reduction proposed by Utkin [6].

The cascade controller proposed in Guenther and De Pieri [5] overcomes the bandwidth limitations when applied to a hydraulic actuator third order linear model and the closed loop is exponentially stable if the parameters are known. Following such methodology some cascade controllers were proposed based on a hydraulic actuator third order nonlinear model [7]. However, as stated in Cunha et al. [8], the servovalve bandwidth limits the hydraulic subsystem control gain. To deal with such limitation and to improve the closed loop performance, in this paper a fixed cascade controller is proposed based on a hydraulic actuator 4th order nonlinear model. The proposed controller combines a Slotine and Li [9] control law for the mechanical subsystem and a fixed control law for the hydraulic subsystem as proposed by Alleyne [10] to the hydraulic actuator force control. Then, by using the Lyapunov's direct method, the exponential stability of the closed loop system is demonstrated under the assumption that the system parameters are known. Experimental results illustrate the main features of the proposed controller.

This paper is organised as follows. In section 2, the hydraulic actuator 4th order model is presented. In section 3, the cascade control strategy is presented and, in section 4 is demonstrated the controller's gain limitation when the controller is designed on a model that does not include the servovalve dynamics. In section 5, a cascade controller based on a nominal 4th order hydraulic actuator nonlinear model including the servovalve dynamics is proposed and the exponential stability is proved by using the Lyapunov's direct method. In section 6 experimental results illustrate the main features of the proposed controller and the conclusions are outlined in section 7.

2 – HYDRAULIC ACTUATOR MATHEMATICAL MODEL

The hydraulic actuator under consideration is shown in figure 1. It consists of a cylinder controlled by a critical centre four-way spool servovalve. The supply fluid pressure P_S comes from a hydraulic power and conditioning unit that provides controlled pressure and temperature.

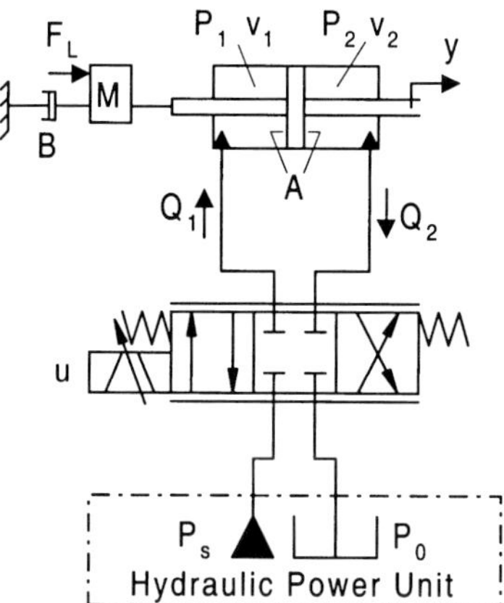

Figure 1 - Hydraulic actuator

In figure 1, P_0 is the return pressure, P_1 is the pressure in the 1st cylinder chamber, P_2 is the pressure in the 2nd cylinder chamber, v_1 is the volume in chamber 1 and line 1, v_2 is the volume in chamber 2 and line 2, Q_1 is the flow rate from the valve to the chamber 1, Q_2 is the flow rate from the chamber 2 to the valve, M represents the total mass of the system, B is the viscous friction coefficient, y is the actuator piston position, F_L represents an external force, and u is the control input.

The modelling of this system is developed applying Bernoulli's law, Newton's second law and the continuity equation to the valve and cylinder. Considering the valve dynamics as a first order system, the hydraulic actuator mathematical model can be written as:

$$M\ddot{y} + B\dot{y} = AP_\Delta + F_L \qquad (1)$$

$$\dot{P}_\Delta = -fA\dot{y} + fK_h g x_v \qquad (2)$$

$$\dot{x}_v = -\omega_v x_v + K_{val}\omega_v u, \qquad (3)$$

where $f = f(y) = \dfrac{\beta v}{\left(\dfrac{v}{2}\right)^2 - (Ay)^2}$, $g = g(x_v, P_\Delta) = \sqrt{P_S - \text{sgn}(x_v)P_\Delta}$, x_v is the valve spool position, A is the cylinder piston cross-sectional area, $P_\Delta = P_1 - P_2$ is the pressure difference in the cylinder, β is the fluid bulk modulus, $v = v_1 + v_2$ is the total fluid volume, sgn( ) is the "signum" function, K_h is a hydraulic constant, K_{val} is a valve constant, and ω_v is the valve bandwidth.

It is important to emphasise that the dynamic model (1)(2)(3) is obtained by making the following assumptions: (i) - The hydraulic power unit delivers a constant supply pressure P_S, irrespective of the oil flow; (ii) - The dynamical behaviour of the pressure in the transmission lines between valves and actuator is assumed to be negligible; (iii) The servovalve is considered a first order system.

Remark 1 – The term "x_v" expresses the measurement result in Volts of spool displacement obtained from an internal valve transducer. In this way, K_{val} is dimensionless. The external force F_L and its first time derivative $\dot{F}_L$ are assumed to be bounded.

3 – THE CASCADE CONTROL STRATEGY

The cascade control strategy applied to the hydraulic actuator consists in dividing the whole model into two subsystems: a mechanical subsystem and a hydraulic one [5].

Inspecting equations (1), (2) and (3) one can see that this system can be interpreted as a mechanical subsystem (1) driven by a hydraulic force AP_Δ, on which a hydraulic subsystem is superimposed to provide a pressure difference P_Δ, when driven by the control input "u". This interpretation enforces the cascade model description. Figure 2 illustrates this interpretation.

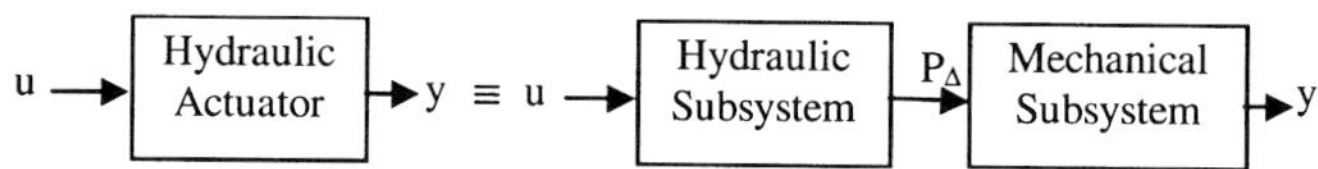

Figure 2 – Cascade control interpretation

To describe (1)(2)(3) as a cascade system $AP_{\Delta d}$ is defined as a desired force so that the mass M achieves a trajectory $y_d(t)$. This force is given by a desired pressure difference $P_{\Delta d}$. Let

$$\tilde{P}_\Delta = P_\Delta - P_{\Delta d} \tag{4}$$

be the pressure difference tracking error. Using (4) and considering $F_L = 0$ (the external force will be considered later) equation (1) can be rewritten as

$$M\ddot{y} + B\dot{y} = AP_{\Delta d} + d(t), \tag{5}$$

where $d(t) = A\tilde{P}_\Delta$ is an input disturbance.

The system (5)(2)(3) is in the cascade form. Equation (5) can be interpreted as a second order mechanical subsystem driven by a desired force $AP_{\Delta d}$ and subjected to an input disturbance d(t). Equations (2)(3) represent the hydraulic subsystem.

The design of the cascade controller for the system (5)(2)(3) can be summarised as:

(i) Compute a control law $AP_{\Delta d}$ for the mechanical subsystem (5) such that the cylinder displacement achieves a desired trajectory $y_d(t)$ taking into account the presence of the

disturbance d(t). With this desired force $AP_{\Delta d}$ one can quantify the desired pressure difference.
(ii) Compute a control law "u" such that P_Δ tracks $P_{\Delta d}$ defined above.

4 – THE IMPORTANCE OF INCLUDING THE SERVOVALVE DYNAMICS

The servovalve dynamics can normally be considered as a first, second or third order linear model depending on the system bandwidth. In Cunha et al. [8], a way to design the cascade controller's gains based on the third order hydraulic actuator model is proposed. From that analysis performed on a system including the servovalve dynamics as a first and a second order system one could verify the limitations imposed by the servovalve dynamics on the cascade controller gain for the hydraulic subsystem. In this section, the limitations imposed by the servovalve dynamics are briefly demonstrated.

Firstly, neglecting the servovalve dynamics, the hydraulic actuator subsystem (2)(3) can be written as

$$\dot{P}_\Delta = -fA\dot{y} + fK_h K_{val} u \sqrt{P_S - \mathrm{sgn}(u) P_\Delta}\ . \tag{6}$$

The fixed hydraulic subsystem control law for this subsystem is given by

$$u = \frac{1}{K_h K_{val} \sqrt{P_S - \mathrm{sgn}(u) P_\Delta}} \left(A\dot{y} + \frac{\dot{P}_{\Delta d} - K_P \tilde{P}_\Delta}{f} \right), \tag{7}$$

where K_P is a positive constant. Substituting (7) into (6) and solving the differential equation gives

$$\tilde{P}_\Delta(t) = \tilde{P}_\Delta(0) e^{-K_P t}\ . \tag{8}$$

From (8) (neglecting the servovalve dynamics) one may conclude that the pressure trajectory tracking error goes to zero at a rate determined by K_P.

Now, consider the servovalve as a first order system. The linearized model for (2) is given by

$$\dot{P}_\Delta = \frac{4\beta}{V}\left(-A\dot{y} - K_C P_\Delta + K_Q x_V \right), \tag{9}$$

where K_C is the pressure-flow rate gain and K_Q is the flow rate gain. Taking the Laplace transform of (3) and (9) with null initial conditions, considering K_C=0 and by combining these equations, gives

$$P_\Delta(s) = \frac{4\beta}{V}\frac{1}{s}\left(-sAY + K_Q K_{val} \frac{\omega_v}{s+\omega_v} U(s) \right). \tag{10}$$

The linearized version for equation (7) is given by

$$u = \frac{1}{\overline{K}_Q}\left[A\dot{y} + \frac{v}{4\beta}\left(\dot{P}_{\Delta d} - K_P\tilde{P}_\Delta\right)\right], \tag{11}$$

where $\overline{K}_Q = K_Q K_{val}$. Taking the Laplace transform with null initial conditions in (11) and substituting into (10) gives

$$P_\Delta(s) = \frac{4\beta}{v}\left(-AY + \frac{\omega_v}{s+\omega_v}AY\right) + \frac{1}{s}\left\{\frac{\omega_v}{s+\omega_v}\left[sP_{\Delta d} - K_P\left(P_\Delta - P_{\Delta d}\right)\right]\right\}. \tag{12}$$

Neglecting the term $\left(-AY + \frac{\omega_v}{s+\omega_v}AY\right)$, one can rewrite (12) as

$$P_\Delta(s) = \frac{\omega_v\left(s + K_P\right)}{s^2 + \omega_v s + K_P\omega_v}P_{\Delta d}(s), \tag{13}$$

which represents the dynamics relation between $P_{\Delta d}$ and P_Δ. Analysing (13) one concludes that for small values of the gain K_P the subsystem has a pole near the origin and another on the left side of the real axis. As the gain K_P is increased the subsystem becomes overdamped, critically damped and, finally, underdamped. This underdamped behaviour limits the K_P value because it may cause vibrations on the actuator. The above analysis when performed considering the servovalve dynamics as a second order system states that K_P must be limited to the relation between $P_{\Delta d}$ and P_Δ to be stable [8]. Then, the value of K_P which was believed to be unbounded by equation (8) is indeed bounded. Furthermore, the relation between $P_{\Delta d}$ and P_Δ depends on the gain K_P and the servovalve bandwidth (equation (13)) instead of being only a function of the gain K_P (equation (8)). Thus the development of control algorithms that include the servovalve dynamics during the design process may improve the closed loop performance.

Remark 2 – The term $\left(-AY + \frac{\omega_v}{s+\omega_v}AY\right)$ was neglected because the goal in this section was to demonstrate the valve influence on the hydraulic subsystem stability. Indeed, it represents an error for the hydraulic subsystem trajectory tracking caused by the valve dynamic.

5 – A NEW FIXED CASCADE CONTROLLER

5.1 – Mechanical subsystem control law

The desired pressure difference ($P_{\Delta d}$) to achieve the trajectory tracking in the mechanical subsystem (5) is calculated by

$$P_{\Delta d} = \frac{1}{A}\left(M\ddot{y}_r + B\dot{y}_r - K_D z\right), \tag{14}$$

where K_D is a positive constant, $\dot{y}_r$ is the reference velocity and z is a measure of the velocity tracking error. In fact, $\dot{y}_r$ can be obtained by modifying the desired velocity $\dot{y}_d$ as follows

$$\dot{y}_r = \dot{y}_d - \lambda\tilde{y} \ , \ \tilde{y} = y - y_d \ , \ z = \dot{y} - \dot{y}_r = \dot{\tilde{y}} + \lambda\tilde{y} \ , \tag{15}$$

where λ is a positive constant.

Expression (14) is based on the control law proposed by Slotine and Li [9] for robot manipulators. It has feedforward terms $(M\ddot{y}_r + B\dot{y}_r)$ and a proportional derivative component $(K_D z)$. Substituting (14) into (5) gives

$$M\dot{z} = A\tilde{P}_\Delta - (B + K_D)z \, . \tag{16}$$

Consider the nonnegative function

$$V_1 = \frac{1}{2}Mz^2 \, , \tag{17}$$

for which, using (16), the time derivative is given by

$$\dot{V}_1 = -(B + K_D)z^2 + A\tilde{P}_\Delta z \, . \tag{18}$$

This expression will be used later.

5.2 – Hydraulic subsystem control law

To control the hydraulic subsystem a fixed control law (based on the system's nominal parameters) proposed by Alleyne [10] for the hydraulic actuators force control problem is used.

Let $\tilde{x}_v = x_v - x_{vd}$ be the spool position error, where x_{vd} is the desired spool position trajectory. Consider two positive constants, ϕ_1 and ϕ_2, and the following nonnegative function:

$$V_2 = \frac{1}{2}\phi_1\tilde{P}_\Delta^2 + \frac{1}{2}\phi_2\tilde{x}_v^2 \, . \tag{19}$$

The time derivative along the hydraulic subsystem trajectories is, from (2), (3) and (4),

$$\begin{aligned} \dot{V}_2 &= \phi_1\tilde{P}_\Delta\left(-fA\dot{y} + fK_h g x_v - \dot{P}_{\Delta d}\right) + \phi_2\tilde{x}_v\left(\dot{x}_v - \dot{x}_{vd}\right) \\ &= \phi_1\tilde{P}_\Delta\left[\left(-fA\dot{y} + fK_h g x_{vd} - \dot{P}_{\Delta d}\right) + fK_h g\tilde{x}_v\right] + \phi_2\tilde{x}_v\left(\dot{x}_v - \dot{x}_{vd}\right). \end{aligned} \tag{20}$$

Using a control law x_{vd} given by

$$x_{vd} = \frac{1}{K_h g}\left[\frac{\left(\dot{P}_{\Delta d} - K_P\tilde{P}_\Delta\right)}{f} + A\dot{y}\right], \tag{21}$$

equation (20) becomes

$$
\begin{aligned}
\dot{V}_2 &= -K_P\phi_1\tilde{P}_\Delta^2 + \phi_1 f K_h g\tilde{P}_\Delta\tilde{x}_v + \phi_2\tilde{x}_v(\dot{x}_v - \dot{x}_{vd}) \\
&= -K_P\phi_1\tilde{P}_\Delta^2 + \phi_2\tilde{x}_v\left(\dot{x}_v + \frac{\phi_1}{\phi_2} f K_h g\tilde{P}_\Delta - \dot{x}_{vd}\right).
\end{aligned}
\tag{22}
$$

Substituting (3) into (22) gives

$$
\dot{V}_2 = -K_P\phi_1\tilde{P}_\Delta^2 + \phi_2\tilde{x}_v\left(-\omega_v x_v + K_{val}\omega_v u + \frac{\phi_1}{\phi_2} f K_h g\tilde{P}_\Delta - \dot{x}_{vd}\right). \tag{23}
$$

Choosing the control law as

$$
u = \frac{1}{\omega_v K_{val}}\left(\dot{x}_{vd} + \omega_v x_v - \frac{\phi_1}{\phi_2} f K_h g\tilde{P}_\Delta - K_V\tilde{x}_v\right), \tag{24}
$$

where K_V is a positive constant, equation (23) gives

$$
\dot{V}_2 = -K_P\phi_1\tilde{P}_\Delta^2 - K_v\phi_2\tilde{x}_v^2. \tag{25}
$$

Equation (25) is used in the sequel. The control laws (21) and (24) were chosen to produce a negative semi-definite function $\dot{V}_2$.

Remark 3 – To calculate $\dot{P}_{\Delta d}$ and $\dot{x}_{vd}$ from the definitions of $P_{\Delta d}$ (14) and x_{vd} (21) it would be necessary to measure the hydraulic actuator acceleration and jerk. To overcome this problem, two ways are possible: one is to calculate these signals using the position, velocity and pressure difference measures and the nominal parameters; another is to use a numeric time derivative with filters and to limit these signals with a saturation function to avoid unbounded signals. It is important to outline that in both cases the estimated signals are bounded.

5.3 – Stability analysis

The fixed cascade control system based on the fourth order hydraulic actuator model is obtained by combining the control law for the mechanical subsystem with the control law for the hydraulic subsystem. This combination is referred to as the New Fixed Cascade Controller (NFCC).

Consider the hydraulic actuator controlled by the NFCC. In this case, the closed loop system is Ω={(1)(2)(3)(14)(21)(24)}. The desired trajectory y_d and its time derivative up to third order are supposed to be bounded. Let $\rho = [\tilde{y} \;\; \dot{\tilde{y}} \;\; \tilde{P}_\Delta \;\; \tilde{x}_v]^T$ be the tracking error vector. In order to prove exponential stability the following convergence lemma will be used.

Lemma 1 – If a real function W(t) satisfies the inequality $\dot{W}(t) + \alpha W(t) \le 0$ where α is a real number, then $W(t) \le W(0)e^{-\alpha t}$ [11].

The following theorem states the closed loop characteristics:

<u>Theorem 1</u> – The Ω system is exponentially stable with respect to the origin of the tracking error vector ρ.

Proof: Consider the following Lyapunov candidate function

$$V = V_1 + V_2 + \frac{1}{2}N\tilde{y}^2, \tag{26}$$

where N is a positive constant to be defined in the sequel. Substituting equations (15)(17)(19) into (26) gives

$$V = \frac{1}{2}\rho^T H_1 \rho, \tag{27}$$

where $H_1 = \begin{bmatrix} \lambda^2 M + N & \lambda M & 0 & 0 \\ \lambda M & M & 0 & 0 \\ 0 & 0 & \phi_1 & 0 \\ 0 & 0 & 0 & \phi_2 \end{bmatrix}$ is a positive definite matrix. The time derivative of V is given by $\dot{V} = \dot{V}_1 + \dot{V}_2 + N\tilde{y}\dot{\tilde{y}}$. Substituting (18) and (25) gives

$$\begin{aligned} \dot{V} &= -(B + K_D)z^2 + N\tilde{y}\dot{\tilde{y}} + Az\tilde{P}_\Delta - \phi_1 K_P \tilde{P}_\Delta^2 - \phi_2 K_v \tilde{x}_v^2 \\ &= -(B + K_D)(\dot{\tilde{y}}^2 + \lambda^2 \tilde{y}^2) + [N - 2\lambda(B + K_D)]\tilde{y}\dot{\tilde{y}} + A\dot{\tilde{y}}\tilde{P}_\Delta + \lambda A\tilde{y}\tilde{P}_\Delta - \phi_1 K_P \tilde{P}_\Delta^2 - \phi_2 K_v \tilde{x}_v^2 \end{aligned}$$

Defining $[N = 2\lambda(B + K_D)]$ the above equation can be written as

$$\dot{V} = -\rho^T H_2 \rho, \tag{28}$$

where $H_2 = \begin{bmatrix} \lambda^2(B + K_D) & 0 & -\frac{1}{2}\lambda A & 0 \\ 0 & (B + K_D) & -\frac{1}{2}A & 0 \\ -\frac{1}{2}\lambda A & -\frac{1}{2}A & \phi_1 K_P & 0 \\ 0 & 0 & 0 & \phi_2 K_v \end{bmatrix}$.

Using the Sylvester's theorem, which states that all principal minors must be strictly positive for a symmetric matrix to be positive definite, shows that if

$$\phi_1 K_D K_P > \frac{1}{2}A^2 \tag{29}$$

then H_2 is a positive definite matrix and, consequently, $\dot{V}$ is negative definite.

Let $\eta_{min}(X)$ and $\eta_{max}(X)$ be the minimum and maximum, respectively, eigenvalues of a matrix X. From matrix theory one can write $\rho^T H_2 \rho \geq \rho^T \eta_{min}(H_2) I \rho$, where I is the identity matrix. Multiplying and dividing the right side of the above equation by $\eta_{max}(H_1)$ results

$$\rho^T H_2 \rho \geq \frac{\eta_{min}(H_2)}{\eta_{max}(H_1)} \rho^T \eta_{max}(H_1) I \rho \geq \gamma V(t) \quad \Rightarrow \quad -\rho^T H_2 \rho = \dot{V}(t) \leq -\gamma V(t),$$

where $\gamma = \dfrac{\eta_{min}(H_2)}{\eta_{max}(H_1)}$. Using the Lemma 1 one can conclude that $V(t) \leq V(0) e^{-\gamma t}$. From this equation one can write $\eta_{min}(H_1)\|\rho(t)\|^2 \leq \eta_{max}(H_1)\|\rho(0)\|^2 e^{-\gamma t}$ which implies that

$$\|\rho(t\| \leq \sqrt{\frac{\eta_{max}(H_1)}{\eta_{min}(H_1)}} \|\rho(0\| e^{-\frac{\gamma}{2}t}. \tag{30}$$

Equation (30) completes the proof and states that the trajectory tracking error vector ρ converges to the origin exponentially with a rate of at least $\gamma / 2$.

Remark 4 – As stated by Vidyasagar and Vannelli [12], the exponential convergence of the closed loop Ω implies that Ω is robust with respect to a limited disturbance such as the load F_L.

6 – EXPERIMENTAL RESULTS

6.1 – Experimental setup[1]

The experimental setup is shown in figure 3. It consists of a double acting cylinder, a proportional valve NG6 BOSCH and its electronic card, position and pressure transducers, temperature transducers in each cylinder chamber, an acquisition and controller board DS1102 – dSPACE, and a computerised hydraulic power and conditioning unit (HPCU) which is responsible for maintaining the fluid conditions.

The description of the hardware system for control is shown in figure 4. It consists of a host P.C. with MATLAB+SIMULINK+C Code Generation software packages and a DSP card with TMS 320 C31. The algorithm of Euler with a sample time of 0.5 milliseconds was used in the experiments.

6.2 – Controller's implementation

The nominal parameters of the system are M=20.66 kg, B=316 Ns/m, A=7.6576x10^{-4} m^2, β=10x10^8 N/m^2, v=9.5583x10^{-4} m^3 and P_s=10 MPa. The cylinder stroke is 1 meter ($|y_{max}| = 0.5$ m). From the valve manufacturer's data, one has that the nominal flow rate is 35 l/min at a pressure differential of 0.8 MPa. Since in this practical application the input "u" is the electronic amplifier input, "u" can be assume values between -10V and +10V, K_{val}=1, and

[1] Laboratory of Hydraulic and Pneumatic Systems - Federal University of Santa Catarina - Brazil

consequently $K_h = 6.5218 \times 10^{-8}\ m^3 / Vs\sqrt{Pa}$. The valve bandwidth $\omega_v = 300\,rad/s$ was estimated from the response time given in the manufacturer's catalogue.

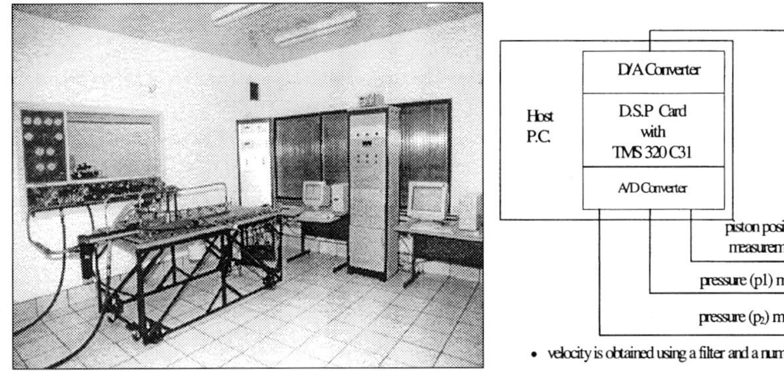

Figure 3 – Hydraulic actuator setup **Figure 4 – Hardware system for control**

As a proportional valve is used instead of a servovalve, a dead zone is expected. By adjusting the electronic card this dead-zone effect can be decreased. However, it is not sufficient for precision applications. Thus, a dead-zone inverse [13] is used as showed in the figure 5. In order to avoid chattering on the control signal and consequently vibration on the hydraulic actuator, in the practical implementation this dead zone compensation is smoothed around the origin.

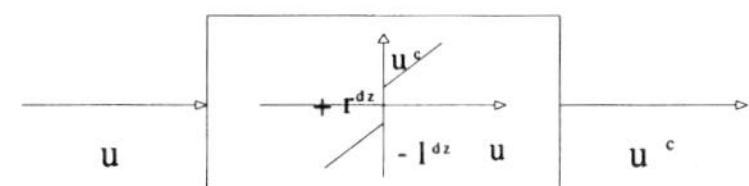

Figure 5 – Dead-zone compensation

To eliminate the dead zone compensation effect on the spool position measure the scheme showed in figure 6 is applied. In these tests: $l_{dz}=r_{dz}=0.7$.

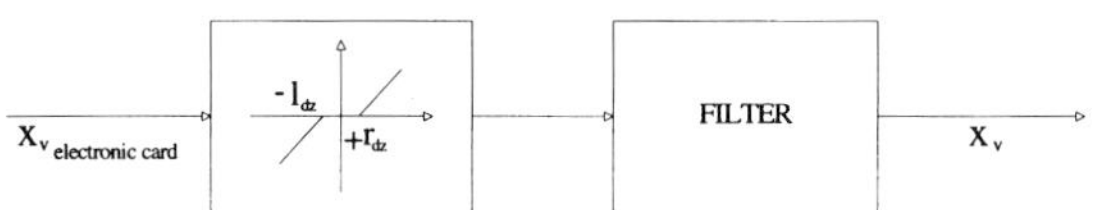

Figure 6 – Spool position measure

In order to obtain a fast response without actuator vibration, the controller gains are chosen as: $\phi_1/\phi_2=5\times10^{-12}$, $K_v=500$, $K_P=500$, $K_D=12000$ and $\lambda=30$. In this section, the New Fixed CC is compared with a fixed cascade controller (named CC) [5] that does not have the spool position in its hydraulic subsystem control law. In order to compare the performance improvement, a desired trajectory which starts with the piston at the centre of the cylinder reaching a steady-state 0.3 [m] at 1s according to a function described by the 7th order polynomial given by equation (31) is used. This position is maintained for 1s. Then the piston returns to the central position in 1s according to a polynomial similar to (31). To complete the task a symmetrical trajectory is developed. A 7th order polynomial was chosen in order to allow the choice of the position, velocity, acceleration and jerk at initial and final time.

$$y_{d1}(t) = -6t^7 + 21t^6 - 25.2t^5 + 10.5t^4 . \tag{31}$$

In figure 7 one can see the response of the CC that can be compared with the results in figure 8 for the NFCC. From this comparison one can conclude that the use of the servovalve spool position in the feedback improves the closed loop performance and in this sense the NFCC produces better results than the CC.

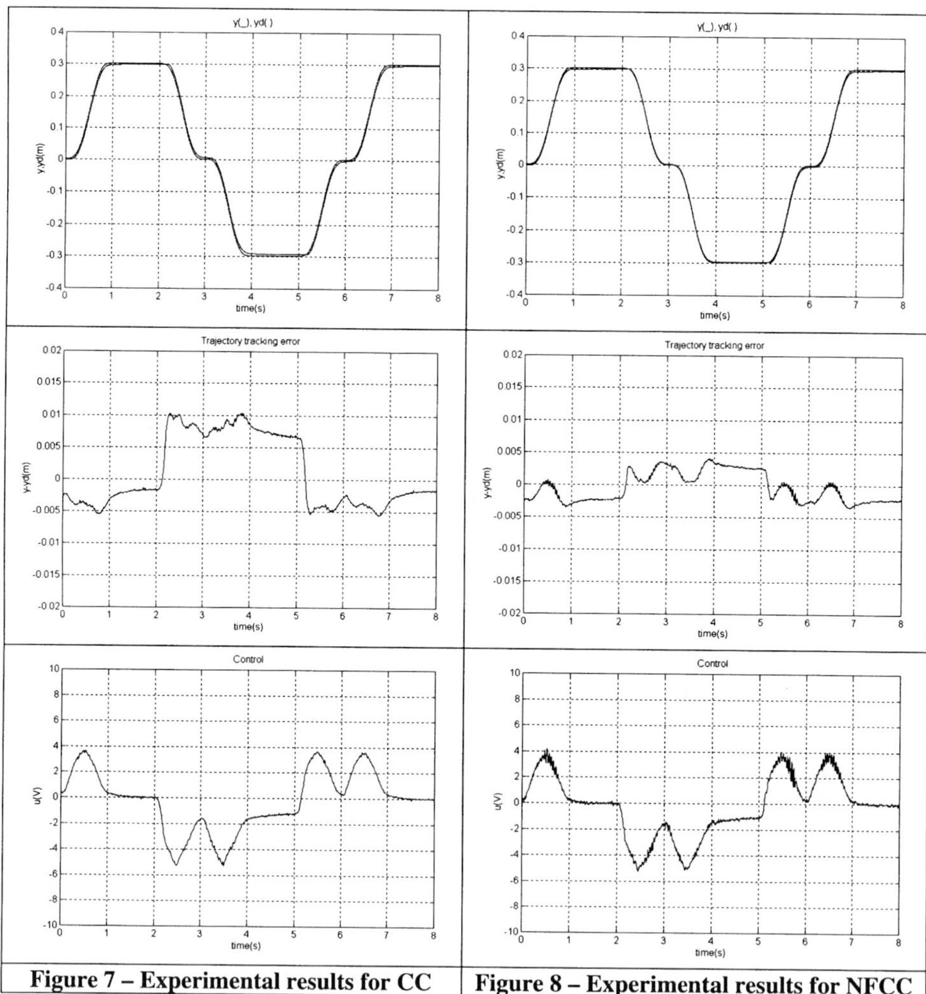

Figure 7 – Experimental results for CC	Figure 8 – Experimental results for NFCC

Now, in order to cause parametric uncertainty the nominal value of β is substituted by $\beta=15\times10^8$ N/m^2. In figures 9 and 10 the trajectory tracking errors for the CC and for the NFCC, respectively, for this value of β are presented. Comparing these two figures one can see that the NFCC also produces better performance than CC when subjected to a parametric uncertainty on β. Comparing the figures 8 and 10, one concludes that the NFCC has low sensibility to a parametric uncertainty on β.

Experimental results have demonstrated that in comparison with the VS-ACC proposed by Guenther et al. [7], the NFCC presents better performance.

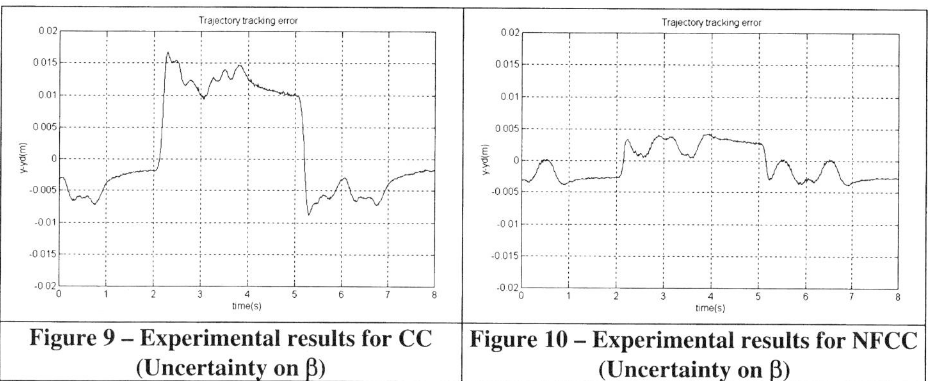

Figure 9 – Experimental results for CC (Uncertainty on β)	Figure 10 – Experimental results for NFCC (Uncertainty on β)

7 – CONCLUSIONS AND PERSPECTIVES

In this work, the limitation on the hydraulic subsystem controller's gains for a cascade controller that is designed based on a hydraulic actuator model that does not include the servovalve dynamics was demonstrated. To overcome this limitation and to improve the closed loop performance in the trajectory tracking a New Fixed Cascade Controller was proposed based on the nominal hydraulic actuator model including the servovalve dynamic as a first order system. Using the Lyapunov's direct method the NFCC was synthesised and the exponential stability was demonstrated in case where the parameters are known. The experimental implementation of the NFCC controller was performed on a test rig and this controller presented a better performance than the CC.

Future work includes the addition of an algorithm to compensate friction.

8 – REFERENCES

[1] Edge, K.A (1997) "The control of fluid power systems - responding to the challenges", Proc. Instn. Mech. Engrs., vol. 211, Part I, 91-110.

[2] Yao, B., Chiu, G.T.C., Reedy, J.T. (1997) "Nonlinear adaptive robust control of electrohydraulic servo systems", In ASME International Mechanical Engineering Congress and Exposition, FPST-vol.4, pp.191-197.

[3] Yao, B., Bu, F., Chiu, G.T.C. (1998) "Nonlinear Adaptive Robust Control of electro-hydraulic actuator servo systems with discontinuous projections", Proceedings of the 37th IEEE CDC, Florida, USA, December, pp. 2265-2270.

[4] Alleyne, A. (1998) "A systematic approach to the control of Electrohydraulic Servosystems", Proceedings of the ACC'98, June, pp. 833-837.

[5] Guenther, R., De Pieri, E.R. (1997) "Cascade position control of hydraulic actuators", Journal of the Brazilian Society of Mechanical Sciences, June, pp. 108-120.

[6] Utkin, V.I. (1987) "Discontinuous Control Systems: State of Art in Theory and Applications", Preprints IFAC 10th World Congress on Automatic Control, Munich, vol1, pp. 75-94.

[7] Guenther, R., Cunha, M.A.B., De Pieri, E.R. (1998) "Experimental implementation of the variable structure adaptive cascade control for hydraulic actuators", Power Transmission and Motion Control, Bath, September, pp. 349-361.

[8] Cunha, M.A.B., Guenther, R., De Pieri, E.R., De Negri, V.J. (2000) "The cascade controller design", Technical Report, Robotics Laboratory, Federal University of Santa Catarina.

[9] Slotine, J.J.E., Li, W. (1988) "Adaptive Manipulator Control: A Case Study", IEEE Transactions on Automatic Control.

[10] Alleyne, A. (1996) "Nonlinear force control of an electro-hydraulic actuator", Japan/USA Symposium on Flexible Automation, Vol. 1, ASME.

[11] Slotine, J.J.E., Li, W. (1991) "Applied Nonlinear Control", Prentice-Hall.

[12] Vidyasagar, M., Vannelli, A. (1982), "New relationship between input-output and Lyapunov Stability", IEEE transaction on Automatic Control, vol. AC-27, no 2, pp. 431-433.

[13] Tao, G., Kokotovic, P.V. (1996) "Adaptive Control of Systems with Actuator and Sensor Nonlinearities", John Wiley & Sons, Inc.

Fuzzy control of a hydraulic servosystem based on practical tracking algorithms

Z RIBAR, R JAVANOVIĆ, and **D SEKULIĆ**
Faculty of Mechanical Engineering, University of Belgrade, Serbia

Abstract

The purpose of this paper is to contribute to the practical applications of fuzzy logic control using practical tracking algorithms. As an object of study one hydraulic servosystem is taken.

In this paper the fuzzy algrithm is based on practical tracking algorithms. This control concept, which is based on the theory of fuzzy logic and the theory of practical tracking, is very applicable for complex and nonlinear control circuits such as hydraulic systems.

Experimental results, as well as simulation results of the nonlinear mathematical models, of the hydraulic servosystem are presented.

1. Notation

$\mathrm{B} \in R^{n\times r}$ - matrix
$\mathrm{D} \in R^{n\times r}$ - arbitrary matrix which satisfies $\det\left(\mathrm{DD}^T\right) \neq 0$
$\mathbf{d}(\cdot) : R \rightarrow R^p$ - the disturbance vector function
$\mathbf{d}(t)$ - the disturbance vector at time t
$\mathbf{d} \in R^p$ - the disturbance vector
$\mathbf{e} \in R^n$ - the output error vector, $\mathbf{e} = \mathbf{y}_d - \mathbf{y}$
e - error (input variable of FLC)
$\mathbf{e}[\cdot; \mathbf{e}_0; \mathbf{u}(\cdot), \mathbf{y}_d(\cdot), \mathbf{d}(\cdot)] : R \rightarrow R^n$ - the output error response, which at time t represents the output error vector $\mathbf{e}(t)$ at the same time $\mathbf{e}[t; \mathbf{e}_0; \mathbf{u}(\cdot), \mathbf{y}_d(\cdot), \mathbf{d}(\cdot)] = \mathbf{e}(t)$
$\mathbf{e}_{M(\cdot)} = \max\{\mathbf{e} : \mathbf{e} \in E_{(\cdot)}\}, \; {(\cdot)} = {A,I}$ - elementwise maximization
$E_A \subset R^n$ - the set of all admitted errors $\mathbf{e}(t)$ on R_τ; closed connected neighbourhood of $\mathbf{0}_e$
$E_I \subset R^n$ - the set of all admitted initial errors $\mathbf{e}(0)$; closed connected neighbourhood of $\mathbf{0}_e$
$E_{(\cdot)i} = \left\{e_i : e_i \in R, e_{im(\cdot)} \leq e_i \leq e_{iM(\cdot)}\right\}, (\cdot) = A, I$

$\mathbf{f}(\cdot) : R^m \times R^p \to R^m$ - the continuous vector function, $\mathbf{f}(\mathbf{x}, \mathbf{d}) \in C(R^m \times R^p)$, which describes internal dynamics of the plant

$\mathbf{g}(\cdot) : R^m \to R^n$ - the output function

$\mathrm{G} = \mathrm{D}^T \left(\mathrm{DD}^T\right)^{-1}$ - matrix, $\mathrm{G} \in R^{r \times n}$; for $n = r = 1 \Rightarrow \mathrm{G} = (\ g\ )$

$I_A(\cdot) : R \times R^n \times 2^{R^n} \to 2^{R^n}$ - the set function of all admitted vector functions $\mathbf{y}(\cdot)$ on R_τ with respect to $\mathbf{y}_d(\cdot)$ and E_A

$I_A(t) = I_A\left[t; \mathbf{y}_d(\cdot); E_A\right]$ - the set value of the set function $I_A(\cdot)$ at time t, with respect to $\mathbf{y}_d(\cdot)$ and E_A

$I_I(\cdot) : R^n \times 2^{R^n} \to 2^{R^n}$ - the set function of all admitted vector functions $\mathbf{y}_0$ with respect to $\mathbf{y}_{d0}$ and E_I

$I_I(\mathbf{y}_{d0}; E_I)$ - the set value of the set function $I_I(\cdot)$ at time t, with respect to $\mathbf{y}_{d0}$ and E_I; if $\mathbf{y}_{d0}$ is chosen $\Rightarrow I_I(\mathbf{y}_{d0}; E_I) = I_I$

$p(\cdot)$ - symbol for $\dfrac{d(\cdot)}{dt}$

$R_+ = [0, +\infty[= \{t : t \in R,\ 0 \leq t < +\infty\}$

$R^+ =]0, +\infty[= \{t : t \in R,\ 0 < t < +\infty\}$

$R_\tau = [0, \tau[\subseteq R_+$

$S_y \subset R^n$ - the set of all admitted $\mathbf{y}_d(\cdot)$ on R_τ; $\mathbf{y}_d(\cdot) \in S_d \Rightarrow \mathbf{y}_d(t) \in C(R_\tau, R^n)$

$S_d \subset R^p$ - the set of all admitted $\mathbf{d}(\cdot)$ on R_τ; $\mathbf{d}(\cdot) \in S_z \Rightarrow \mathbf{d}(t) \in C(R_\tau, R^p)$

$\mathrm{sign}(\cdot)$ - the signum scalar function; $\mathrm{sign}(\cdot) : R \to R$; $\mathrm{sign}\,\xi = |\xi|^{-1}\xi$ iff $\xi \neq 0$ and $\mathrm{sign}\,\xi = 0$ iff $\xi = 0$

$\mathbf{s}(\cdot)$ - the signum vector function; $\mathbf{s}[\mathbf{e}(t)] = (\mathrm{sign} e_1(t)\ \mathrm{sign} e_2(t)\ \ldots\ \mathrm{sign} e_n(t))^T$

$\mathrm{sgn}(\cdot) : R \to R$; $\mathrm{sgn}\,\xi = |\xi|^{-1}\xi$ iff $\xi \neq 0$ and $\mathrm{sgn}\,\xi \in [-1, 1]$ iff $\xi = 0$

S_e - scaling factor for error

S_{u^-} - scaling factor for output variable in previous instance

S_u - scaling factor for output variable

$S_{\Delta e}$ - scaling factor for change of error

$\mathbf{u}(\cdot) : R \times \ldots \to R^r$ - the vector function which describes evolution of the control vector

$\mathbf{u} \in R^r$ - the control vector

$\mathbf{u}(t)$ - the value of the function $\mathbf{u}(\cdot)$ at time t

$\mathbf{u}(t^-) = \mathbf{u}(t - \varepsilon)$, $\varepsilon \to 0^+$ - a control vector of the just realised control at time t

$\mathbf{x} \in R^m$ - the state vector

$\mathbf{y}\left[\cdot; \mathbf{y}_0; \mathbf{u}(\cdot), \mathbf{y}_d(\cdot), \mathbf{d}(\cdot)\right]$ - the real output response, which at time t equals the real output vector at same time, $\mathbf{y}\left[t; \mathbf{y}_0; \mathbf{u}(\cdot), \mathbf{y}_d(\cdot), \mathbf{d}(\cdot)\right] = \mathbf{y}(t)$

$\mathbf{y} \in R^n$ - the real output vector

$\mathbf{y}_d(\cdot) : R \to R^n$ - the desired output vector function

$\mathbf{y}_d(t)$ - the desired output vector at time t

$\mathbf{y}_d \in R^n$ - the desired output vector

$\beta \in R^+$

$\gamma \in [\beta, +\infty[$

δe - sum-of errors (input varible of FLC)

Δe - change of error (input variable of FLC)

$\Lambda \in R^{n \times n}$, $\Lambda = \mathrm{diag}\{e_{1MA}\ e_{2MA}\ \ldots\ e_{nMA}\}$

2. Introduction

Traditionally, controllers are designed on the basis of a mathematical description using a linearized model and in this way of controller design is sufficient for many control systems. Therefore, it is difficult to implement these model-based controllers to a real system, especially to a system which is complex and nonlinear. An alternative to model-based control is the fuzzy logic control. With the help of fuzzy logic it is possible to use the knowledge of process operators to synthesize the closed loop controllers for a given process. Classical fuzzy controllers are obtained from conventional controllers by fuzzification of all input variables of the controller. These controllers possess nonlinear characteristics, and they are suitable for control of nonlinear systems.

The main purpose of this paper is the design of a fuzzy logic controller (FLC) based on the concept practical tracking, developed by Grujić in 1984., [1]. This nonconventional control concept has further been developed by Grujić [2, 6] and Lazić [13]. FLC based on the concept practical tracking has been introduced by Jovanović [14].

3. Problem statement

The electrohydraulic servosystem, which is considered, consists of an electrohydraulic servovalve with a hydraulic cylinder. This hydraulic servosystem can be described by a mathematical model expressed by the state space and output equations:

$$\begin{aligned} \dot{\mathbf{x}}(t) &= \mathbf{f}\left[\mathbf{x}(t), \mathbf{d}(t)\right] + \mathbf{B}\mathbf{u}(t), \\ \mathbf{y}(t) &= \mathbf{g}\left[\mathbf{x}(t)\right]. \end{aligned} \tag{1}$$

It is required that real dynamic behaviour of system (1) tracks its desired dynamic behaviour with prescribed quality as long as initial real output values are allowable and real dynamic behaviour is realizable.

4. Fuzzy logic controllers (FLC)

A typical fuzzy logic controller is composed of five basic parts and its general structure is shown in Fig.1. In most cases, for simplicity, the membership functions of the input and output variables are defined on a common normalized domain. This means that the physical values of the actual inputs and outputs of the controller are mapped on the predetermined normalized domain. An input scaling factor transforms a physical signal into normalized domain of controller input, whereas the output denormalization factor transforms a defuzzified output signal from normalized domain into the physical output domain. The membership functions for the inputs and outputs of controller can

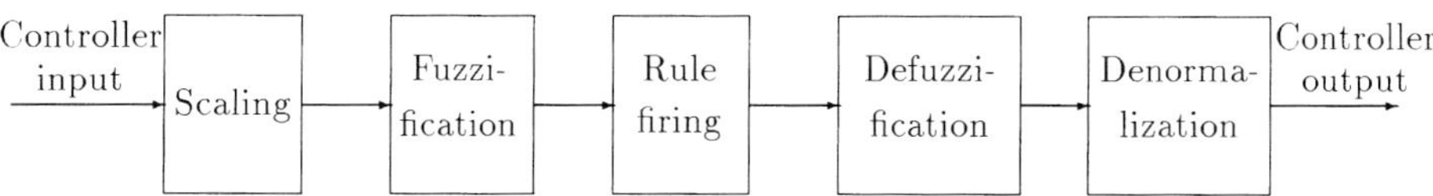

Fig. 1. Structure of FLC.

be defined on their physical domains and then the inputs and outputs of the controller are processed only using fuzzification, rule firing and defuzification. The fuzzification block transforms the continuous input signal into linguistic fuzzy variables such as *Small, Medium, Large, ...* . The fuzzy engine (or rule firing) carries out rule inference where human experience can easily be injected through linguistic rules. The defuzzification block converts denormalized fuzzy output into a continuous signal.

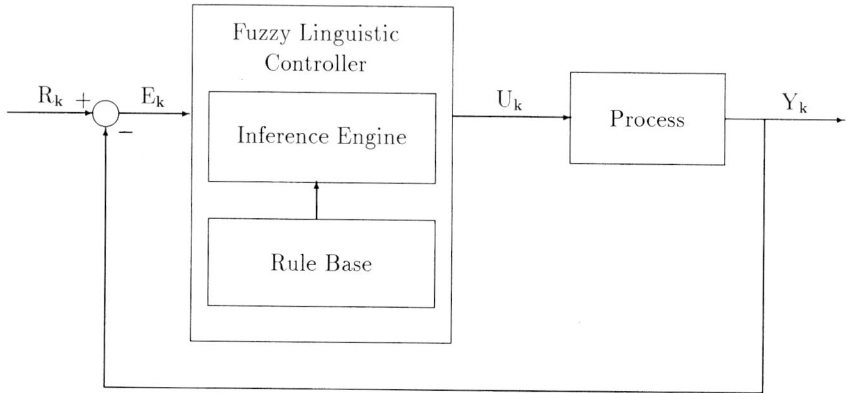

Fig. 2. FLC for feedback Error Control.

Classification of systems which use FLC as essential system components is given in [26]. The most popular type is FLC for feedback Error Control, shown on Fig.2. This is called a fuzzy linguistic control system and its basic characteristic is to generate control action based upon the error information resulting from the difference between the input signal and feedback signal. The input to this system R_k usually is a desired or reference input.

The most popular fuzzy controllers of this type are fuzzy PD, PI and PID controllers, which are the fuzzy equivalents of the appropriate conventional controllers.

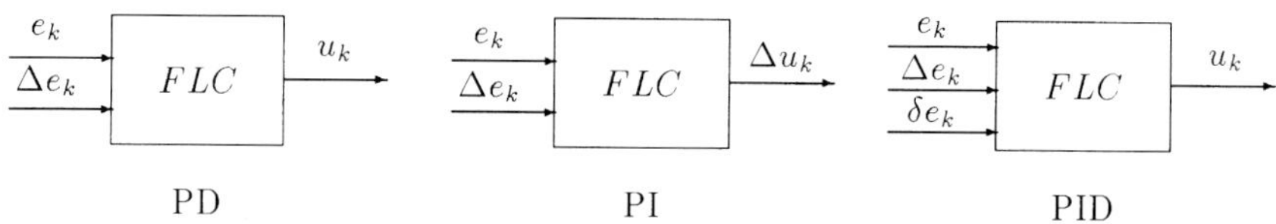

Fig. 3. Types of FLC.

Behaviour of these controllers, in discrete forms, shown in Fig.3, can be represented by functional expressions

$$u_k = FLC(e_k, \Delta e_k), \tag{2}$$

$$\Delta u_k = FLC(e_k, \Delta e_k), \tag{3}$$

$$u_k = FLC(e_k, \Delta e_k, \delta e_k). \tag{4}$$

The e_k, Δe_k and δe_k are fuzzy variables defined as follows:

$$\begin{aligned} e_k &= r_k - y_k \\ \Delta e_k &= e_k - e_{k-1} \\ \delta e_k &= \sum_{i=1}^{k-1} e_i \end{aligned} \tag{5}$$

and they represent the instantaneous values of the process error at the k-th sampling instance, its rate of change, and sum-of errors,whereas u_k is the fuzzy output variable of FLC. FLC denotes the functional characteristic of FLC. Design and development of these controllers can be found,for example, in [24].

5. Practical exponentional tracking

From the technical viewpoint, the concept of practical tracking has a great importance. Consideration of dynamic behaviour of technical plants on limited time intervals, with pre-specified quality of that behaviour, requires constraints to be placed on any technical plant. For many technical plants the most adequate tracking concept is practical tracking. This concept most completely satisfies the practical and technical requirements on dynamic behaviour and on quality of dynamic behaviour.

In this paper we proceed from practical exponential tracking algorithm [13]. A definition and algorithm are now given.

Definition

DEFINITION 1 The system (1) controlled by $\mathbf{u}(\cdot) \in S_u$ exhibits *practical exponential tracking with respect to* $\{\tau, \Lambda, \beta, I_I(\cdot), I_A(\cdot), S_d, S_y\}$ (fig. 4) if and only if $[\mathbf{y}_0, \mathbf{y}_d(\cdot), \mathbf{d}(\cdot)] \in I_I(\mathbf{y}_{d0}) \times S_d \times S_y$ implies

$$\mathbf{y}\,[t; \mathbf{y}_0; \mathbf{u}(\cdot), \mathbf{y}_d(\cdot), \mathbf{d}(\cdot)] \in I_A(t),\ \forall\, t \in R_\tau, \tag{6}$$

and $\forall\, i \in \{1, 2, \ldots, n\}$ and $\forall\, t \in R_\tau$ holds

$$y_i\,[t; y_{i0}; \mathbf{u}(\cdot), \mathbf{y}_d(\cdot), \mathbf{d}(\cdot)] \geq y_{di}(t) - \alpha_i(y_{di0} - y_{i0})e^{-\beta t},\ \ y_{i0} \leq y_{di0}, \tag{7}$$

$$y_i\,[t; y_{i0}; \mathbf{u}(\cdot), \mathbf{y}_d(\cdot), \mathbf{d}(\cdot)] \leq y_{di}(t) - \alpha_i(y_{di0} - y_{i0})e^{-\beta t},\ \ y_{i0} \geq y_{di0}. \blacksquare \tag{8}$$

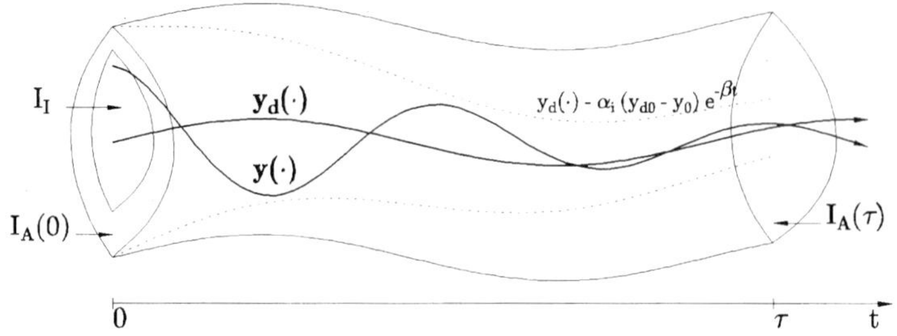

Fig. 4. Practical exponential tracking.

Algorithm

The algorithm is based on the natural tracking control concept, Grujić [5, 6, 7].

The main characteristic of this principle is the existence of the local positive feedback in the control **u** which compensates influences of the disturbances and internal dynamics of the controlled object. The information about disturbances and internal dynamics of the controlled object are not used when the control signal is calculated.

The character of the change output error value $e(t)$ can be defined exactly by means of this algorithm, as well as the time instant when output error value reaches zero.

ASSUMPTION 1 All elements of vectors $\mathbf{y}(t)$ and $\dot{\mathbf{y}}(t)$ are measurable in any time instant $t \in R_\tau$. ∎

THEOREM 1 Let Assumption 1 hold and let $S_u = \{\mathbf{u}(\cdot)\}$ with control $\mathbf{u}(\cdot)$:

$$\mathbf{u}(t) = \mathbf{u}(t^-) + \mathrm{D}^T(\mathrm{DD}^T)^{-1}\left[\dot{\mathbf{e}}(t) + \gamma(e(t); \boldsymbol{\tau}_r)\mathbf{s}[\mathbf{e}(t)]\right], \quad \forall\, [t, \mathbf{e}_0, \mathbf{y}_d(\cdot), \mathbf{d}(\cdot)] \in R_\tau \times E_I \times S_y \times S_d, \tag{9}$$

where $\mathrm{D} \in R^{n \times r}$ is an arbitrary matrix satisfied $\det(\mathrm{DD}^T) \neq 0$, and $\gamma \in [\beta, +\infty[$.

The system (1) controlled by $\mathbf{u}(\cdot)$, (9), exhibits practical exponential tracking with respect to $\{\tau, I, \beta, I_I(\cdot), I_A(\cdot), S_y, S_d\}$. ∎

PROOF 1 If there is not delay in the local positive feedback loop than $\mathrm{u(t)} = \mathrm{u(t^-)}$ and following vector equation [13] one gets equation (9) one obtains:

$$\mathrm{D}^T(\mathrm{DD}^T)^{-1}\left[\dot{\mathbf{e}}(t) + \gamma e(t)\right] = \mathbf{0}, \quad \forall\, [t, \mathbf{e}_0, \mathbf{y}_d(\cdot), \mathbf{d}(\cdot)] \in R_\tau \times E_I \times S_y \times S_d. \tag{10}$$

Multiplying equation (10) by matrix D from the left side gives:

$$\dot{\mathbf{e}}(t) = -\gamma e(t), \quad \forall\ [t, \mathbf{e}_0, \mathbf{y}_d(\cdot), \mathbf{d}(\cdot)] \in R_\tau \times E_I \times S_y \times S_d, \tag{11}$$

which is based on (3) and the Theorem from [13].This proves the theorem. ∎

6. Fuzzy algorithm based on practical exponential tracking

Condition (9) of Theorem 1 for the considered electrohydraulic servosystem can be represented as follows:

$$\begin{aligned} u(t) &= u^-(t) + g\dot{e}(t) + \lambda e(t) \\ \lambda &= g\gamma \end{aligned} \tag{12}$$

On the base of (12) the system's symbolic block diagram with inserted FLC is shown in Fig.5, [14].

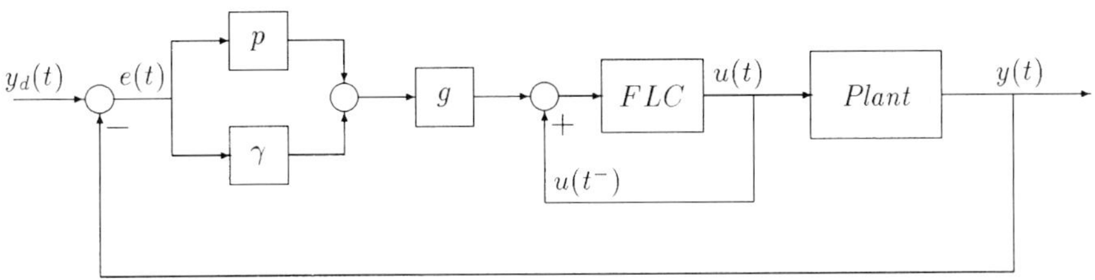

Fig. 5. Symbolic block diagram of the system.

The relationship between input variables, error, change of error and output control action in the previous instance, and output control action can be expressed approximately as follows:

$$\frac{u(t)}{S_u} = \frac{u^-(t)}{S_{u^-}} + \frac{e(t)}{S_e} + \frac{\Delta e(t)}{S_{\Delta e}} \tag{13}$$

or

$$\begin{aligned} u(t) &= \frac{S_u}{S_{u^-}} u^-(t) + \frac{S_u}{S_e} e(t) + \frac{S_u}{S_{\Delta e}} \Delta e(t) \\ &= \frac{S_u}{S_{u^-}} u^-(t) + \frac{S_u}{S_e} e(t) + \frac{S_u}{S_{\Delta e}} T\dot{e}(t) \\ &= K_u u(t) + K_\lambda e(t) + K_g \dot{e}(t). \end{aligned} \tag{14}$$

In discrete form, behaviour of the controller, shown in Fig.6, can be represented by the functional expression

$$u_k = FLC(e_k, \Delta e_k, u_{k-1}) \tag{15}$$

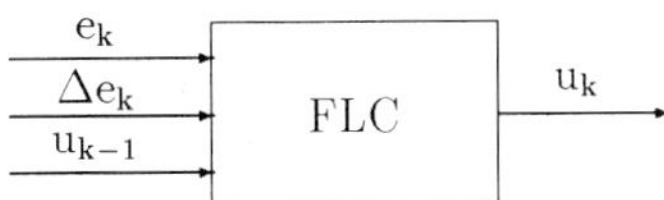

Fig. 6. FLC based on practical exponential tracking.

The e_k and Δe_k are fuzzy variables defined in (5), u_{k-1} represent the fuzzy output variable of FLC in the previous, (k-1) sampling instance. The control law, given by

$FLC(\cdot)$, where FLC denotes the functional characteristic of FLC, can be rewritten as

$$u_k = K_u u_{k-1} + K_\lambda e_k + K_g \frac{\Delta e_k}{T} \tag{16}$$

where T is the sampling time, and $S_e, S_{\Delta e}, S_{u^-}$ and S_u are the scaling factors of error, change of error, output variable in previous instance and output variable, respectively:

$$K_u = \frac{S_u}{S_{u^-}},\ K_\lambda = \frac{S_u}{S_e},\ K_g = \frac{S_u}{S_{\Delta e}} T \tag{17}$$

7. System description

The hydraulic servosystem is shown in Fig.7.

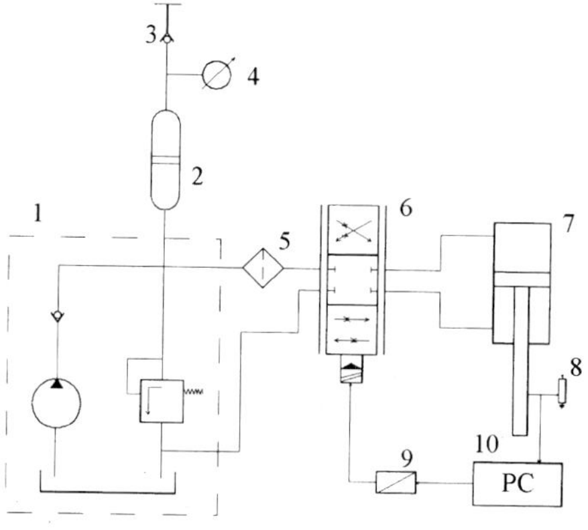

Fig. 7. Electrohydraulic servosystem.

The electrohydraulic servovalve is two-stage with force feedback.

Hydraulic cylinder characteristics are:

- piston diameter, 50 mm
- rod diameter, 32 mm
- piston pitch, 200 mm.

The basic parts of the electrohydraulic servosystem shown in Fig.7 are: 1. hydraulic power supply; 2. accumulator 747-0200; 3. charge valve JUR-743-0000; 4. pressure gauge device 1070; 5. filter 748-8300; 6. electrohydraulic servovalve D.210.040.0500.K2; 7. hydraulic cylinder 12-9830; 8. potentiometer; 9. personal computer - PC; 10. U/I converter.

All enumerated components except the potentiometer, personal computer and U/I converter have been made by "PRVA PETOLETKA", Trstenik, Yugoslavia.

The real output of the electrohydraulic servosystem is measured and is compared to the desired output value. Based on the output error value calculated in the PC, in accordance with the fuzzificated practical tracking control algorithm, the control signal is computed and forwarded to the U/I converter.

The nonlinear mathematical model, in absolute values, of the hydraulic servosystem, which include influences of Coulomb friction and linear viscous friction, in the form of the state and output equations are given in Appendix 1. The mathematical model is derived in [22], with all assumptions.

8. Simulation and experimental results

On the basis of defined algorithm in the fuzzification step we employ three inputs: the error signal, the change of the error signal and control output in previous instance, and one control output. A set of fuzzy membership functions for e, Δe, u^- and u defined on normalized domain, are given in Fig.8 and Fig.9 and they are symmetric from -1 to 1. Here, symbols $NS, Z, PS, \ldots$ have common meanings "negative small", "zero", "positive small", respectively. Membership functions for control action u are chosen with higher resolution then other variables because the controller must be able to make aggressive and less aggressive control moves based on process nonlinearity. Based on these membership functions, the fuzzy control rules are given in Fig.10. In the defuzzification procedure we use the Center-of-Sums method. Digital computer simulation and experiment are carried out with parameters:

- Fuzzy algorithm: $S_e = 0.003, \quad S_{\Delta e} = 0.002, \quad S_{u^-} = 5, \quad S_u = 6$;
- Exponential tracking algorithm: $g = 8, \quad \gamma = 1$.

Simulation and experimental results are shown in Figs. 11, 12 and 13.

As can be seen from Fig.13 the reachability time for fuzzy control is smaller then for exponential tracking. The character of change output error value $e(t)$ can be exactly defined by means of exponential tracking algorithm contrary to fuzzy control algorithm, shown here.

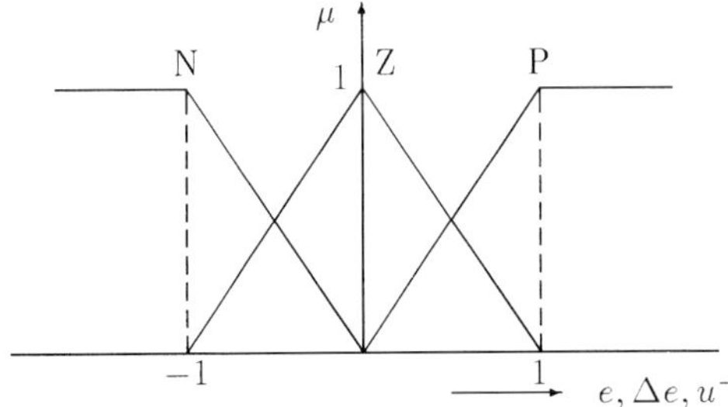

Fig. 8. Membership functions for input variables.

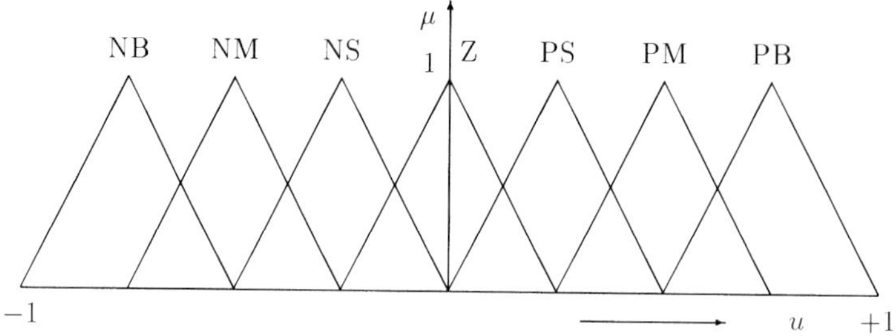

Fig. 9. Membership functions for output variable.

$u^{-}=N$		$\dot{e}$ N	$\dot{e}$ Z	$\dot{e}$ P
e	N	NB	NM	NS
e	Z	NM	NS	Z
e	P	NS	Z	PS

$u^{-}=Z$		$\dot{e}$ N	$\dot{e}$ Z	$\dot{e}$ P
e	N	NM	NS	Z
e	Z	NS	Z	PS
e	P	Z	PS	PM

$u^{-}=P$		$\dot{e}$ N	$\dot{e}$ Z	$\dot{e}$ P
e	N	NS	Z	PS
e	Z	Z	PS	PM
e	P	PS	PM	PB

Fig. 10. Fuzzy rules.

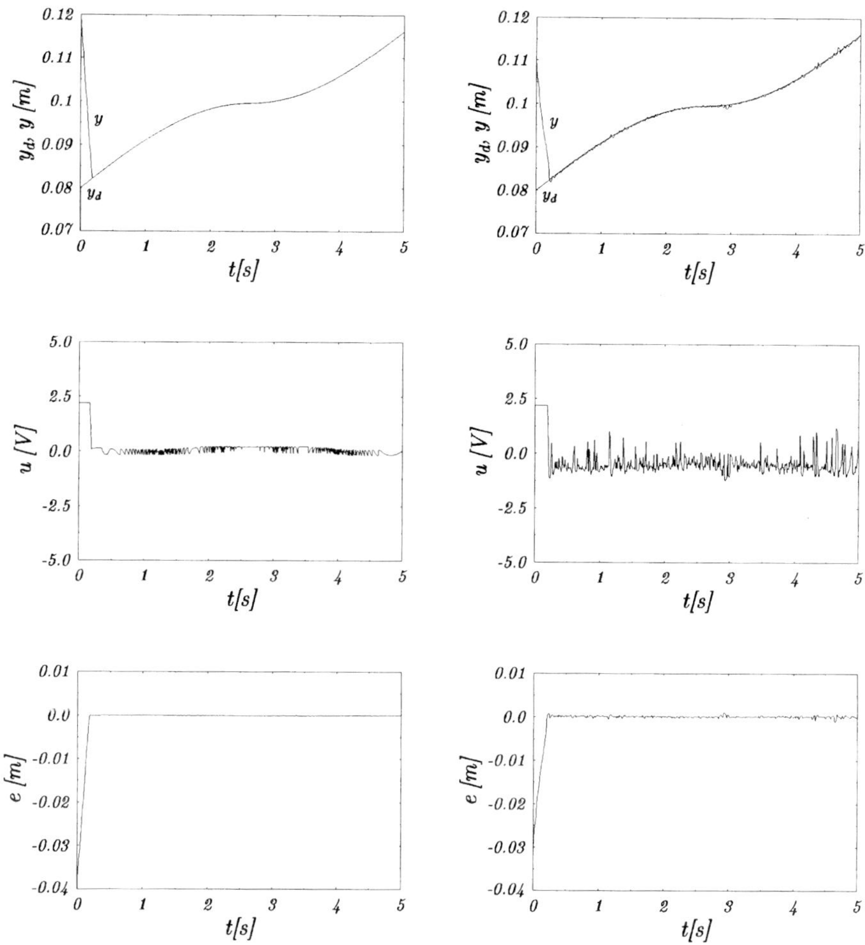

Fig. 11. Fuzzy control based on exponentional tracking - simulation (left column) and experimental (right column) results.

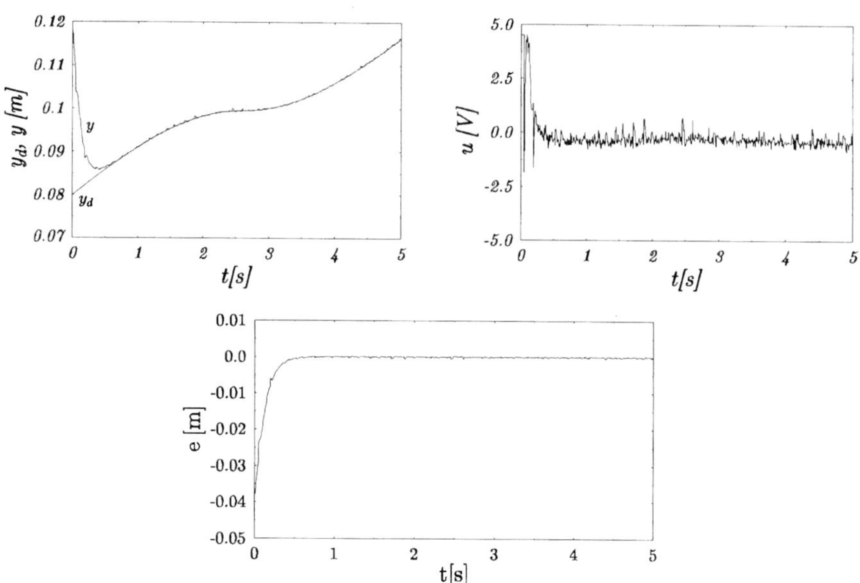

Fig. 12. Practical exponential tracking - experimental results.

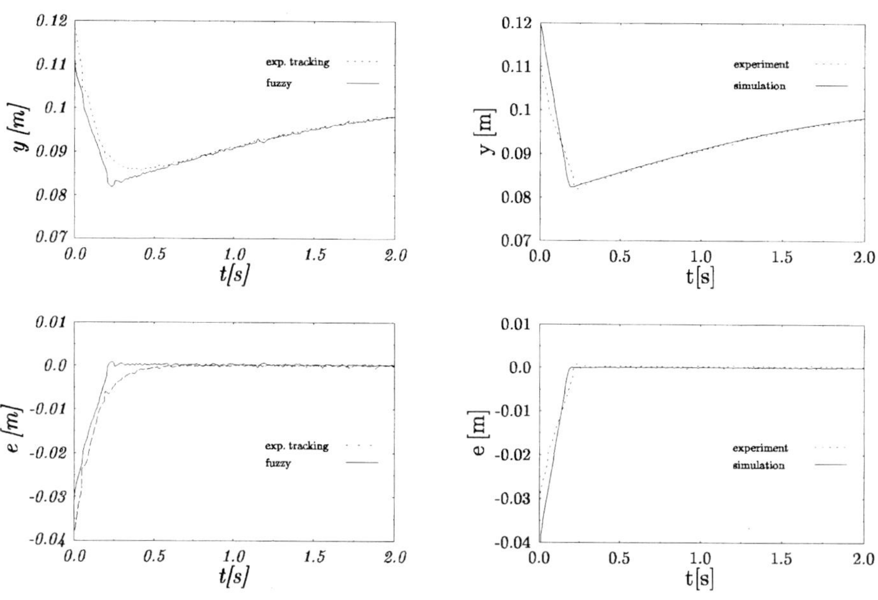

Fig. 13. Comparation of the fuzzy control and exponential tracking experimental results (left column) and fuzzy control simulation and experimental results (right column).

9. Conclusions

In this paper we proposed a fuzzy controller which was based on the exponential tracking algorithm.

The experimental and simulation results show the high quality of fuzzy control. For control design, the internal dynamics of the controlled object need not be known and measurement of real output values only are required.

The resulting controller has the same linear structure as a controller based on the exponential tracking algorithm, but with nonconstant parameters g and γ. These parameters can be adjusted by means of scaling factors change.

10. Appendix

Nonlinear mathematical model of the hydraulic position servosystem is given as,

$$\begin{aligned}
\dot{x}_1 &= x_2,\\
\dot{x}_2 &= \frac{1}{M_t}\Big\{A_1 x_3 - A_2 x_4 - \delta_0 x_2 - \delta_1 \,\mathrm{sgn} x_2 - F_L\Big\},\\
\dot{x}_3 &= \frac{\beta_e}{V_{10} + A_1 x_1}\Big\{C_d x_5 \sqrt{\frac{1}{\rho}\Big((P_S - x_3)(1 + \mathrm{sign} x_5) + x_3(1 - \mathrm{sign} x_5)\Big)} - A_1 x_2 - C_{im}(x_3 - x_4)\Big\},\\
\dot{x}_4 &= \frac{\beta_e}{V_{20} - A_2 x_1}\Big\{A_2 x_2 + C_{im}(x_3 - x_4) - C_d x_5 \sqrt{\frac{1}{\rho}\Big(x_4(1 + \mathrm{sign} x_5) + (P_S - x_4)(1 - \mathrm{sign} x_5)\Big)}\Big\},\\
\dot{x}_5 &= \frac{1}{T_r}\Big\{-x_5 + \frac{K_r}{K_q} u\Big\},\\
y &= x_1,
\end{aligned} \tag{18}$$

where,

$x_1 = X_p$ - displacement of piston $[m]$,
$x_2 = \dot{X}_p$ - velocity of piston $[\frac{m}{s}]$,
$x_3 = P_1$ - pressure in the cylinder $[Pa]$,
$x_4 = P_2$ - pressure in the cylinder $[Pa]$,
$x_5 = X_v$ - valve displacement $[m]$,
are elements of the state vector for hydraulic position servosystem and
$M_t = 2.6\,kg$ - total mass of piston,
$A_1 = 1.96 \cdot 10^{-3}\,m^2$ - large area of piston, $A_2 = 1.16 \cdot 10^{-3}\,m^2$ - small area of piston,
$\delta_0 = 770\,\frac{Ns}{m}$ - viscous damping coefficient, $\delta_1 = 16\,N$ - internal friction coefficient,
$V_{10} = V_{20} = 1.75 \cdot 10^{-4} m^3$ - initial volume of both chamber,
$\beta_e = 1.4 \cdot 10^9\,Pa$ - oil bulk modulus, $C_d = 0.61$ - dimensionless damping coefficient,
$C_{im} = 1.69 \cdot 10^{-11}\,\frac{m^3}{Pa \cdot s}$ - internal or cross-port leakage coefficient,
$\rho = 850\,\frac{kg}{m^3}$ - oil density, $P_S = 10^7\,Pa$ - supply pressure,
$T_r = 0.01\,s$ - valve time constant, $K_r = 0.001588\,\frac{m^3}{sV}$ - valve gain, $K_q = 1.44\,\frac{m^2}{s}$ - valve flow gain.

REFERENCES

[1] Ljubomir T. Grujić, **"Stability Versus Tracking in Automatic Control Systems", (in Serbian),** *Proceedings JUREMA 29.*, Zagreb, 1984, part 1, pp. 1-4.

[2] L. T. Grujić, **"Phenomena Concepts and Problems of Automatic Tracking: Continuous-time Stationary Non-linear Systems with Variable Inputs", (In Serbian),** *Proceedings of the First International Seminar Automaton and Robot* Belgrade, 1985, pp. 307-330.

[3] Ljubomir T. Grujić, **"On The Non-Linear Tracking Systems Theory: I-Phenomena, Concepts and Problems via Output-Space",** *Automatika 27,* Zagreb, No. 1-2, 1986, pp. 3-8.

[4] Ljubomir T. Grujić, **"On The Non-Linear Tracking Systems Theory: II-Phenomena, Concepts and Problems via State-Space",** *Automatika 27,* Zagreb, No. 1-2, 1986, pp. 9-16.

[5] Ljubomir T. Grujić, **"Tracking with Prespecified Performance Index Limits: Control Synthesis for Non-linear Objects",** *Proceedings of II International Seminar and Symposium: Automaton and Robot,* SAUM and ZZEE, Belgrade, October 27-29, 1987, pp. S-20 to S-52.

[6] Ljubomir T. Grujić, **"Tracking Versus Stability: Theory",** *Proceedings of 12 th World Congress on Scientific Computation,* IMACS, Paris, July 18-22, 1988, Vol. 1, pp. 319-327.

[7] Ljubomir T. Grujić, **"Tracking Control Obeying Prespecified Performance Index",** *Proceedings of 12th World Congress on Scientific Computation,* IMACS, Paris, July 18-22, 1988, Vol. 1, pp. 332-336.

[8] Ljubomir T. Grujić, **"On the Theory of Nonlinear Systems Tracking With Guaranteed Performance Index Bounds: Application To Robot Control",** *Proceedings of IEEE International Conference on Robotics and Automation,* Arizona, May 14-19, 1989, Vol. 3, pp. 1486-1490.

[9] Ljubomir T. Grujić and William Pratt Mounfield, **"Natural Tracking Control of Linear Systems",** *Proceedings of 13th World Congress on Computation and Applied Mathematics,* IMACS, Dublin, July 22-26, 1991, Vol. 3, pp. 1269-1270.

[10] Ljubomir T. Grujić and William Pratt Mounfield, **"Natural Tracking Control of Time Invariant Linear Systems Described by IO Differential Equations",** *Proceedings of 30th Conference on Decision and Control,* IEEE, Brighton, December 11-13, 1991, Vol. 3, pp. 2441-2446.

[11] Ljubomir T. Grujić and William Pratt Mounfield, Jr, **"Natural PD Exponential Tracking Control: State-Space Approach to Linear Systems",** *Proceedings of The Cairo Third IASTED International Conference,* Cairo, Egypt, December 26-29, 1994, pp. 432-435.

[12] Lj.T. Grujić, P. Borne, J.C. Gentina and J.P. Richard, **"Stability Domains: Time-Invariant Continuous-Time Systems",** *Ecole Centrale de Lille,* 1994.

[13] Dragan V. Lazić, **"Analysis and Synthesis of Automatic Practical Tracking Control", (in Serbian)** Ph.D. Dissertation, Belgrade, 1995.

[14] Radiša Ž. Jovanović, **"Application of the fuzzy logic in control systems", (in Serbian),** M.Sc. Thesis, Belgrade, 1999.

[15] William Pratt Mounfield and Ljubomir T. Grujić, **"High-Gain Natural Tracking Control of Linear Systems",** *Proceedings of 13th World Congress on Computation and Applied Mathematics,* IMACS, Dublin, July 22-26, 1991, Vol. 3, pp. 1271-1272.

[16] William Pratt Mounfield and Ljubomir T. Grujić, **"High-Gain Natural Tracking Control of Time-Invariant Systems Described by IO Differential Equations",** *Proceedings of 30th Conference on Decision and Control,* IEEE, Brighton, December 11-13, 1991, Vol. 3, pp. 2447-2452.

[17] W. P. Mounfield and L. T. Grujić, **"High-gain PI Natural Control for Exponential Tracking of Linear Single-Output Systems with State-Space Description",** *RAIRO-APII,* 1992, Vol. 26, pp. 125-146.

[18] William Pratt Mounfield and Ljubomir T. Grujić, **"Natural Tracking Control for Exponential Tracking: Lateral High-Gain PI Control of an Aircraft System with State-Space Description",** *Neural, Parallel & Scientific Computations,* 1993, Vol. 1, No. 3, pp. 357-370.

[19] William Pratt Mounfield and Ljubomir T. Grujić, **"High-Gain PI Control of an Aircraft Lateral Control System",** *Neural, Parallel & Scientific Computations,* 1993, Vol. 2, pp. 736-741.

[20] W.P. Mounfield Jr and L.T. Grujić, **"High-Gain PI Natural Tracking Control for Exponential Tracking of Linear MIMO Systems With State-Space Description",** *Int. J. Systems Sci.,* 1994, Vol. 25, No. 11, pp. 1793-1817.

[21] William Pratt Mounfield and Ljubomir T. Grujić, **"Natural PD Exponential Tracking Control Part II: Computational Resolutions",** *Proceedings of The Cairo Third IASTED International Conference,* Cairo, Egypt, December 26-29, 1994, pp. 436-439.

[22] Nenad A. Čebašek, **"Control of Pneumatic and Hydraulic Cylinders Using Unconventional Control Algorithms", (in Serbian),** M.Sc. Thesis, Belgrade, July, 1996.

[23] D.Driankov, H.Hellendoorn and M. Reinfrank, **"An Introduction to Fuzzy Control",** *Springer-Verlag,* Berlin, 1993.

[24] Ching-Yu Tyan, "**Intelligent control based on Fuzzy logic and constraint processing**", *Department of Electrical Engineering Duke Universtity,* Ph.D. Thesis, 1995.

[25] S. Joe Qin and Guy Borders, "**A Multiregion Fuzzy Logic Controller for Nonlinear Process Control,**" *IEEE Transaction on Fuzzy Systems,* 1994, Vol. 2, No. 1, pp. 74-81.

[26] Paul P. Wang and Ching-Yu Tyan, "**Fuzzy Dynamic System and Fuzzy Linguistic Controller Classification,**" *Automatica,* 1994, Vol. 30, No. 11, pp. 1769-1774.

[27] Young-Moon Park, Un-chul Moon and Kwang Y. Lee, "**A Self-Organizing Fuzzy Logic Controller for Dinamic Systems Using a Fuzzy Auto-Regresive Moving Average (FARMA) Model,**" *IEEE Transaction on Fuzzy Systems,* 1995, Vol. 3, No. 1, pp. 75-82.

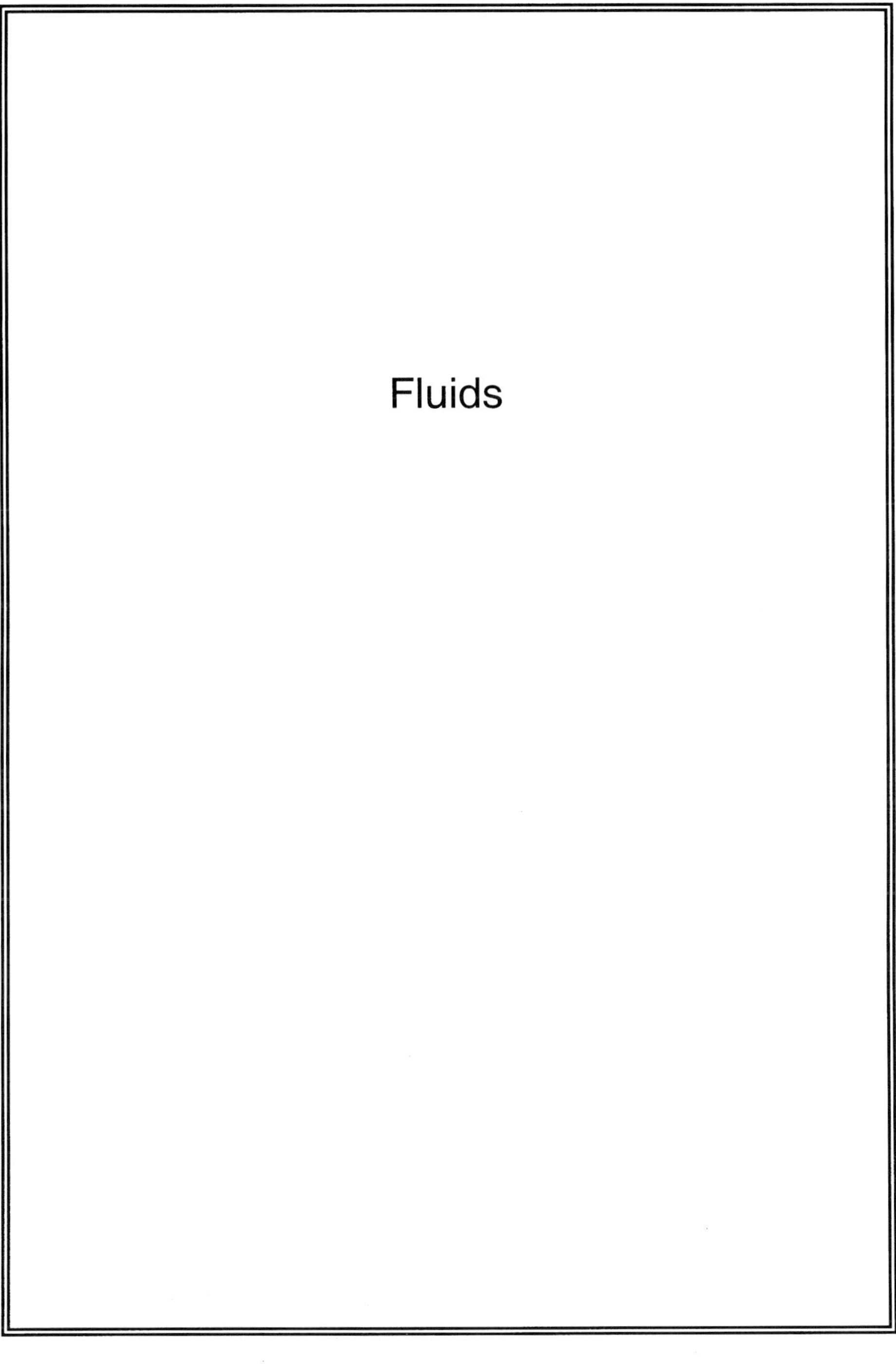

Fluids

Investigations on the ageing of ecologically acceptable fluids

M WERNER and **H MURRENHOFF**
Institute of Fluidpower Transmission and Control, RWTH Aachen, Germany

1. INTRODUCTION

Ecologically acceptable fluids based on renewable resources are an important step towards an unpolluted environment. Due to the difference in the chemical structures of these fluids compared with mineral based oils there will be changes in the properties of the fluids by use in a hydraulic system. The main effects of ageing are oxidation and hydrolyses of the fluid. This paper deals with the influence of temperature and especially catalysts on the ageing of various fluid families.
Hydraulic components consist of several metals or alloys of metals. These materials influence the ageing of fluids. Wolf (5) reports on the most frequent materials for steam turbines. He worked at the influence of these materials on the oxidation of steam turbine oils using the Baader-Test. Remmelmann (6) examined the influence of coatings to the oxidation of synthetic esters. A complete examination of the influence of pure metal elements on the ageing of ecologically acceptable fluid is still missing.
This paper deals with investigations on the oxidative as well as the hydrolytic stability of three typical base oils for hydraulic fluids. The first fluid is a native ester, rape seed oil. Next there is TMPO, a partly saturated synthetic ester, followed by a saturated synthetic ester as the third one. These fluids were tested in combination with seven typical metals used in hydraulic systems.

2. OXIDATION STABILITY

The oxidation stability was tested by a modified rotating-bomb-test. The test set up is shown in **figure 1**. 35 g of the fluid are weigh out into a glass which is placed in a metal bomb. The atmosphere inside the bomb is of pure oxygen with a start pressure of 6.25 bar at 25 °C. The closed bomb is fixed in a heating device, where it starts rotating. By heating the bomb there is an increase in pressure reaching a maximum when the whole bomb achieves its temperature. The reaction of oxygen and the fluid causes a decrease in pressure. The pressure inside the bomb is measured by a sensor which is observed by a computer. A pressure drop of 1.75 bar is defined as test end and the achieved time is a measure of the fluids lifetime.

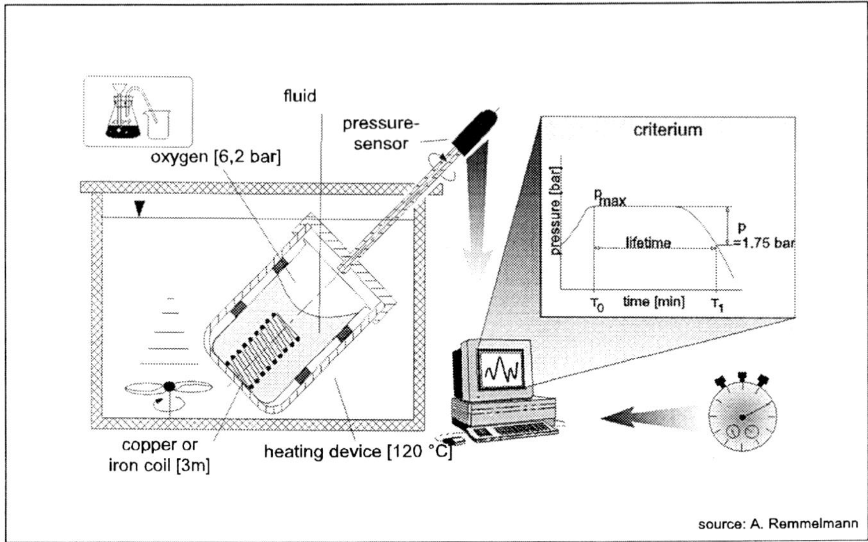

figure 1: Oxidation test

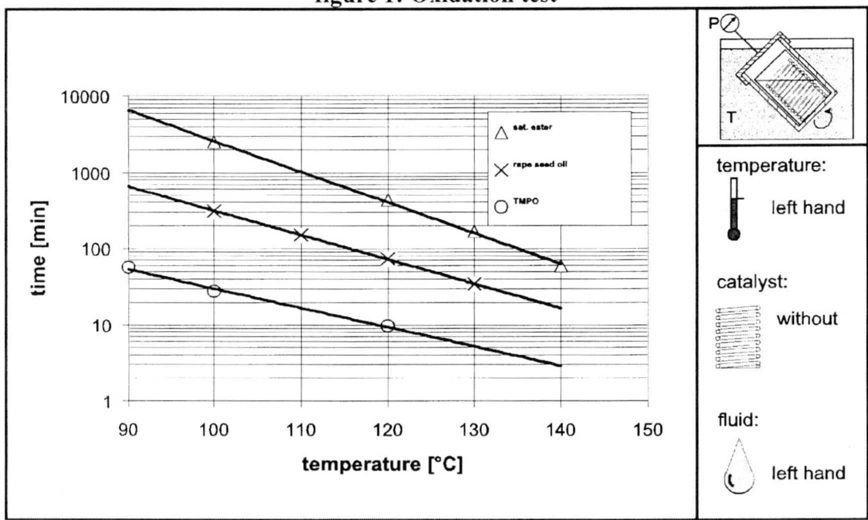

figure 2: Lifetime of different base oils in oxidation test without catalyst

First the oxidation stability of the fluids was tested at several temperatures without any catalysts. The selected temperatures ensure a lifetime of at least 10 minutes, shorter lifetimes can't be measured exactly enough. Checking out the lifetime at several temperatures increases the exactness again.

The test results of the pure fluids without any catalysts are shown in **figure 2**. The lifetime of the fluids decreases about 50 % by an increase in temperature of 10 K. Total lifetime is best for saturated ester. Native rape seed oil is second and the TMPO third. The difference in lifetime is about one decade between each fluid.

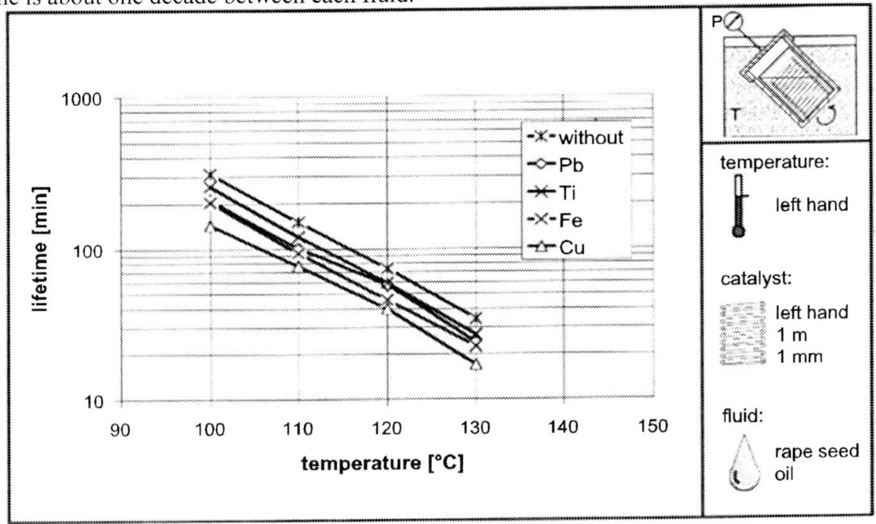

figure 3: Lifetime of rape seed oil vs. metal catalysts

Next the oxidation stability of the rape seed oil was tested in presence of metal catalysts. The catalysts are copper, iron, lead, titanium, aluminium, nickel and molybdenum. The tests were worked out at four temperatures from 100 °C up to 130 °C. For each test a new metal coil was used. The measured lifetime varies from about 15 up to 350 minutes, **figure 3**.

In order to compare the influence of different catalysts on the fluid without any influence of temperature a *relative lifetime* was defined. This means the quotient of lifetime tested with catalyst to the lifetime tested without catalyst. The average of the quotients of all four tested temperatures led to the results. This average lifetime quotient is a measure for the catalytic influence of each metal. The results for the influence of the tested metals on the oxidation of rape seed oil are shown in **figure 4**. In presence of copper there is a relative lifetime of 0.5. This means copper increases reaction speed by about 100 %, or the lifetime with presence of copper is 50 % of the lifetime without copper. Iron is the next important catalyst for oxidation of rape seed oil. It reduces the lifetime in the test by about 35 %. Lead is the third one with a relative lifetime of 0.7. Titanium, molybdenum, nickel and aluminium reduce the lifetime by less than 20 %.

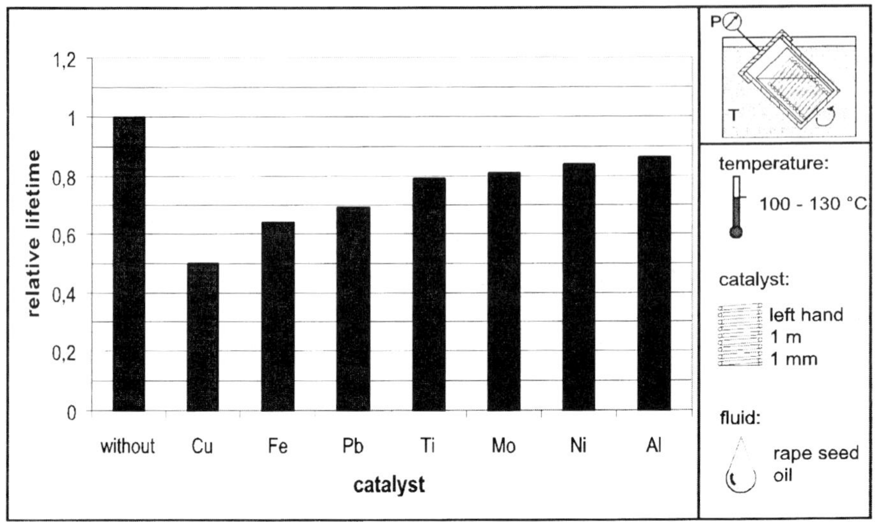

figure 4: Relative lifetime of rape seed oil vs. catalyst

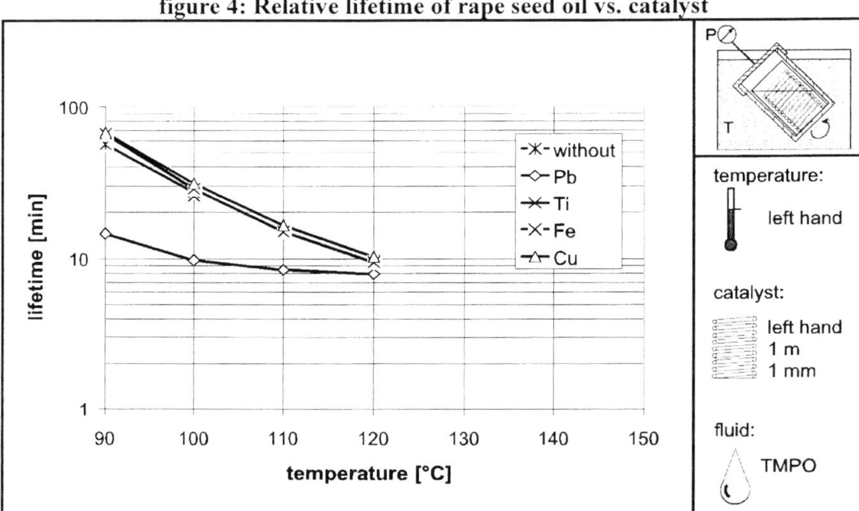

figure 5: Lifetime of TMPO vs. catalyst

TMPO is a partly saturated synthetic ester. Lifetime of the pure oil is less than lifetime of the rape seed oil. It's a chemical product and the characteristics can vary in a wide range, depending on the fluid it is made of. The absolute lifetime of TMPO in presence of the catalysts is shown in **figure 5**, the relative lifetime in **figure 6**.

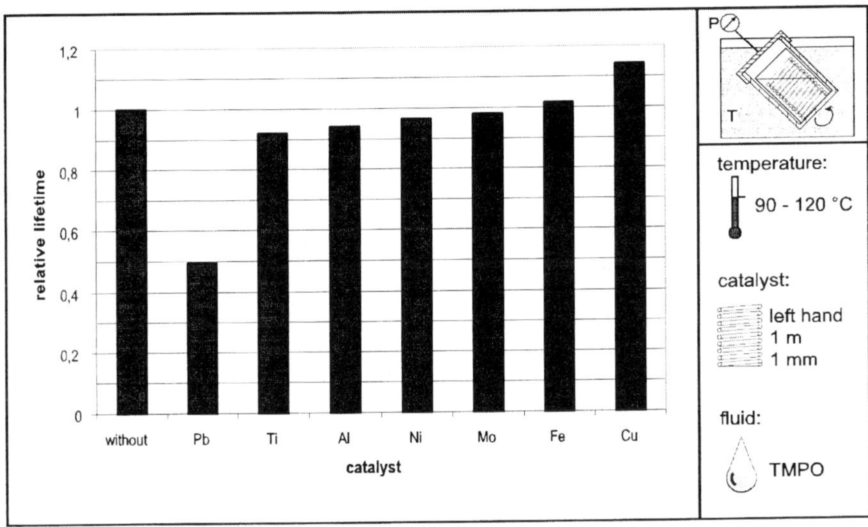

figure 6: Relative lifetime of TMPO vs. catalyst

In presence of lead the lifetime is reduced to 50%. There is no significant influence of titan, aluminium, nickel, molybdenum and iron. In presence of copper lifetime in this oxidation test is about 15 % higher than without any catalyst.

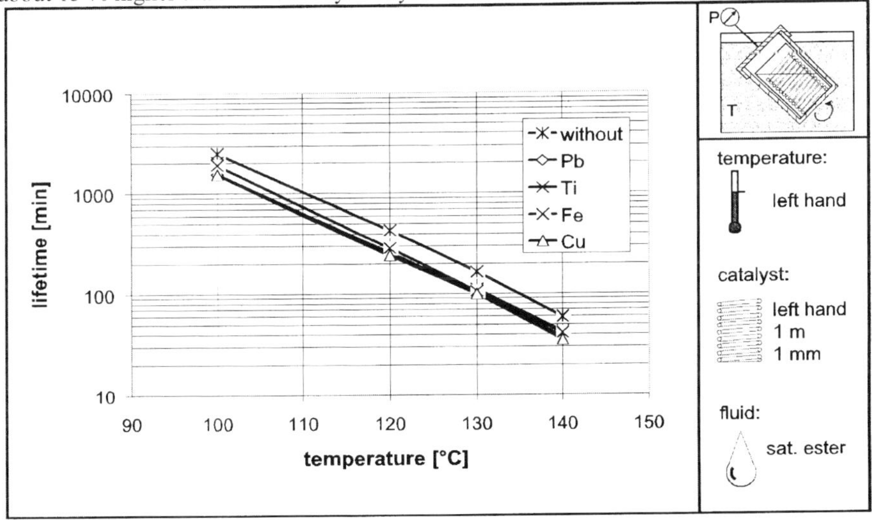

figure 7: Lifetime of a saturated ester vs. catalyst

The third fluid tested is a synthetic ester as well but it's a saturated one. Absolute lifetime of this fluid in the oxidation test in presence of the different catalysts is shown in **figure 7**. Absolute lifetime of the fluid in presence of all tested catalysts is less than without.

The relative lifetime for all seven catalysts is shown in **figure 8**.

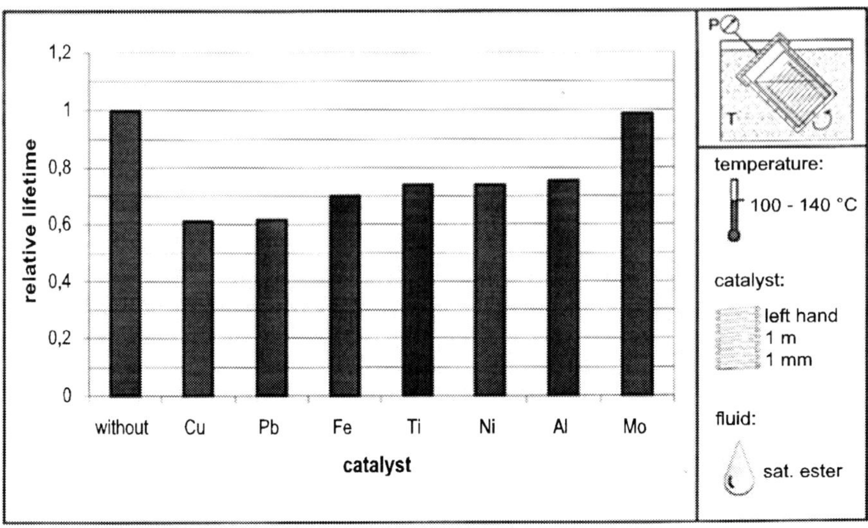

figure 8: Relative lifetime of a saturated ester vs. catalyst

Copper and Lead reduce the lifetime by about 40 %. The influence of iron, titanium, nickel and aluminium is about 25 to 30 %. Molybdenum has no influence on the lifetime at all.

To analyse the differences in catalytic influence of the seven metals on these three fluids, all the results are summarised in **figure 9**.

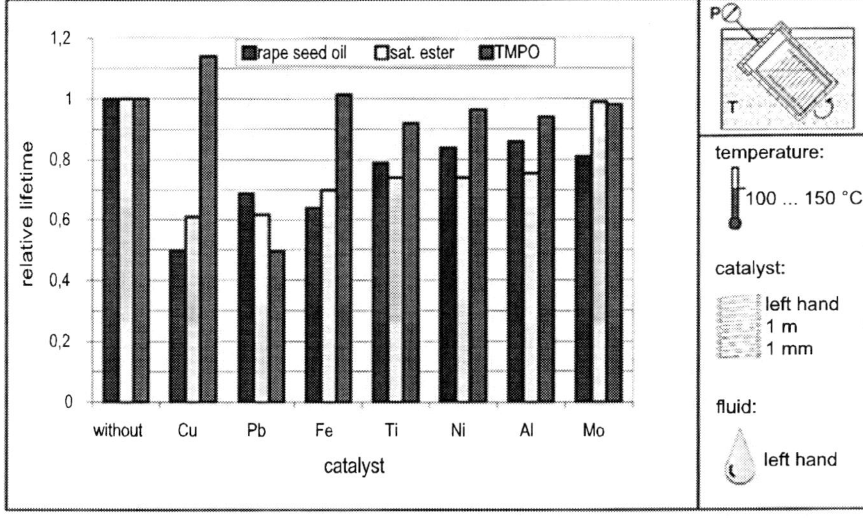

figure 9: Relative lifetime of different fluids vs. catalysts

A similar behaviour of the rape seed oil and the saturated ester can be seen. All metals reduce the lifetime in the oxidation test. This means an increase in reaction speed. The catalytic influence is strongest for copper and lead. Molybdenum has no significant influence.

The behaviour of TMPO is totally different. Copper, which is the strongest catalyst for all the other fluids, reduces the reaction speed about 15 %.

Looking at the chemical structure of the three fluids there is no way for an explanation. All fluids are esters, and the fluids with the greatest difference in structure (rape seed oil and saturated ester) show a similar behaviour while the fluids with the same acids (rape seed oil and TMPO) completely differ. To understand this behaviour it would be helpful to analyse the produced ageing products.

3. HYDROLYTIC STABILITY

Hydrolysis is the other important way of ageing an ecologically acceptable hydraulic fluid in a laboratory test. Hydrolysis means the splitting of an ester molecule into an acid and an alcoholic part. The progress of this reaction can be measured by determination of the TAN.
Experiments for evaluation of hydrolytic stability of the fluids were done using the following experimental set up. A mixture of 70 % fluid and 30 % water is filled into a round-bottomed flask. The metal wire is coiled along a magnetic mixing stick and placed into the flask as well. The flask is closed by a cork, with a long capillary for pressure exchange in it.

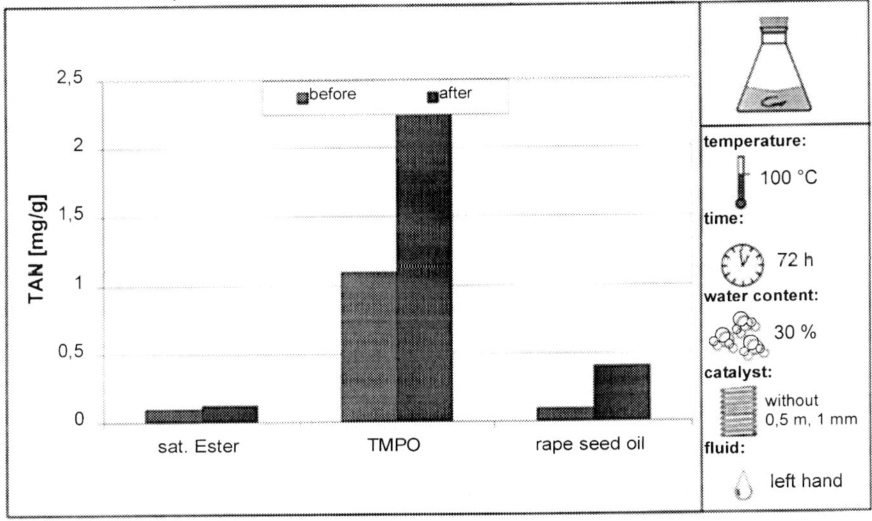

figure 10: Change of the fluids TAN during hydrolytic test

First of all hydrolytic stability of the fluid is examined without any catalyst. The experiments last 72 hours each and work at a temperature of about 100 °C. The change in TAN is shown in **figure 10**.

TAN of the saturated ester is on a very low level, before the test as well as after it. This stands for a good hydrolytic stability of this ester. TAN of the TMPO doubles from about 1 to 2 mg/g. On rape seed oil a TAN four times than before (from 0.1 to 0.4 mg/g) can be measured after the test.

To compare catalytic influence of the different metals, a factor for relative change in the TAN had to be defined. This means TAN after the test with catalyst is divided by TAN after the test without any catalyst. A higher relative change factor indicates a faster hydrolysis of the ester. A factor of 1 indicates no influence of the metal catalyst.

Figure 11 depicts the relative change factors of six tested metals for the hydrolysis of rape seed oil. In presence of copper or iron there is a lower TAN-level. Nickel is without any influence on hydrolysis. Aluminium, molybdenum and especially titanium increase the speed of hydrolysis up to 30 %.

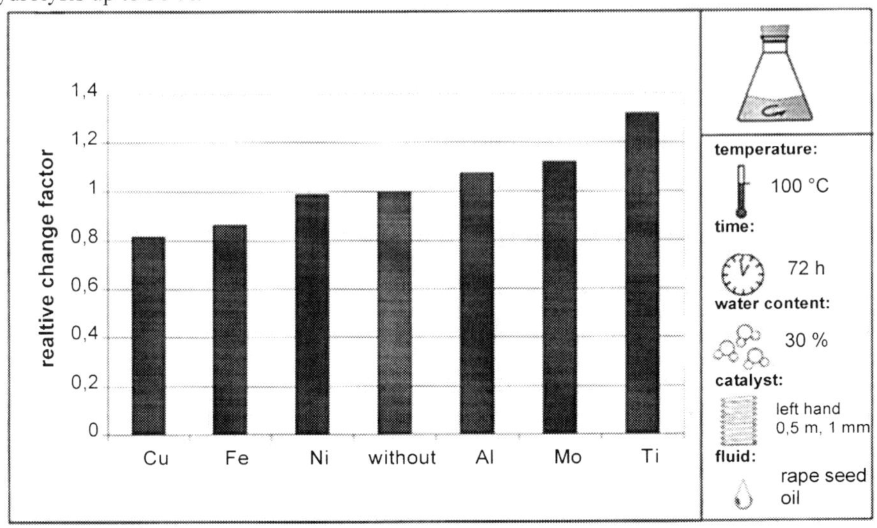

figure 11: Relative change factor of rape seed oil vs. catalysts

Analysing catalytic influence on hydrolysis of TMPO a completely changed influence of titanium can be recognised, **figure 12**. In contrast to rape seed oil speed of hydrolysis of TMPO decreased down to 60%. The influence of copper, iron, molybdenum and nickel is mostly the same compared to rape seed oil. The relative change factor of aluminium is increased from about 1.1 for rape seed oil up to 1.3 for the TMPO.

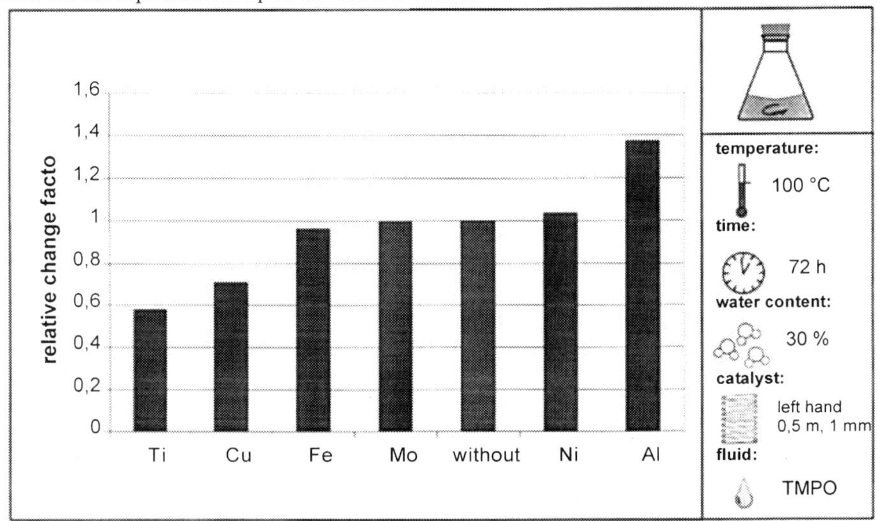

figure 12: Hydrolytic stability of TMPO vs. catalysts

Relative change factor for hydrolysis of saturated ester are presented in **figure 13**.

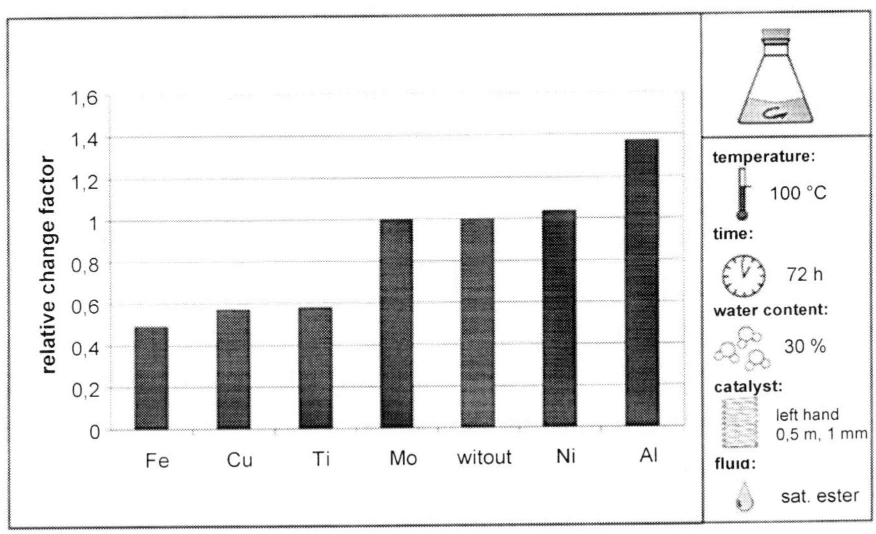

figure 13: Hydrolytic stability of saturated ester vs. catalysts

Again aluminium represents the metal with the greatest influence on speed of hydrolysis. Nickel and molybdenum don't show a significant influence. Iron, copper and titanium reduce the level of hydrolysis down to 50%.

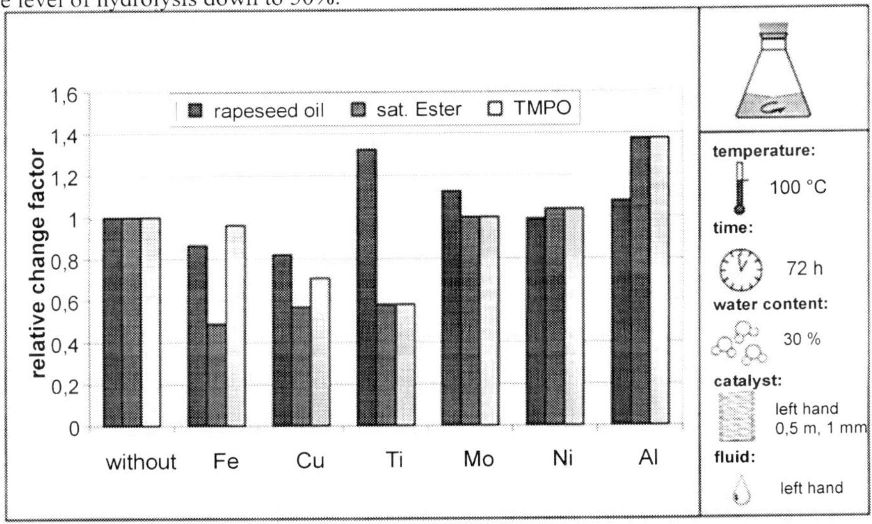

figure 14: Relative change factors of different fluids vs. catalysts

From **figure 14** the relative factor for all fluids and catalysts can be taken in a single bar chart. Aluminium forces an increase in TAN for all tested fluids. Nickel and molybdenum are of almost no effect. The influence of titanium varies by the fluid. Speed of hydrolysis is increased for rape seed oil but decreased for the saturated ester and TMPO. Iron and copper reduce the TAN on all fluids.

Analysing these hydrolysis tests it is important to recognise that there may be a reaction of the acids with metal coils. So the measured TAN is not really an exact measure for the hydrolysis of the fluid. But on the other hand a reaction of acids with metallic coils can take place for all fluids so there is no significant misinterpretation comparing all these results achieved under comparable and predefined conditions.

4. SUMMARY

Finally the conclusion can be drawn, that there is no general difference in oxidation stability between native and synthetic esters, however there is a difference between partly saturated and saturated esters. Oxidation reaction of the tested fluids evaluated by the achieved lifetime is influenced by presence of metal catalysts. But there is no simple and easy way explaining the results. Neither the chemical structure of the fluids nor the position of the metals in the period table of elements gives a good reason for explanation. To achieve a real understanding of these reactions some more chemical analysis of the ageing products needs to be done.

Concerning hydrolytic stability of esters again there are some differences between partly saturated and saturated esters. But again no general correlation between chemical structure of the fluids and the metal catalysts can be found.

Ageing of ecologically acceptable fluids is a very complex problem influenced by many parameters. Only by an interdisciplinary work of engineers and chemists it seams to be possible to understand the complete chemistry of hydraulic fluids in regard to its complex ageing behaviour. To understand the ageing process of the fluids will be the best way to develop better products in the future providing preconditions for a cleaner and less polluted environment.

5. REFERENCES

/1/ Busch, Ch.: Untersuchung und Analyse der Eigenschaften und Eigenschaftsänderungen einer rapsölbasischen Flüssigkeit in ihrer Funktion als hydraulisches Druckübertragungsmedium
Dissertation, RWTH Aachen, 1995

/2/ N.N.: VDMA-Einheitsblatt 24568: Biologisch schnell abbaubare Druckflüssigkeiten – Technische Mindestanforderungen
Beuth Verlag, March 1994

/3/ NN: Standard Test Method for Oxidation Stability of Inhibited Mineral Insulating Oil by Rotating Bomb
ASTM D 2112 – 93

/4/ Werner, M.; Das Alterungsverhalten biologisch schnell abbaubarer Hydraulikflüssigkeiten
Ölhydraulik und Pneumatik, No 2, 2000

/5/ Wolf, K.: Die Schmierung von Dampfturbinen
Springer-Verlag Berlin, Göttingen, Heidelberg 1951

/6/ Remmelmann, A.: Die Entwicklung und Untersuchung von biologisch schnell abbaubaren Druckübertragungsmedien auf Basis von synthetischen Estern
Dissertation, RWTH Aachen, 1999

Comparison of different fluid models

T H A NYKÄNEN, S ESQUÉ, and **A U ELLMAN**
Institute of Hydraulics and Automation, Tampere University of Technology, Finland

ABSTRACT

In many simulation studies fluid properties such as bulk modulus and viscosity are considered as a constant component. However bulk modulus of fluid exerts a strong influence on systems performance and it depends on air content of fluid. In more advanced fluid power simulation packages fluid is considered by the means of approximation for state equation of fluid. However, this leads to different models for pressure generation in a volume. In this paper comparison with two different fluid models and those behaviour in simple simulation tests is discussed. Temperature is assumed as a constant and its influence is neglected.

INTRODUCTION

In most simulation studies fluid properties such as bulk modulus and viscosity are considered as constant quantities. However these properties have a significant influence on system performance. Bulk modulus is a function of pressure and temperature. In this paper the influence of temperature variations is neglected. The equation of state for a liquid cannot be mathematically derived from physical principles. Therefore Taylor's series first term may be used as an approximation (1). However this leads to different models for pressure generation in a volume, depending on how density is defined. Air is always present in a fluid and it appears either as a free air or dissolved air. Because the density of air is small compared to the density of a liquid it have a significant influence on density of the air-liquid mixture. A two-phase model for an air-liquid mixture is needed when considering significant pressure changes. Also in advanced fluid power simulation packages the fluid is considered as a two-phase mixture. Dissolved air has no influence on the density of a liquid but it is potential to generate free air and must be taken into consideration in cavitation studies. This model is based on the mass continuity law. In references (3) and (4) the equations presented for bulk modulus are good static models. However usage of these equations leads complicated equations for pressure generation. A two-phase fluid model is needed when cavitation, large pressure fluctuations are under consideration.

NOMENCLATURE

B_e = effective bulk modulus
B_{eff} = effective bulk modulus of liquid-gas mixture
B_{gas} = bulk modulus of gas
B_{liq} = bulk modulus of liquid
k = polytropic gas constant
m_{gas} = mass of gas
m_{lg} = mass of liquid-gas mixture
m_{liq} = mass of the liquid
p = pressure
p_{lg} = pressure of liquid-gas mixture
p_0 = pressure at initial state
T = current temperature
T_0 = temperature at initial state
T_1 = temperature at first definition point
T_2 = temperature at second definition point
V_{gas} = gas volume
V_{gas0} = gas volume at initial state
V_{liq} = liquid volume
V_{liq0} = liquid volume at initial state
V_T = geometric volume
x = volumetric fraction of entrapped gas at (p_0, T_0)
α = coefficient of thermal expansion
ν_1 = viscosity at temperature T_1
ν_2 = viscosity at temperature T_2
ρ = density
ρ_0 = gas density at initial state
ρ_{gas0} = gas density at (p_0,T_0)
ρ_{lg} = density of liquid-gas mixture
ρ_{liq} = liquid density
ρ_{liq0} = liquid density at (p_0,T_0)

1. APPROXIMATION OF THE EQUATION OF STATE FOR A LIQUID

The density of a liquid is a function of pressure and temperature. Changes in density as a function of pressure and temperature are small. Therefore an approximation may be used (1).

$$\rho = \rho_0 + \left(\frac{\partial \rho}{\partial p}\right)_T (p - p_0) + \left(\frac{\partial \rho}{\partial T}\right)_P (T - T_0) \tag{1}$$

Where ρ_0, p_0 and T_0 are initial values for density, pressure and temperature. The normally used linearized form is:

$$\rho = \rho_0\left[1+\frac{1}{B_{liq}}(p-p_0)-\alpha(T-T_0)\right] \tag{2}$$

, where

$$B_{liq} \equiv \rho_0\left(\frac{\partial p}{\partial \rho}\right)_T \quad \text{and} \quad \alpha \equiv -\frac{1}{\rho_0}\left(\frac{\partial \rho}{\partial T}\right)_p$$

Later an approximation is used where α is neglected and density change as a function of pressure and air content is taken into consideration. Influence of the temperature is usually neglected because including heat convection leads to complicated equations.

2. FLUID MODEL

The density for a liquid is defined as mass per unit of volume,

$$\rho_{liq} = \frac{m_{liq}}{V_{liq}} \tag{3}$$

Air is always present in a liquid and when the percentage of air is small density varies only slightly. With high percentages of air, density is dependent on pressure and at low-pressure, density collapses. At high-pressure level air volumetric percentage is significantly small. Therefore air influence on density is low at high-pressure level.

The gas content of a liquid-gas mixture at initial state:

$$x = \frac{V_{gas0}}{V_{liq0}+V_{gas0}} \tag{4}$$

Let us consider a liquid volume, which has a unit volume at initial state (p_0, T_0).

Volume of the gas at initial state is:

$$V_{gas0} = x \tag{5}$$

The mass of the gas is constant, only the volume is altering as a function of pressure:

$$m_{gas} = x \cdot \rho_{gas0} \tag{6}$$

Assuming a polytropic gas process, the volume of the gas is:

$$p_0 V_{gas0}^k = p V_{gas}^k \tag{7}$$

$$V_{gas} = \left(\frac{p_0}{p}\right)^{1/k} \frac{x}{1-x} \tag{8}$$

Volume of liquid at initial state:

$$V_{liq0} = 1-x \tag{9}$$

Mass of liquid:

$$m_{liq} = \rho_{liq0} \cdot V_{liq0} \tag{10}$$

Volume of liquid:

$$V_{liq} = \frac{V_{liq0}}{1+\frac{1}{B_{liq}}(p-p_0)} = \frac{1-x}{1+\frac{1}{B_{liq}}(p-p_0)} \tag{11}$$

Density of liquid-gas mixture is:

$$\rho_{lg} = \frac{m_{gas}+m_{liq}}{V_{gas}+V_{liq}} \tag{12}$$

$$\rho_{lg} = \frac{x \cdot \rho_{gas0} + \rho_{liq0}(1-x)}{\left(\frac{p_0}{p}\right)^{1/k} x + \frac{1-x}{1+\frac{1}{B_{liq}}(p-p_0)}} \tag{13}$$

3. CONTINUITY EQUATION

The pressure generation equation for the control volume is derived using the continuity equation based on mass conservation. In the control volume is a mass, which is the difference between incoming and outgoing mass flow.

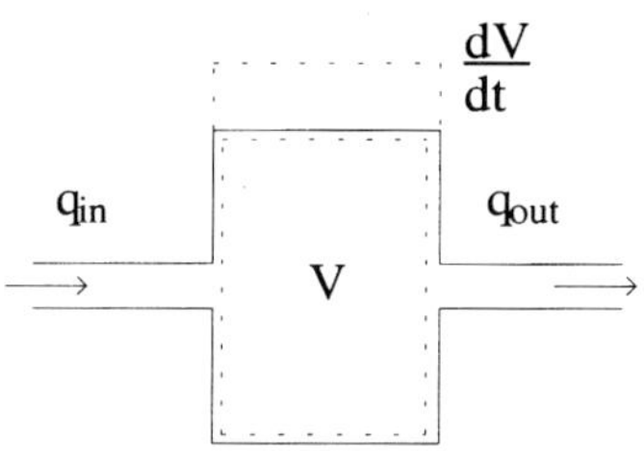

Figure 1. Control volume of mass flow.

$$\sum \frac{dm_{in}}{dt} - \sum \frac{dm_{out}}{dt} = \frac{dm}{dt} = \frac{d(\rho \cdot V)}{dt} \tag{14}$$

On the other hand mass flow may be derived by using volume flow:

$$\frac{dm}{dt} = \rho \cdot q \tag{15}$$

Because the densities of the incoming and the outgoing liquid are the same, it can be shown that:

$$\sum \rho_{in} \cdot q_{in} - \sum \rho_{out} \cdot q_{out} = \rho \cdot \sum q = \rho_0 \cdot \frac{\partial V}{\partial t} + V \cdot \frac{\partial \rho}{\partial t} \tag{16}$$

Using the definition of density according to the equation of state approximation, the continuity equation can be expressed as:

$$\rho = \rho_0 + \frac{\rho_0}{B} \cdot p \tag{17}$$

$$\left(\rho_0 + \frac{\rho_0}{B} p\right) \sum q = \rho_0 \frac{\partial V}{\partial t} + V \frac{\partial \rho}{\partial t}$$

$$\rho_0 \left[\left(1 + \frac{1}{B} p\right) \sum q - \frac{\partial V}{\partial t}\right] = V \frac{\partial \rho}{\partial p} \frac{\partial p}{\partial t}$$

$$\rho_0 \left[\left(1 + \frac{1}{B} p\right) \sum q - \frac{\partial V}{\partial t}\right] = V \rho_0 \frac{1}{B_{eff}(p)} \frac{\partial p}{\partial t}$$

$$\left(1 + \frac{1}{B} p\right) \sum q - \frac{\partial V}{\partial t} = \frac{V}{B_{eff}(p)} \frac{\partial p}{\partial t}$$

$$\frac{\partial p}{\partial t} = \frac{B_{eff}(p)}{V} \left[\left(1 + \frac{1}{B} p\right) \sum q - \frac{\partial V}{\partial t}\right] \tag{18}$$

Equation (18) can be simplified

1. Assuming $p << B$. In that case approximately 2 % error at 200 bar pressure level is introduced.
2. Assuming Beff constant. Above 100 bar pressure level the error is acceptable, but at lower pressure level this assumption leads to great errors, especially if the percentage of air in the liquid is significant.

In a cavitating flow the bulk modulus collapses totally. Using these assumptions the equation for pressure generation that is commonly used is established. However this equation does not completely satisfy the law of mass conservation.

$$\frac{\partial p}{\partial t} = \frac{B_e}{V} \left[\sum q - \frac{\partial V}{\partial t}\right] \tag{19}$$

4. COMPRESSIBILITY

Bulk modulus is defined as:

$$B = -V_0 \cdot \frac{dp}{dV} = -\frac{m_0}{\rho_0}\frac{\partial p}{\partial \rho}\frac{\partial \rho}{\partial V} = -\frac{m_0}{\rho_0}\frac{\partial p}{\partial \rho}\frac{-m_0}{V_0^2} = \rho_0 \frac{1}{\frac{\partial \rho}{\partial p}} \tag{20}$$

To evaluate the density of the liquid-gas mixture at the initial stage Eq (13) is used substituting $p = p_0$. Equation (13) for the density of the liquid-gas mixture and equation (20) yields an equation for effective bulk modulus,

$$\rho_{lg0} = x \cdot \rho_{gas0} + (1-x) \cdot \rho_{liq0} \tag{21}$$

$$B_{eff} = \rho_{lg0} \frac{1}{\frac{\partial \rho_{lg}}{\partial p}} \tag{22}$$

$$B_{eff} = -\frac{\left(\left(\frac{p_0}{p}\right)^{\frac{1}{k}} x + \frac{1-x}{1+\frac{p-p_0}{B_{liq}}}\right)^2}{-\frac{\left(\frac{p_0}{p}\right)^{\frac{1}{k}} x}{k \cdot p} - \frac{1-x}{\left(1+\frac{p-p_0}{B_{liq}}\right)^2 B_{liq}}} \tag{23}$$

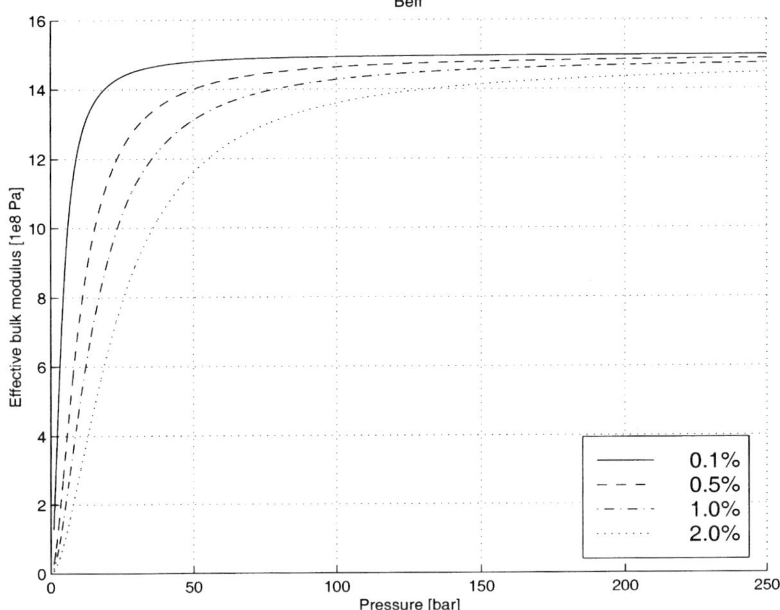

Figure 2. Effective bulk modulus as a function of pressure level.

The percentage free air content remains constant as the pressure increase. The equation (23) includes compression of the free air as the pressure rises. In practise some of the free air will dissolve, but that effect is not included. However usage of the equation (23) gives conservative results, especially when pressure decreases and the free air expands.

5. VISCOSITY

Viscosity alters considerably with temperature and has a strong effect on flow mechanics of various fluid components. Therefore viscosity changes due to the temperature are included in the fluid model. However it is assumed that a system operates at certain temperature level and temperature changes within the system are neglected. Viscosity dependence on temperature is described according to the ASTM D 341-43 standard,

$$\nu = e^{e^{y_3}} - 0.7 \quad \text{, where} \tag{24}$$

$$y_3 = a + bx_3 \qquad a = y_1 - bx_1$$

$$b = \frac{y_1 - y_2}{x_1 - x_2} \qquad x_3 = \log(T)$$

$$x_1 = \log(T_1) \qquad y_1 = \log(\log(\nu_1 + 0.7))$$

$$x_2 = \log(T_2) \qquad y_2 = \log(\log(\nu_2 + 0.7))$$

6. TEST SYSTEM AND SIMULATION RESULTS

In simulation tests two different fluid models, model 1 eq. (19) and model 2 eq. (23) are compared. In the simulation the volume of the test chamber is altered. This results pressure increase in the measuring chamber.

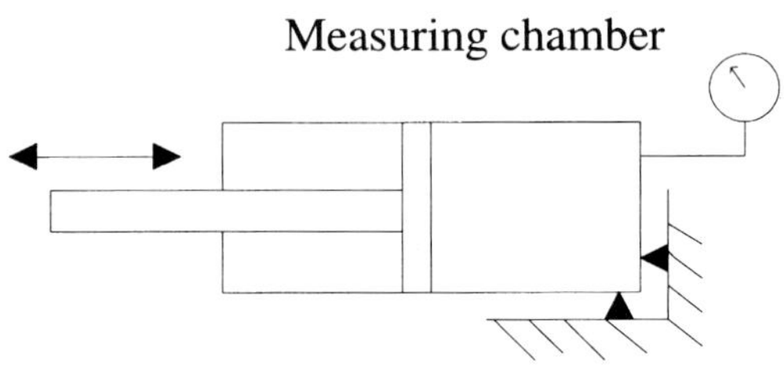

Figure 3. Test circuit.

- Volume of measuring chamber 2 dm^3
- Maximum $\frac{\Delta V}{V}$ 4 %

Figure 4 shows model 1 and model 2 results at different air percentages when the volume of the measuring chamber is compressed. Air percentages of the model 2 varies from 0.5% to 2.0%. The influence of air is significant especially with high percentages and low pressure levels. The numeric value for the air percentage depends on the system type. It is basic engineering knowledge to estimate the magnitude of the air percentage.

Figure 5 shows simulation results where model 1 and model 2 are compared when the volume of the measuring chamber is altered sinusoidally. Model 1 allows pressure decrease to negative values. The used air percentage in model 2 varies from 0% to 2.0%. Model 2 reduces the pressure above zero without any added limiter. The maximum pressure of model 2 is a little bit higher than the maximum pressure of model 1, however the difference is only few bars and it has no significance in practice.

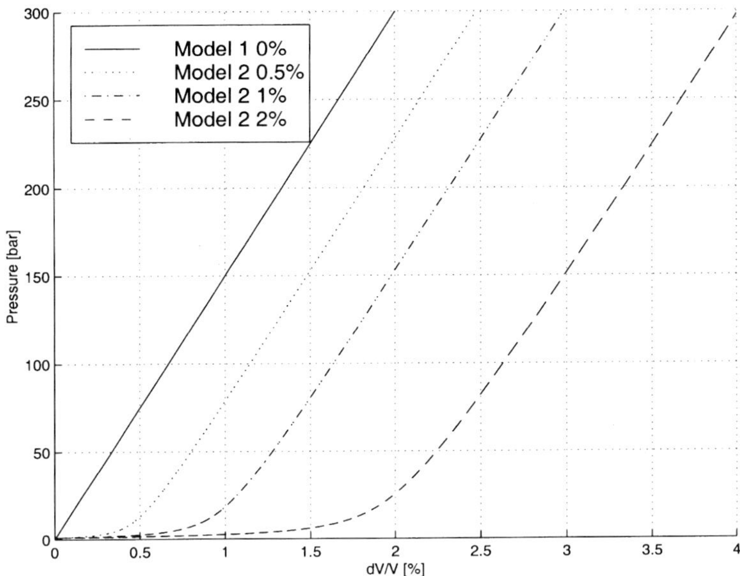

Figure 4. Pressure increase as a function of relative volume change.

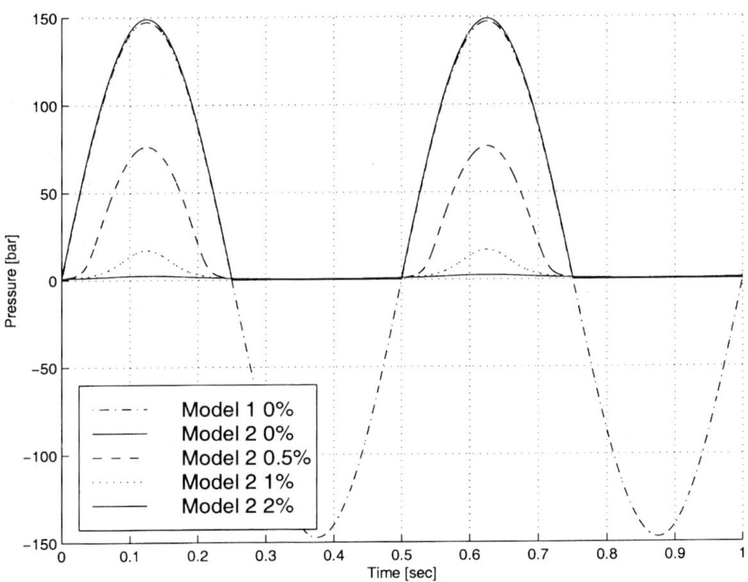

Figure 5. Pressure variation.

7. SUMMARY

A one-phase fluid model is widely used in fluid power simulation. However lately two-phase fluid models are coming into use especially in more advanced simulation packages. It is clear that with a two-phase fluid model more accurate results can be achieved. Two-phase fluid modelling is needed when large pressure changes are under consideration. It also includes effect of free air in fluid which is considerable at low-pressure levels. Use of a two-phase model is particularly necessary under cavitating conditions.

8. REFERENCES

[1] Merritt, HE., Hydraulic Control Systems, John Wiley & Sons Inc, 1967

[2] Watton, J., Fluid power systems, Prentice Hall, 1989

[3] Hayward, ATJ. Aearation in hydraulic systems – its assessment and control. Paper 17. Oil Hydraulics Conference, 1961. Institution of Mechanical Engineers.

[4] Feldmann DG. Der Einfluß nicht gelöster Luft auf die Kompressibilität des Übertragungsmediums hydrostatischer Kreise. Ölhydraulik und pneumatik 14(1970) Nr 8.

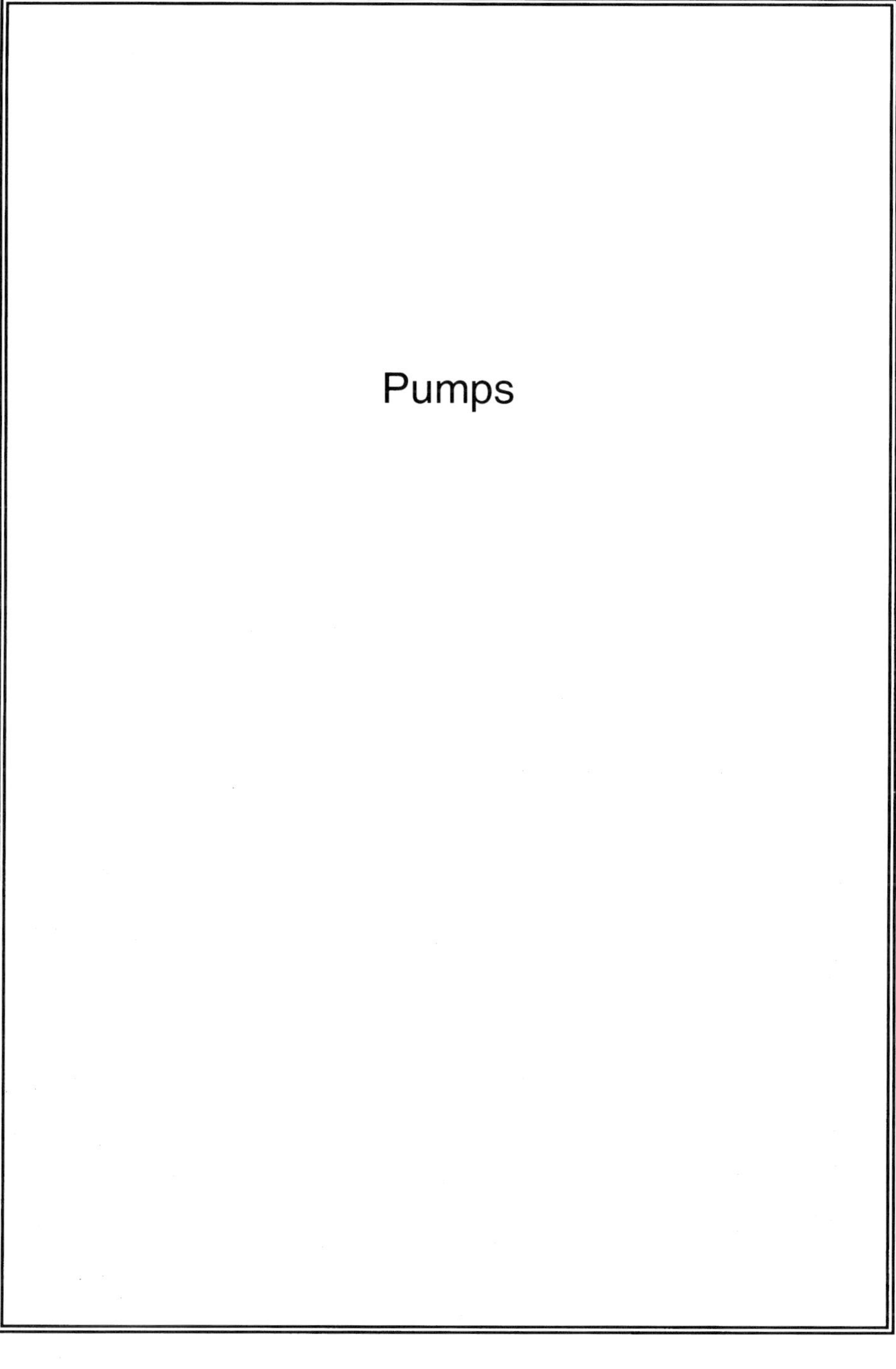

Pumps

CASPAR – a computer-aided design tool for axial piston machines

U WIECZORECK and **M IVANTYSYNOVA**
Institute for Aircraft Systems Engineering, Technical University Hamburg-Harburg, Germany

ABSTRACT

This Paper presents the simulation tool CASPAR (Calculation of swash plate type axial piston pump and motor). A method for the simulation of the non-isothermal gap flow for self-adjusting gaps is presented. Based on the simulation of the flow through the lubrication gaps in swash plate type axial piston machines the calculation of the main losses due to friction and leakage flow is possible. The gap micro and macro geometry as well as the material determine the friction and leakage flow. To verify the simulation model, the instantaneous cylinder bore pressure and the drainage flow rate have been measured.

Nomenclature

a_p	Piston acceleration	[m/s²]
A_{rHD}	Valve plate opening to high pressure side	[m²]
A_{rND}	Valve plate opening to low pressure side	[m²]
c_p	Specific heat capacity	[J/(kg K)
e	Absolute elastic deformation	[m]
$F_{\omega S}$	Centrifugal force of the slipper	[N]
F_{AP}	Axial piston force	[N]
F_{aP}	Inertia piston force	[N]
F_c	Contact force	[N]
F_e	External force	[N]
F_f	Fluid force	[N]
F_N	Normal force applied on slipper	[N]
F_{PP}	Pressure force applied on piston	[N]
F_{TP}	Friction force applied on piston	[N]
h	Gap height	[m]
h_{min}	Minimal gap height	[m]
J	Jacobian matrix (matrix of partial derivatives)	
K	Bulk modulus	[N/m²]

k_T	Temperature coefficient viscosity	[1/K]
M	Torque	[Nm]
m_P	Mass of piston and slipper	[kg]
m_S	Mass of slipper	[kg]
n	Shaft speed	[rpm]
p	Pressure	[Pa]
p_D	Pressure displacement chamber	[Pa]
p_e	Pressure case	[Pa]
p_{rHD}	Pressure pressure port	[Pa]
P_{rND}	Pressure suction port	[Pa]
Q_{HD}	Total flow pressure port	[m³/s]
Q_{out}	Flow through orifice valve (load)	[m³/s]
Q_r	Flow through valve plate	[m³/s]
Q_{rHDi}	Flow from cylinder to high pressure side	[m³/s]
Q_{rHDi}	Flow from cylinder to low pressure side	[m³/s]
Q_{SB}	Leakage flow cylinder block	[m³/s]
Q_{SG}	Leakage flow slipper	[m³/s]
Q_{SK}	Leakage flow piston	[m³/s]
R_B	Piston pitch radius	[m]
R_S	Outer radius slipper	[m]
T	Temperature	[°C]
t	Time	[sec]
V	Volume	[m³]
Vg	Geometric displacement	[m³]
v_{px}, v_{py}	Velocity piston relative to cylinder	[m/s]
v_x, v_y	Fluid velocity	[m/s]
x_s, y_s, z_s	Cartesian coordinates slipper	
α_D	Flow coefficient	
α_P	Pressure coefficient viscosity	[1/Pa]
β	Swash plate angle	[°]
ε	Relative elastic deformation	
γ, r, z	Cylindrical ccordinates	
ρ	Density	[kg/m³]
λ	Thermal conductivity	[W/(mK)]
τ	Shearing stress	[N/m²]
σ	Normal stress	[N/m²]
ω	Angle velocity	[rad/s]
Φ_D	Dissipation function	
μ	Viscosity	[Pa s]

1 INTRODUCTION

Swash plate type axial piston machines are prevalent in hydraulic systems because of their simple structure and fast response. The variable displacement pumps and motors which are currently available, usually achieve a relatively long operating life and sufficient reliability, but very often at low efficiency and top much noise emission. Thus to optimize the efficiency makes these machines even more competitive, especially for pump controlled systems instead of valve controlled solutions [7]. The losses in hydrostatic machines are mainly influenced by friction and leakage flow in the gaps. In order to reduce friction and abrasion in the gaps, it is desirable to maintain fluid lubrication. The gaps in an axial piston pump are self adjusting gaps, which means the gap geometry is not only determined by the dimensions of the parts. It depends on the operating conditions due to the mobility of the components.

The design of sealing and bearing gaps in hydrostatic machines is normally done by experience of former products. Considering this it is desirable to support the design engineer by means of computer aided tools. The enormous increase in the performance of computer hardware and the simultaneous decrease of the price makes the use of CFD (computational fluid dynamics) possible for many applications in almost all branches of engineering. Different simulation models for the gap flow have been developed (e.g. [3][6][8][14][15]). In CASPAR a non-isothermal approach for the gap flow is realized [6][14]. The structure of the simulation tool CASPAR is shown in Figure 1. Predefined parameters such as the geometry and operating conditions are the inputs of the program. The flow field and the load bearing in the gap can be calculated numerically. The simulation of the non-isothermal gap flow is applied for the gaps between piston and cylinder, between slipper and swash plate and between cylinder block and valve plate. The equilibrium of all forces exerted on the moving part of the rotating group defines the gap geometry. Since the gap geometry determines the hydrodynamic pressure field in the gap, an iteration has to be done to solve the motion equation. This interaction between the fluid flow and the moving parts of the rotating group is the essential element of the simulation. The leakage flow, friction and the pressure in the displacement chamber are obtained as results.

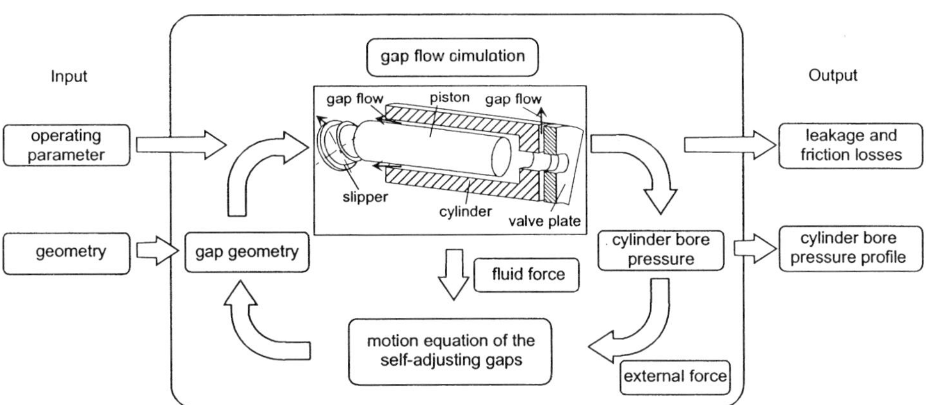

Figure 1 Simulation Tool CASPAR

2 GAP FLOW SIMULATION

Sealing and bearing gaps are essential basic elements of hydrostatic machines and they are used to separate chambers of different pressure and to bear high forces as hydrostatic or hydrodynamic bearings. Due to the very small gap height laminar flow can be assumed in case of viscous friction condition. The pressure distribution for a hydrodynamic bearing can be calculated based on the Reynolds equation [13]. In the following it is explained for the example of the gap between piston and cylinder. Since the gap-diameter ratio is very small, the gap can be expanded on the x-y plane, where x denotes the gap circumference and y denotes the co-ordinate in direction of the piston axis.

$$\frac{\partial}{\partial x}\left(\frac{\partial p}{\partial x}\frac{h^3}{\mu}\right)+\frac{\partial}{\partial y}\left(\frac{\partial p}{\partial y}\frac{h^3}{\mu}\right)=6\left((v_{px})\frac{\partial h}{\partial x}+(v_{py})\frac{\partial h}{\partial y}+2\frac{\partial h}{\partial t}\right) \tag{1}$$

To avoid negative pressure values, a negative value is set to zero when it occurs in the iteration [9]. Energy dissipation occurs due to friction of the viscous fluid in the gap. The fluid and the whole unit is heated by the dissipated energy. Depending on the thermodynamic state of the liquid, local changes of viscosity can occur. This influence can not be neglected, thus with decreasing viscosity the flow increase and the load bearing capacity decrease. The temperature distribution is described by the energy equation (2).

$$c_p\rho\left(\frac{\partial T}{\partial t}+\mathbf{v}\ \ \mathrm{grad}\,T\right)=\lambda\,\mathrm{div}\ (\mathrm{grad}\,T)+\mu\Phi_D(\mathbf{v}) \tag{2}$$

The simulation model assumes steady state conditions, thus local changes do not occur $\partial T/\partial t$=0. Assuming a cartesian system of co-ordinates the energy dissipation due to fluid friction is given by [12]

$$\Phi_D(\mathbf{v})=\left(\frac{\partial v_x}{\partial z}\right)^2+\left(\frac{\partial v_y}{\partial z}\right)^2 \tag{3}$$

with the velocity components v_x and v_y. Once the pressure and temperature distribution is obtained the change of the viscosity can be evaluated. The pressure and temperature behaviour of the viscosity can be calculated as described in [4]

$$\mu=\mu_0\exp(\alpha_p p-k_T(T-T_0)). \tag{4}$$

The pressure distribution is again computed based on these results. The iteration stops if the required precision is achieved. In this manner Reynolds and energy equation are solved simultaneously. Both equations are partial differential equations, which can only be solved numerically. Therefore the finite volume method is used [11]. By integrating the differential equation over a control volume the differential equation can be instantly turned into the discretization equation, which means a system of linear equations. To solve this algebraic equations the Gauss-Seidel iteration method with over relaxation is used [10]. The velocity distribution in the gap can be calculated based on the pressure field. The velocity v_y of the fluid in direction of the piston axis depending on the gap height z is given by

$$v_y=\frac{1}{2\mu}\frac{\partial p}{\partial y}(z^2-h\ \ z)+v_{Py}\frac{z}{h} \tag{5}$$

The load bearing capacity can be calculated by integrating the pressure distribution. The leakage flow is given by integrating the velocity distribution. Based on the shear stress the viscous friction can be calculated.

3 PRESSURE IN THE DISPLACEMENT CHAMBER

For the simulation of the gap flow the instantaneous pressure in the displacement chamber or in the cylinder volume respectively in each time step of the simulation has to be concerned. It defines the boundary conditions for the pressure distribution in the gap and it determined the external forces. The bore pressure profile is essential for an optimization of the valve plate and has strong influence on the noise emission. The differential equation for the pressure behaviour in the displacement chamber is given by

$$\frac{dp_{Di}}{dt} = \frac{K}{V_i}\left(Q_{ri} - Q_{SKi} - Q_{SBi} - Q_{SGi} - \frac{dV_i}{dt}\right) \tag{6}$$

where dV_i/dt is given by the product of piston velocity v_P and piston face area A_P.

$$\frac{dV}{dt} = v_P A_P \tag{7}$$

The leakage flow Q_{SK}, Q_{SG} and Q_{SB} correspond to the gap flow as shown in Figure 2. Since the leakage flow through the gap influence the bore pressure profile, the gap flow in each gap influence each other. The leakage flow is obtained from the solution of the Reynolds equation.

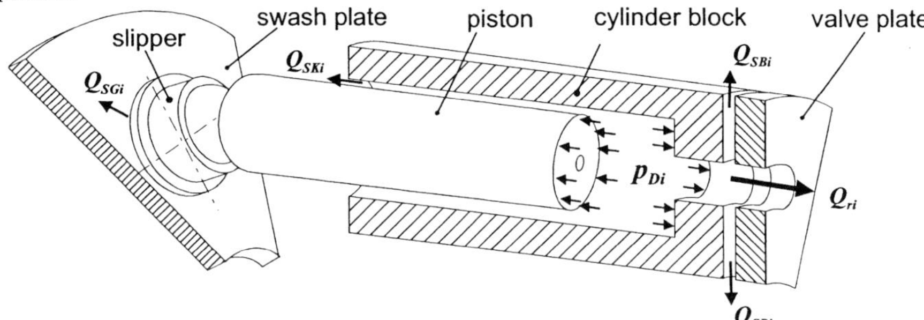

Figure 2 Pressure in the displacement chamber

In equation (6) Q_{ri} denotes the flow through the valve plate into the suction and pressure port from the cylinder i. To calculated the flow Q_{ri} turbulent flow is assumed and the opening to flow in the valve plate has to be determined for each time step. For some valve plates the cylinder is connected to both suction and pressure ports during the change over, to achieve a more gradual pressure increase. In order to describe the behaviour of this so called cross port, the flow to high and low pressure side Q_{rHDi} and Q_{rNDi} are considered separately. The turbulent flow through the valve plate can be expressed as

$$Q_{rHDi} = \alpha_D A_{rHD}\sqrt{\frac{2}{\rho}\left|p_{Di} - p_{rHD}\right|}\,\mathrm{sgn}(p_{rHD} - p_{Di}) \tag{8}$$

$$Q_{rNDi} = \alpha_D A_{rND}\sqrt{\frac{2}{\rho}\left|p_{Di} - p_{rND}\right|}\,\mathrm{sgn}(p_{rND} - p_{Di}) \tag{9}$$

where A_{rHD} denotes the opening to flow on the pressure port side and A_{rND} the opening to flow on the suction port side. The flow Q_{ri} is given by

$$Q_{ri} = Q_{rHDi} + Q_{rNDi} \tag{10}$$

The pressure in the displacement chamber is mainly influenced by the pressure in the pressure and suction port. The pulsation of the flow, that is determined by the number of pistons, causes a pulsation of the pressure in the high pressure line, that effects the bore pressure profile. Figure 3 shows the calculated flow pulsation on the pump outlet.

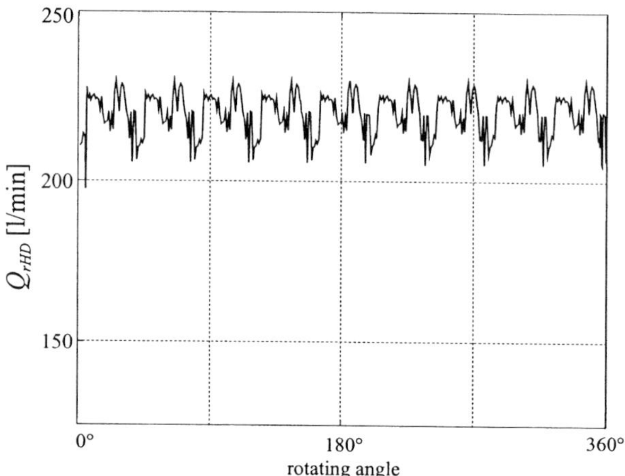

Figure 3 Flow rate on the pump outlet

To determine the pressure pulsation caused by the flow pulsation in the high pressure line the simulation model is extended by the following model:

$$\dot{p}_{rHD} = \frac{K}{V_H}\left(-Q_{HD} + Q_{out}\right) \tag{11}$$

where Q_{HD} denotes the flow into pressure port and Q_{out} the flow through the orifice as shown in Figure 4.

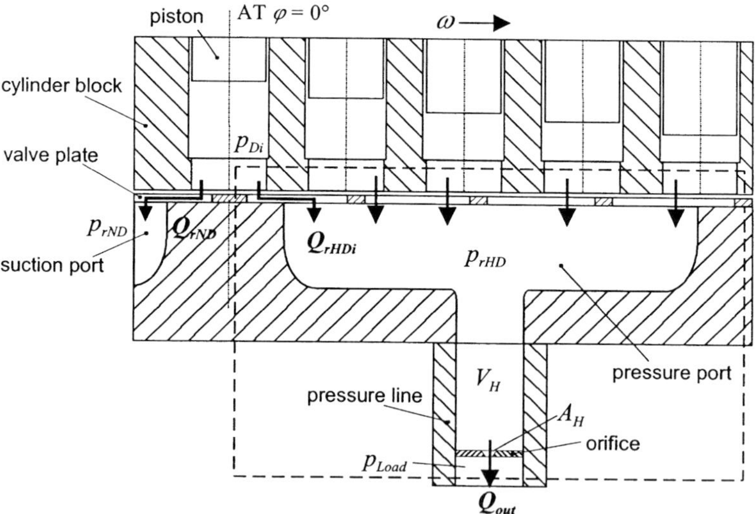

Figure 4 Pressure in the pressure port

The length of the pipe influences the volume V_H and the pressure pulsation on the pump outlet. The total flow through the pressure port is obtained by

$$Q_{HD} = \sum_i Q_{rHD_i} \tag{12}$$

For the load simulation the program CASPAR uses a simple variable orifice. According to the turbulent flow model of an orifice for the effective flow rate Q_{out} follows

$$Q_{out} = \alpha_D A_H \sqrt{\frac{2}{\rho}\left|p_{rHD} - p_{Load}\right|}\,\mathrm{sgn}(p_{Load} - p_{rHD}) \tag{13}$$

Through variation of the opening of the orifice valve A_H different pressure levels can be adjusted, p_{Load} denotes the pressure behind the orifice valve. In this case p_{Load} is set to tank pressure. This approach corresponds to the test assembly that is realized for the measurement of the instantaneous pressure in the cylinder volume. In chapter 5 the test assembly is explained in detail.

4 MOTION EQUATION

As already mentioned the gaps in an axial piston pump are self adjusting. To determined the gap geometry the balance of forces between all external forces exerted on the components and the load bearing capacity of the lubricating film has to be considered. The motion equation will be explained for the example of the slipper. The calculation is carried out similarly for the gaps between the piston and cylinder and between the cylinder block and valve plate. The function of the slipper is to bear the piston on the swash plate. As shown in Figure 5 high pressure is fed by a drilling under the slip plane to obtain hydrostatic bearing of the slipper. Additional hydrodynamic forces occur due to relative movement.

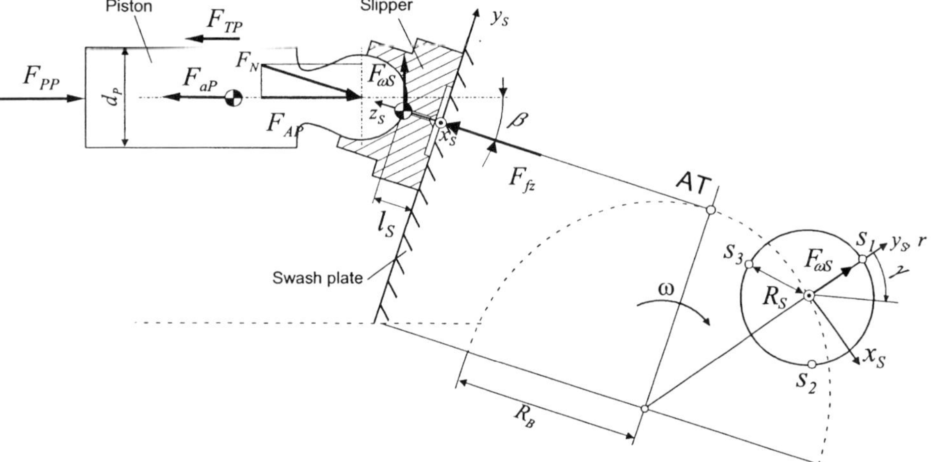

Figure 5 External forces on the slipper

The piston is loaded with a force in axial direction F_{AP}, that can be expressed by

$$F_{AP} = F_{PP} + F_{TP} + F_{aP} \tag{14}$$

where F_{PP} denotes the force due to the pressure in the displacement chamber, F_{TP} denotes the friction forces and F_{aP} the inertia forces.

$$F_{PP} = \frac{\pi d_p^2}{4}(p_D - p_e) \tag{15}$$

$$F_{aP} = -m_P a_P = m_P \omega^2 R_B \tan\beta \cos\varphi \tag{16}$$

The friction force is calculated based on the shearing stress μ.

$$\tau = \mu \frac{\partial v}{\partial z} \tag{17}$$

Integrating the shearing stress gives the axial friction force on the piston.

$$F_{TP} = \int_A \tau dA \tag{18}$$

Due to the sliding bearing of the slipper the swash plate can only incorporate forces vertical to the surface. The normal force is given by

$$F_N = \frac{F_{AP}}{\cos\beta} \tag{19}$$

Additional the centrifugal force is exerted on the slipper. The pitching torque is obtained by

$$M_x = F_{\omega S} l_S \cos\beta = m_S R_B \omega^2 \cos\beta \tag{20}$$

The gap height depends on the fluid forces of the fluid film under the slipper. The position of the slipper can be described by the clearance between slipper and swash plate in the points s_1, s_2 and s_3 given by the vector $\mathbf{h} = (h_1, h_2, h_3)$. The gap height can be expressed as

$$h_S(r,\gamma) = \frac{\sqrt{\frac{1}{3}}}{R_S}(h_2 - h_3) r \sin\gamma + \frac{1}{3R_S}(2h_1 - h_2 - h_3) r \cos\gamma + \frac{1}{3}(h_1 + h_2 + h_3) \tag{21}$$

For the motion equation the balance of forces of the fluid force, external force and contact force in this points are considered. The normal force F_N and the pitching torque M_x can be expressed by three forces in the points s_1, s_2 and s_3 represent by the vector $\mathbf{F}_e = (F_{e1}, F_{e2}, F_{e3})$

$$\begin{pmatrix} 1 & 1 & 1 \\ R_S & -0.5R_S & -0.5R_S \\ 0 & -\frac{\sqrt{3}}{2}R_S & \frac{\sqrt{3}}{2}R_S \end{pmatrix} \begin{pmatrix} F_{e1} \\ F_{e2} \\ F_{e3} \end{pmatrix} = \begin{pmatrix} -F_N \\ -M_x \\ 0 \end{pmatrix} \tag{22}$$

The fluid force and the torque due to the pressure field under the slipper is obtained by

$$F_{fz} = \sum_{i,j} p_{ij} r_j \Delta\gamma \Delta r \tag{23}$$

$$M_{fx} = \sum_{i,j} p_{ij} r_j^2 \Delta\gamma \Delta r \cos\gamma_i \tag{24}$$

$$M_{fy} = \sum_{i,j} - p_{ij} r_j^2 \Delta\gamma \Delta r \sin\gamma_i \tag{25}$$

By solving the linear system (26) the fluid force $\mathbf{F}_f = (F_{f1}, F_{f2}, F_{f2})$ in point s_1, s_2 and s_3 is obtained.

$$\begin{pmatrix} 1 & 1 & 1 \\ R_S & -0.5R_S & -0.5R_S \\ 0 & -\frac{\sqrt{3}}{2}R_S & \frac{\sqrt{3}}{2}R_S \end{pmatrix} \begin{pmatrix} F_{f1} \\ F_{f2} \\ F_{f3} \end{pmatrix} = \begin{pmatrix} -F_{fz} \\ -M_{fx} \\ -M_{fy} \end{pmatrix} \tag{26}$$

If the external load can not be incorporated by the fluid force, the slipper will move in the direction of the external force. The movement of the slipper is described by the shifting velocity $\dot{\mathbf{h}}$. Due to the squeeze effect the pressure and the load bearing capacity of the fluid film under the slipper increases. The squeeze film term in the Reynolds equation $\partial h / \partial t$ can be calculated based on the shifting velocity.

$$\frac{\partial h(r,\gamma)}{\partial t}=\frac{\sqrt{\frac{1}{3}}}{R_S}(\dot{h}_2-\dot{h}_3)r\sin\gamma+\frac{1}{3R_S}(2\dot{h}_1-\dot{h}_2-\dot{h}_3)r\cos\gamma+\frac{1}{3}(\dot{h}_1+\dot{h}_2+\dot{h}_3) \tag{27}$$

If the lubrication film is interrupted, solid to solid contact can occur. In this case a minimal gap height of 0.1μm is assumed, so that the Reynolds equation can still be applied for the calculation of the pressure field. Where the gap height is smaller then the minimal gap height elastic deformation of the components is assumed. The elastic deformation can be expressed as

$$\varepsilon=\begin{cases}0 & \text{with } h \geq h_{\min} \\ \dfrac{h_{\min}-h}{l_S} & \text{with } h < h_{\min}\end{cases} \tag{28}$$

With the relative deformation ε and the Young's modulus E of the component the contact force can be estimated. The normal stress is given by

$$\sigma=\varepsilon E \tag{29}$$

Similar to the fluid force the contact force is obtained by integrating the strain.

$$F_{Cz}=\sum_{i,j}\sigma_{ij}r_j\Delta\gamma\Delta r \tag{30}$$

$$M_{Cx}=\sum_{i,j}\cos\gamma_i\sigma_{ij}r_j^2\Delta\gamma\Delta r \tag{31}$$

$$M_{Cy}=\sum_{i,j}-\sin\gamma_i\sigma_{ij}r_j^2\Delta\gamma\Delta r \tag{32}$$

The contact force $\mathbf{F}_C = (F_{C1}, F_{C2}, F_{C2})$ in point s_1, s_2 and s_3 is obtained by solving the linear system

$$\begin{pmatrix}1 & 1 & 1\\ R_S & -0.5R_S & -0.5R_S\\ 0 & -\frac{\sqrt{3}}{2}R_S & \frac{\sqrt{3}}{2}R_S\end{pmatrix}\begin{pmatrix}F_{C1}\\ F_{C2}\\ F_{C3}\end{pmatrix}=\begin{pmatrix}-F_{Cz}\\ -M_{Cx}\\ -M_{Cy}\end{pmatrix} \tag{33}$$

Note that the contact force is needed only for the reason of stability of the numerical procedure. Under normal operating conditions no contact force occurs. For a certain shifting velocity of the slipper the equilibrium of forces can be achieved. In this manner equation (34) denotes a system of non linear equations with the shifting velocity $\dot{\mathbf{h}}$ as the unknown.

$$\mathrm{f}=-\mathbf{F}_f(\mathbf{h},\dot{\mathbf{h}},t)-\mathbf{F}_e(t)-\mathbf{F}_c(\mathbf{h},t)=0 \tag{34}$$

To solve this system the damped newton iterative method is used [2]. The iterative rule is.

$$\dot{\mathbf{h}}^{(k)}=\dot{\mathbf{h}}^{(k-1)}+\Delta\dot{\mathbf{h}}^{(k)}2^{-k} \text{ with} \tag{35}$$

$$\Delta\dot{\mathbf{h}}^{(k)}=-\mathbf{J}^{-1}(\dot{\mathbf{h}}^{(k)})\mathbf{f}(\dot{\mathbf{h}}^{(k)}) \tag{36}$$

The newton method requires the evaluation of the Jacobian $\mathbf{J}$ and the solution of linear systems in every time step. The computation of the Jacobian is a time consuming task, the Reynolds - and energy equations have to be solved several times to obtain the differentiation.

Note that the Newton method already lost its property of quadratic convergence when the Jacobian is approximated [1]. To reduce the approximation error, the Jacobian is approximated by central difference scheme.

$$\frac{\partial \mathrm{f}_j}{\partial \dot{h}_i} = \frac{\mathrm{f}(\mathbf{h}, \dot{\mathbf{h}} + \Delta \dot{h}_i) - \mathrm{f}(\mathbf{h}, \dot{\mathbf{h}} - \Delta \dot{h}_i)}{2\Delta \dot{h}_i} \tag{37}$$

The aim of the evaluation of the motion equation is to modify the position of the components for given external forces, so that with the modified position the system is in an equilibrium state. The position of the components is obtained by integrating the shifting velocity. The positions converge to a cyclical movement with the same period time as the shaft speed. The position of the slipper over three revolutions is shown in Figure 6.

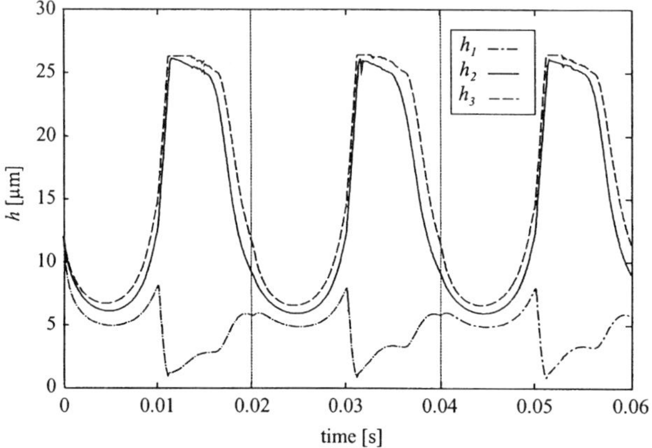

Figure 6 Position of the slipper

In Figure 7 the simulation flow for all gaps is shown.

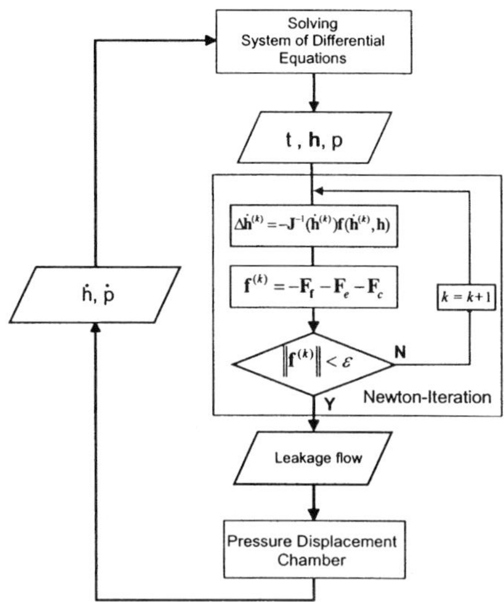

Figure 7 Simulation flow

Starting with the initial values for the position and the pressure in the displacement chamber the Newton iteration is done to determined the gap height of the considered gaps. The shifting velocity is varied until the pressure field can bear the external forces. With the leakage flow the differential equation for the pressure in the displacement chamber is solved. The shifting velocity and the derivation of the pressure denotes a system of ordinary differential equations that is solved numerically.

5 VALIDATION OF THE SIMULATION MODEL

With the simulation tool CASPAR the leakage flow, the viscous friction in the gap and the pressure in the displacement chamber can be calculated. In order to verify the new simulation approach different methods have been used. The simulated external leakage flow was compared with the measured drainage flow rate. Since the leakage flow occurs only in the concerned gaps the external leakage flow can be calculated exactly based on the gap flow, and the measurement of the leakage flow is very easy to realize. Table 1 shows the simulated and measured external leakage flow for different operating parameters.

Shaft Speed [RPM]	Pressure Difference [bar]	Swash Plate Angle [°]	External Leakage Simulation [l/min]	Drainage Flow Rate Measurement [l/min]
2000	100	17	1.10	1.21
2000	200	17	1.32	1.41
3000	200	17	1.70	1.67

Table 1 Simulated and measured leakage flow

It can be seen that the simulated external leakage flow corresponds to the measurements. The slight difference between simulation and measurement can be explained by tolerances of the dimensions of the components and the accuracy of the flow meter. To prove the computed friction force a special test device for the measurement of the friction between piston and cylinder has been used [5][7]. To use CASPAR for the optimization of valve plates it is essential to know about the pressure behaviour in the displacement chamber. Therefore another test assembly was built up that allows the measurement of the instantaneous pressure in displacement chamber.

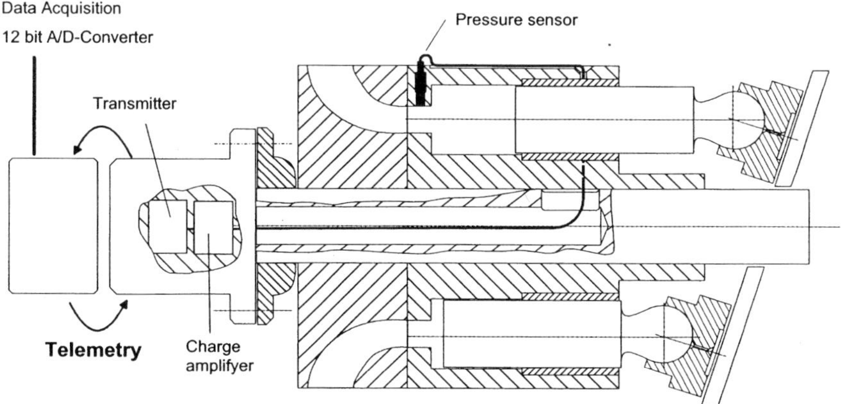

Figure 8 Measurement system for the pressure in the displacement chamber

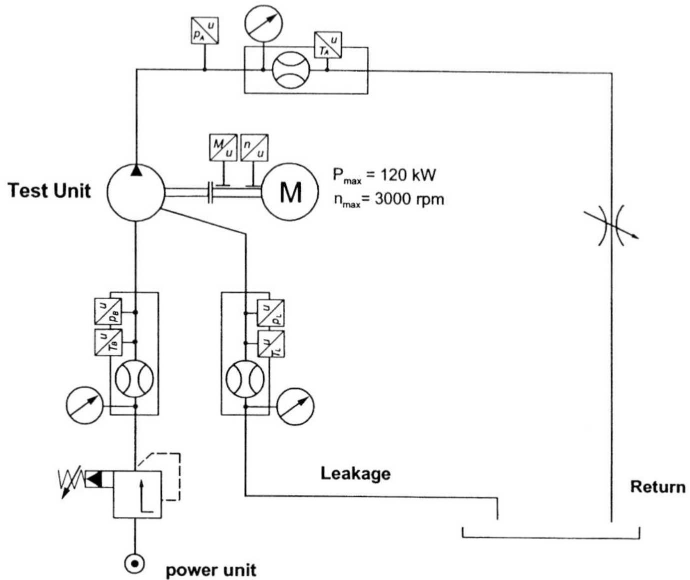

Figure 9 Test assembly for experiments

Figure 9 shows the measurement of the instantaneous pressure in the displacement chamber at 3000 rpm shaft speed and the simulation for the same operating parameter.

An axial piston pump of 75ccm displacement shown in Figure 8 was equipped with a piezoelectric pressure sensor. The measurement data are transmitted to the data acquisition system by means of a telemetry and are recorded over one revolution of the cylinder block. As already mentioned the pressure on the pump outlet was adjust by means of a variable orifice. The test assembly shown in Figure 9 was also used to measure power losses and drainage flow rate as shown in Table 1.

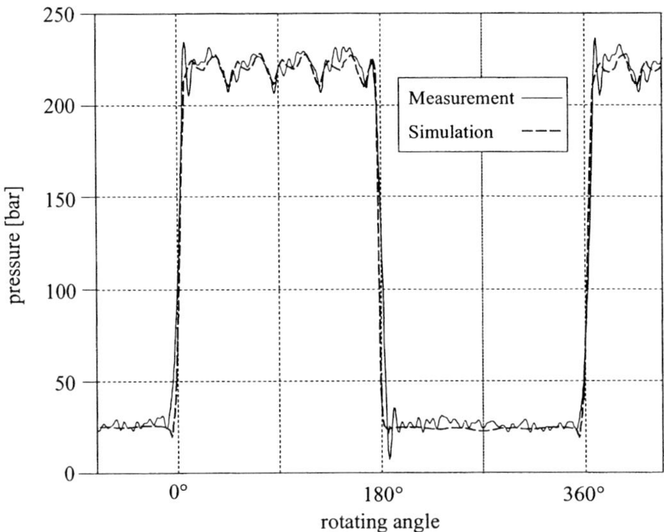

Figure 10 Pressure in the displacement chamber - measurement and simulation

The simulation provides a very good approximation of the pressure profile on the high pressure side, although the flow through the valve plate is approximated by a very simple model. Since the inductivity of the oil in the pipe has not been considered the simulation could differ from the measurement. The difference between measurement and simulation on the low pressure side can be easily explained. For simplification purposes the pressure in the suction port was set to a constant value in the simulation model. It is possible to simulate also the pressure pulsation in the low pressure line by an extension of the simulation model.

6 CONCLUSION

A method for the calculation of the non-isothermal gap flow in the main gaps of swash plate type axial piston machines has been developed. By solving the motion equation for the moving parts the gap geometry can be calculated. With the calculated pressure and velocity field the leakage flow and the friction in the gap can be determined. As mentioned before leakage flow can only occur in the concerned gaps, so the internal and external leakage flow can be determined precisely. The comparison of the simulation data with the measurements especially the external leakage flow has shown a very good fit. So an estimation of the volumetric efficiency is possible with CASPAR. The simulation model only considers viscous friction but this is not a big disadvantage, since viscous friction conditions are the desirable operating conditions. The results for the instantaneous pressure in the displacement chamber correspond to the measurement, so CASPAR can be used for optimization of the valve plate as well as for the design of the bearing and sealing gaps. The simulation tool CASPAR enables the design engineer to calculated the losses due to the gap flow and their dependence on design and operating parameters. With the simulation tool CASPAR a lot of different variants can be analysed without long testing and the development time can be reduced.

Reference

[1] Eich-Soellner E., Führer C.: Numerical Methods in Multibody Dynamics. Stuttgart: Teubner, 1998.

[2] Engeln-Müllges G., Uhlig f.: Numerical Algorithms with C. Berlin, Heidelberg: Springer-Verlag, 1996.

[3] Fang Y., Shirakashi M.: Mixed Lubrication Characteristics between the Piston and Cylinder in Hydraulic Piston Pump-Motor. Journal of Tribology, Trans. ASME, Vol. 117, pp. 80-85. 1995.

[4] Ivantysyn J., Ivantysynova M.: Hydrostatische Pumpen und Motoren. Würzburg: Vogel Verlag, 1993.

[5] Ivantysynova M.: A New Approach to the Design of Sealing and Bearing Gaps of Displacement Machines. 4th JHPS International Symposium on Fluid Power. Tokyo Japan, 1999.

[6] Ivantysynova M.: Temperaturfeld im Schmierspalt zwischen Kolben und Zylinder einer Axialkolbenmaschine. Maschinenbautechnik 34. S. 532 - 535. 1985.

[7] Ivantysynova M., Lasaar R.: Ein Versuchträger zur Messung der Reibkräfte zwischen Kolben und Zylinder in Axialkolbenmaschinen. Konstruktion 6, pp. 57 – 65. 2000

[8] Kleist A.: Design of Hydrostatic Static Bearing and Sealing Gaps in Hydraulic Machines. 5th Scandinavian International Conference on Fluid Power, Lingköping, 1997.

[9] Krasser J., Laback O., Loinbennegger B., Priebsch H.: Anwendung eines elastohydrodynamischen Verfahrens zur Berechnung von Kurbeltriebslagern. Motortechnische Zeitschrift 55, pp. 656-663. 1994.

[10] Noll B.: Numerische Strömungsmechanik. Berlin Heidelberg: Springer-Verlag, 1993.

[11] Patankar S.V.: Numerical Heat Transfer and Fluid Flow. New York, Washington: Hemisphere Publishing Corporation, 1980.

[12] Vogelpohl G.: Beiträge zur Kenntnis der Gleitlagerreibung. VDI-Forschungsheft 386. Berlin: VDI-Verlag, 1937.

[13] Vogelpohl G.: Betriebssichere Gleitlager. Band 1, 2. Aufl. Berlin, Heidelberg: Springer-Verlag, 1967.

[14] Wieczorek U., Ivantysynova M.: Computer Aided Optimization of Bearing and Sealing Gaps in Hydrostatic Machines. Proceedings 1st Bratislavian Fluid Power Symposium, Bratislava, Slavakia, 1998.

[15] Zhu D., Cheng H.S. , Arai T. and Hamai K.: A numerical analysis for piston skirt in mixed librication. Journal of Tribology Part I and II., 114 ,115, pp. 553- 562, 125 – 133. 1992

Innovative pump design to reduce pressure pulsations of axial piston pumps

D BECHER and **S HELDUSER**
Institute of Fluid Power and Motion Control, Dresden University of Technology, Germany

ABSTRACT

The paper deals with the description of an innovative pump design to reduce the flow ripple of an axial piston pump. The new system operates with an elastic ring which realises a pressure depending cross-sectional area between a high pressure ring groove and the cylinder chamber.
Starting with an analysis of requirements in function, load, geometry and assembly, a constructive design of the ring system was done developing an FE-model of the ring and numerical simulation models of the pump. Using simulation models reduces time and cost intensive investigations on the test rig. To prove the efficiency in reducing pulsation, shown in the numerical calculations, an axial piston pump was modified. The measured values of pulsation conform to the calculated ones. Depending on the operating point, a reduction of the pressure ripple is possible in a range between 5 and 60 %.

0 SYMBOLS

$A_{Th,Ring}$	Throttling area between ring chamber and piston bore	mm^2
d_a, d_b, b	Outside, inside diameter, width of elastic ring	mm
$d_{control}$	Diameter of control bore	mm
F_{Flex}	Restoring force	N
F_p	Pressure force	N
M_1	Angular momentum	Nm
p_1	Delivery pressure	bar
p_T	Intake pressure	bar
p_P	Pressure in cylinder bore	bar
s_1, s_2	Parameters descriping the opening gap of elastic ring	mm
Q_1	Flow at pump outlet port	l/min
Q_{Le}	External leakage	l/min
Q_{ZHD}	Flow between the cylinder chamber and the high pressure kidney	l/min
V_1	Volumetric displacement	cm^3
V_{1max}	Maximum volumetric displacement	cm^3
α	Swing angle	deg
φ	Rotation angle	deg
η_t	Overall efficiency	%
η_{vol}	Volumetric efficiency	%
$\eta_{h/m}$	Hydraulic/mechenical efficiency	%
Δp_1	Pressure pulsation at pump outlet port (peak to peak)	bar
$\Delta p_{1,standard}$	Pressure pulsation of the standard pump	bar
$\Delta p_{1,modified}$	Pressure pulsation of the modified pump	bar

1 INTRODUCTION

Low noise levels play an important role for the selection of hydraulic drives. An active noise source is the hydrostatic pump. Due to the pressure changes during the transition of the pistons from the intake side to the delivery side or vice versa, a dynamically alternating force is transmitted to the swash plate and the pump casing through the piston face so that the pump itself emits noise. In addition, the kinematic volume flow pulsation and, above all, the compressible oil in the cylinders produce a pulsating volume flow at the pump outlet. This pulsating volume flow flows into the hydraulic system and produces pressure pulsations due to flow resistances which may cause even those components which are far away from the actual pump to emit noise. The pressure pulsation also acts on the pump outlet through the pressure medium and therefore reacts on the pump casing by way of a vibration and noise stimulation mechanism. In commercial pumps, the rate of pressure change during the piston transition phase is controlled by control slots or bores. However, this measure shows optimum results only for a particular operating point or range of the pump. To reduce pressure pulsation, it seems therefore reasonable to adapt the transition phase according to the instantaneous operating parameters operating pressure and, if required, swing angle.

2 DESIGN AND WORKING PRINCIPLE OF THE RING VALVE

The ring valve /1/ realizes a change in the cross-sectional area between the piston chamber and the high-pressure kidney during the transition of the piston from the intake side to the delivery side. This change depends on the pressure p_l. The basic design of the system is shown in **Fig. 1**.

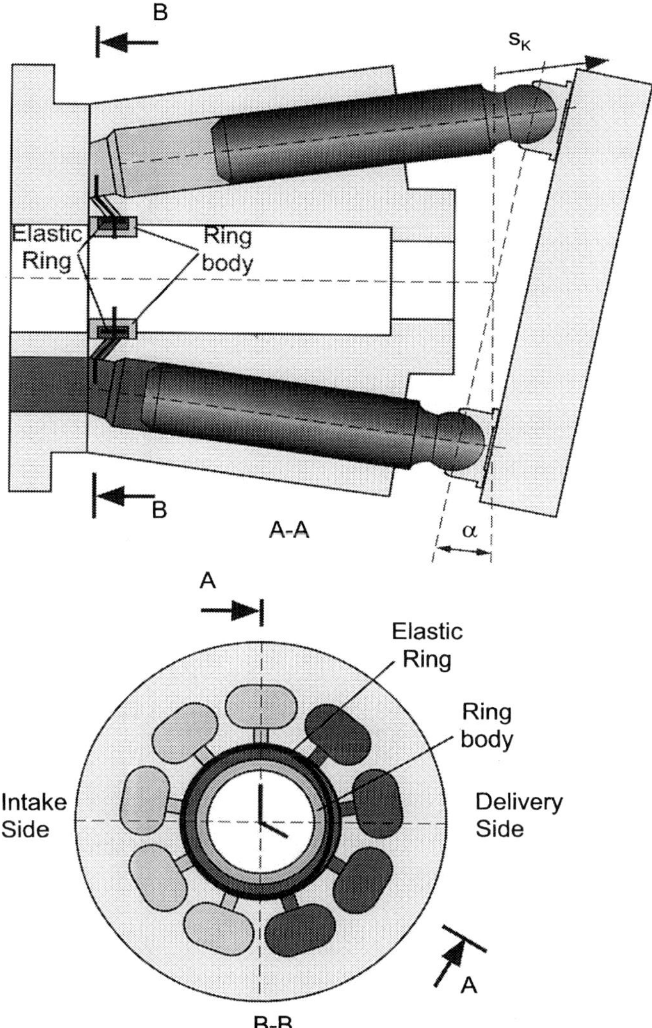

Fig. 1: Axial piston machine with ring valve

Every piston chamber is connected to a ring channel via a control bore. The ring channel is formed by a bore shoulder in the cylinder block and a ring body. A floating elastic ring with radial play whose outer diameter is smaller than the diameter of the bore shoulder is arranged in the ring channel. The inner diameter of the ring is larger than the diameter of the groove base. During operation, the ring channel is connected to the high-pressure-carrying piston via the control bore. This ensures that the ring channel always has system pressure. In addition, the throttling ports connected to the intake side are closed by the elastic ring due to the pressure differential across the control bore. A force that changes according to the pressure p_l and hence on the pressure differential p_l - p_T acts on the ring via the control bores and thus has an effect on the deformation behaviour and therefore on the opening behaviour between ring chamber and control bore in dependence on the pressure. Moreover, the ring prevents the build-up of pressure peaks in this area during the transition from the intake side to the delivery side.

3 DESIGN OF THE RING SYSTEM

3.1 FE-analyses of ring deformation

The pulsation at the pump outlet is substantially influenced by the change with time of the volume flow between piston chamber and high-pressure kidney while the piston is communicating with the high-pressure side. The size of the volume flow in the ring system studied depends, in turn, on the pressure differential and the relevant cross-sectional area between piston chamber and high-pressure kidney. To evaluate the pulsation behaviour, it is therefore necessary to know the change in the throttling area between control bore and ring chamber and its dependence on the operating point and swing angle.

The ring deformation for different loads is computed by means of a Finite Element program. The computation of the ring deformation under operating load is a static structural analysis, that is, node shifts and stresses are computed according to the load acting on the ring. The actual ring is modelled from beam elements which allow the computation of deformations and stresses. To model the contact face between ring and ring chamber, so-called "contact elements" are used. When both faces come into contact, the stiffness in this point is increased in such a away that both faces cannot penetrate each other. If the distance between both contact faces exceeds zero, the nodes of both faces can move freely according to the load. The contact element is a non-linear element, as stiffness changes in dependence on the system state. To stabilize the computation in case of contact problems, additional spring elements are used which solely serve to stabilize the computation algorithm. In this case, the stiffness of the spring elements is smaller than that of the material by about an order of magnitude.

Fig. 2 shows the loading forces acting on the ring. When a piston is on the intake side, the relevant control bore is closed due to the pressure differential between the system pressure p_l in the ring channel and the tank pressure p_T in the piston chamber and the resulting force across the bore area. The deformation of the ring produces a restoring force that tries to restore the original circular ring shape. During the transition from the intake side to the delivery side, an increasing cylinder pressure p_P is built up in the piston chamber due to the compression movement of the piston. The cylinder pressure causes the ring to detach from the ring chamber to open a throttling cross section between control bore and ring chamber. The size of this throttling cross section determines the size of the volume flow that additionally flows from the high-pressure side into the piston chamber due to the pressure differential between ring chamber (p_l) and piston chamber (p_P).

The ring deformation is computed for a constant pressure p_l but different cylinder pressures p_P for the piston which is just traversing from the intake side to the delivery side. As a result, the change of the throttling cross section between control bore and ring chamber is obtained which is depending on the cylinder pressure p_P.

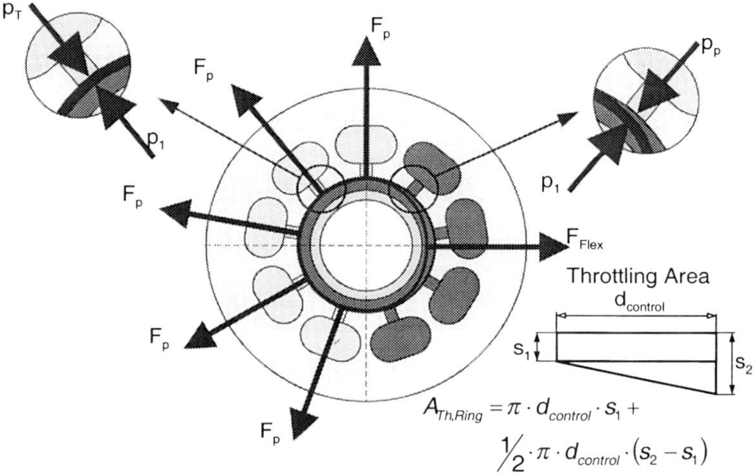

Fig. 2: Determination of the throttling area between control bore and ring chamber in dependence on pressure p_1 and cylinder pressure p_P

3.2 Combining the results of the FE analysis with the numerical pump model

The use of a non-linear numerical computation model allows an effective investigation of the pulsation behaviour of axial piston pumps /2/. The analysis of the dynamic behaviour of a pump with a ring system requires that the pressure-dependent deformation of the elastic ring computed by the FE program should be integrated into the numerical simulation model of the pump. Usually, interfaces at the program level must be used for an exact coupling of both models. This requires, however, a revision of the source text of the software. Because such a software coupling would be beyond the scope of the project, a simplified method is used.

The ring deformation for a constant pressure p_1 but different cylinder pressures p_P for the piston that is moving from the intake side to the delivery side is computed by means of the FE model. As a result, the change in the throttling cross section between control bore and ring chamber in dependence on the cylinder pressure p_P is obtained. The cylinder pressure p_P is then computed for a constant pressure p_1 in dependence on the swing angle α by means of the numerical simulation model of a single piston. After the inner dead centre of the piston has been passed, an increasing pressure p_P is first generated solely by the compression movement of the piston. When during the simulation a cylinder pressure p_P is reached which coincides with the load pressure in the FE computation, the simulation is stopped at this point and the throttling cross section determined by the FE model is used as the cross section of the flow between the high-pressure side and the piston chamber. The change in the throttling cross section between control bore and ring chamber in dependence on the swing angle is obtained by step-by-step computation. This change is transferred as a characteristic field into the overall model of the pump (**Fig. 3**). This allows a dynamic computation of the hydraulic parameters of the pump with a combined ring valve according to the operating point.

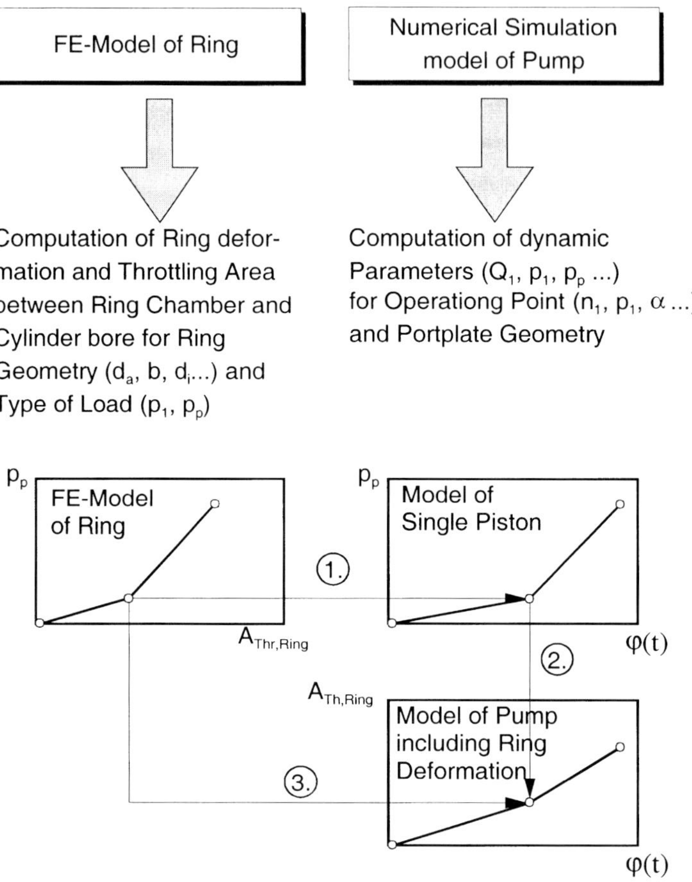

Fig. 3: Combining the results of the FE analysis with the numerical simulation model of the pump

3.3 Design

For the realization of the ring system, various designs are conceivable. The design shown in **Fig. 4** was used for the fundamental investigations on the function of the ring system. The ring chamber is created during the assembly of the ring body. The advantage of this design is the ease of assembly and disassembly of the elastic ring. This saves time, particularly at the experimental stage where different geometric designs are studied.

In the design shown in **Fig. 4**, the pressure p_l acts on the inner circular ring face of the ring body. To prevent the ring body from being forced out in the axial direction, the cylinder block has one cross hole between two cylinder bores through which a straight grooved pin is driven in up to the ring body. These pins absorb the relevant axial forces acting on the ring body. Two inserted O-rings seal the ring channel towards the casing.

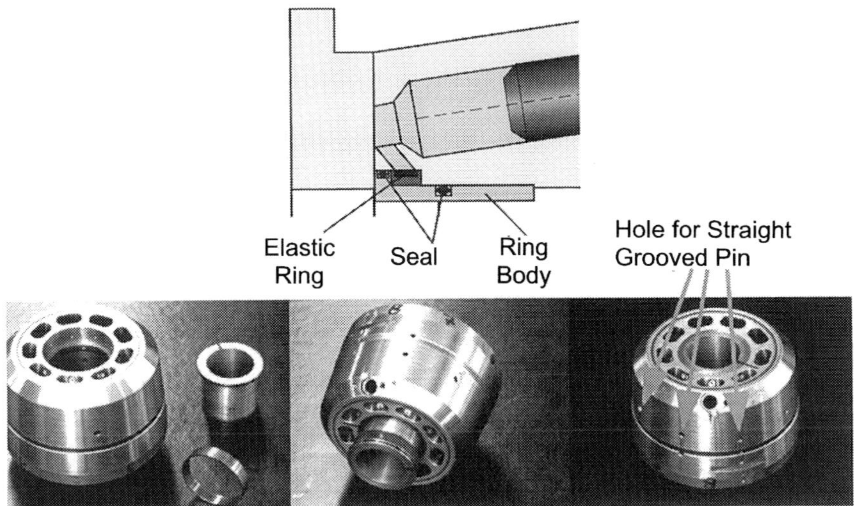

Fig. 4: Design of the ring system for investigation

Fig. 4 also shows how this design was realized in terms of hardware. The ring system is composed of a cylinder block, elastic ring, and ring body. When the ring system is assembled, the ring is put over the ring body and both components are forced into the cylinder block that was machined appropriately. The cylinder block is pinned to the ring body via holes running perpendicular to the drive shaft to prevent an axial movement caused by the pressure acting on the circular ring face at the front.

4 RESULTS

4.1 Determination of pulsation

Due to the limited dynamics of today's known principles of flow measurement (gear flowmeter, turbine) a direct measurement of the flow pulsation at the pump outlet port is not possible. However, the flow pulsation leads to a proportional pressure pulsation in the connected hydraulic system. This pressure ripple can be simply measured with dynamic pressure sensors. The pressure ripple is not only a function of the pump design but also of the circuit in which it is measured. To determine the pump generated pulsation different measurement techniques, for instance the "secondary-source method /3/ or measurement by using a "reflection loss end of pipe" /4/ were developed. A drawback of both techniques is a relative high demand in measurement hard- and software. A simpler method is described in /5/. The pump outlet port is connected to a termination orifice via a short test line. The measurement of the pressure ripple is made with a pressure transducer at the pump outlet port. According to ISO/FDIS 10767-2, Part 2 a test rig for measurement of pulsation was built up

(**Fig. 5**). The amplified sensor signal is lead to a FFT-Analyzer which allows a transformation in the frequency range.

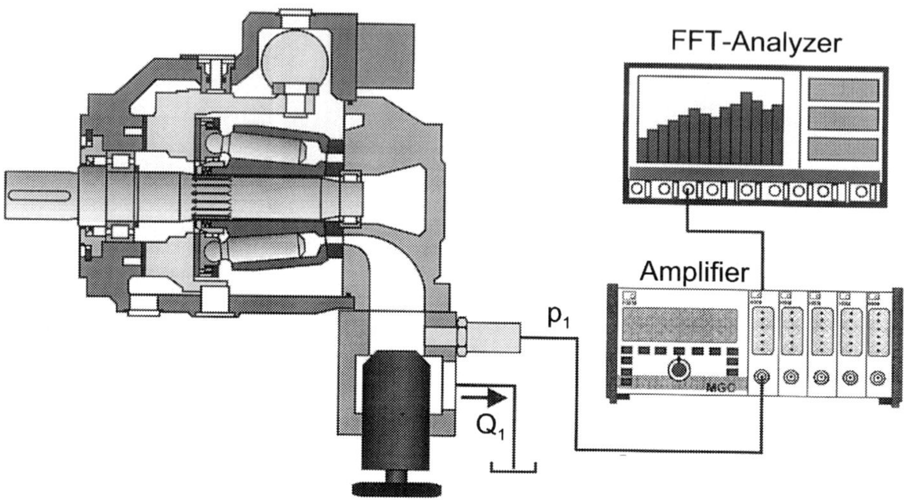

Fig. 5: Test rig for measurement of pulsation

Fig. 6 shows the measured pressure signal for the unmodified (9 pistons) pump on the left site. As expected the shape of graph is determined from an oscillation of the fundamental pump frequency:

$$f_f = \frac{n_1 \left[\min^{-1}\right]}{60} \cdot z \tag{1}$$

and its period:

$$T_f = \frac{1}{f_n} \tag{2}$$

respectively. The value of the fundamental period of a 9 piston pump at $n_1 = 2{,}000\ \min^{-1}$ is $T_f = 3.3$ ms. The natural pump frequency is superimposed by oscillations with higher frequency that result from the harmonics and the inductance between the piston chamber and the high-pressure kidney during the transition of the piston from the intake side to the delivery side /4/.

The right site of **Fig. 6** shows the calculated pressure ripple using a mathematical model of the pump. Both shape and amplitude Δp_1 of pressure pulsation agree well with the measurement.

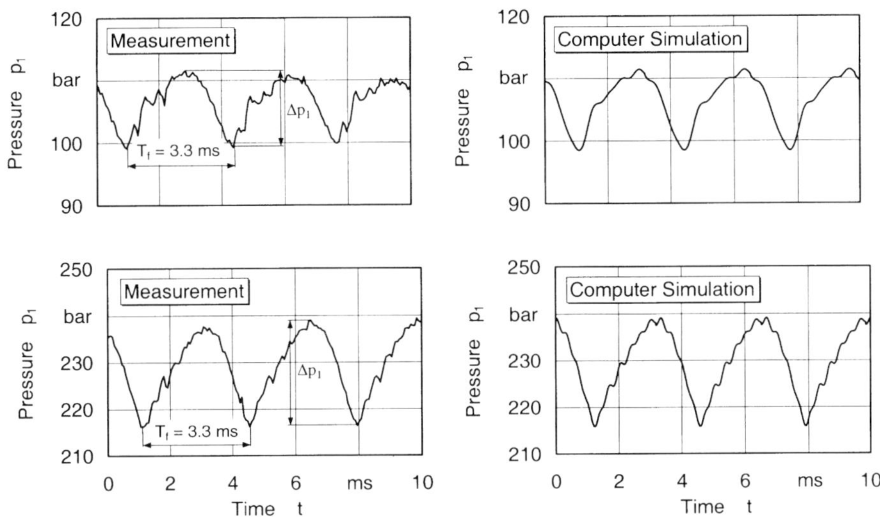

Fig. 6: Comparison of measured and simulated pressure ripple (Unmodified pump, $V_1/V_{1max} = 1$, $n_1 = 2.000\ min^{-1}$)

4.2 Results of the investigations

By means of the simulation model of a pump, a control system for a pump with a ring system was developed which meets the demands for a reduction of the pulsation at the pump outlet. To prove the pulsation-reducing action, a standard pump was appropriately modified. The modifications involve an altered cylinder block as well a new designed portplate. To measure the pulsation, a constant speed n_1 and a constant volumetric displacement is adjust. With the variable orifice at the end of the test line, the average pressure p_1 is set to the desired value and the pressure ripple is stored with a PC-based measuring data recording program.

For the evaluation, the difference between the maximum and the minimum measured pressure value (peak to peak level) is calculated. **Fig. 7** shows the ratio of the pressure pulsation of the modified and the standard pump $\frac{\Delta p_{1,\text{modified}}}{\Delta p_{1,\text{standard}}}$. A value of for instance 40% means that the pressure amplitude of the modified pump is reduced to 40% of the value of the standard pump.

The amount of reduction strongly depends on the operating point. For both maximum volumetric displacement V_{1max} and a displacement volume of $0.25 \bullet V_{1max}$, a reduction of the pulsation at the pump outlet by about 5 to 60 % in comparison with the standard design is achieved.

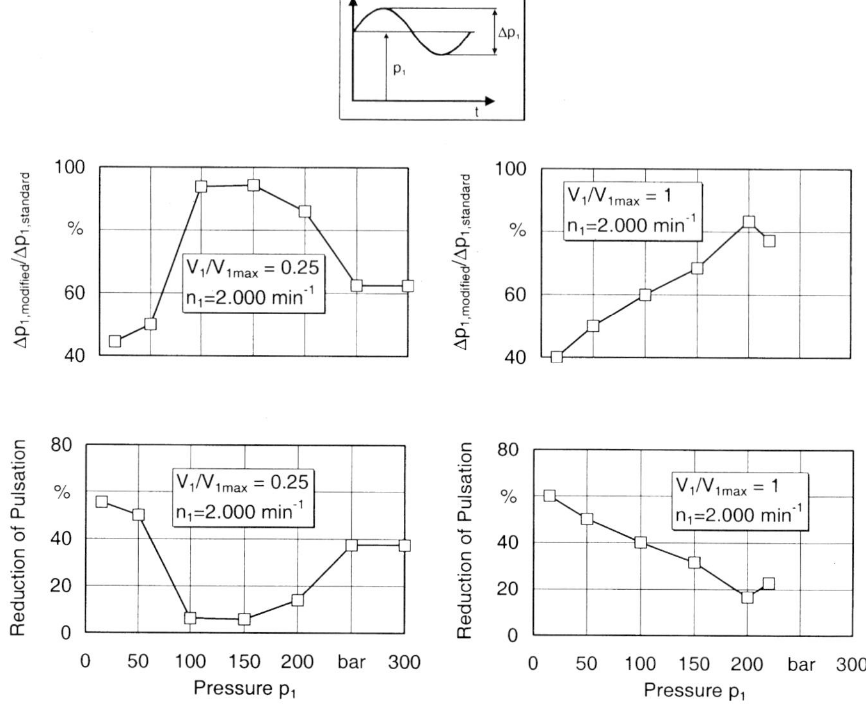

Fig. 7: Ratio of pressure pulsation of the modified and the standard design and reducing of pulsation (measurement)

Apart from pulsation, the efficiency of the pump is an important criterion of selection. To determine the effect that might have been caused by the modified design, comparative measurements of efficiency and of power loss were carried out in the standard and the modified pump. **Fig. 8** shows the difference in overall efficiency, volumetric and hydraulic-mechanical efficiency for the maximum volumetric displacement at $p_1 = 220$ bar in dependence on speed n_1. The maximum difference is for all efficiencies approximately 0.6 % and is therefore within the range of measurement accuracy of the arrangement. The measurements show that the efficiency of the pump is not affected by the modifications.

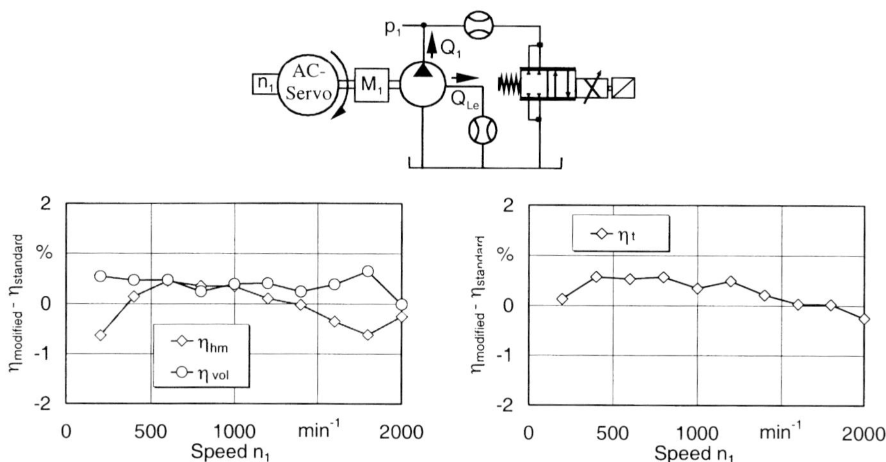

Fig. 8: Difference in the efficiencies between the modified and the standard pump ($V_1/V_{1max} = 1$, $p_1 = 220$ bar)

5 SUMMARY

We investigated an innovative system for the reduction of the pulsation at the pump outlet of axial piston pumps in which an elastic ring varies the cross-sectional area between the high-pressure side and the cylinder bore according to the operating pressure p_1 while the cylinder is communicating with the high-pressure kidney.
By means of a simulation program, a computation model for a swashplate axial piston pump was prepared. The combination with the results of the deformation analysis of the elastic ring allows the dynamic behaviour of such a pump and the effects of parameter changes to be investigated using computer simulations.
By means of the simulation model of a pump, a ring system was developed which meets the demands for a reduction of the pulsation at the pump outlet. The subsequent measurement of the modified pump shows that the pulsation is reduced throughout the operating range. Measurements also show that the efficiency of such a pump is not affected.

6 REFERENCES

/1/ Hydrostatische Maschine, Offenlegungsschrift DE 44 41 449 A1, 1996

/2/ Helduser S, Becher D: Simulation des dynamischen Verhaltens von Axialkolbenpumpen, Dresdner Tagung “Simulation im Maschinenbau 2000”, Band 2, 24. und 25. Februar 2000

/3/ ISO/FDIS 10767-2, Hydraulic fluid power – Determination of pressure ripple generated in systems and components – Part1: Precision method for pumps, 1996

/4/ Jarchow, M.: Maßnahmen zur Minderung hochdruckseitiger Pulsationen hydrostatischer
Schrägscheibeneinheiten, Dissertation, RWTH Aachen, 1997

/5/ ISO/FDIS 10767-2, Hydraulic fluid power – Determination of pressure ripple generated in systems and components – Part2: Simplified method for pumps, 1999

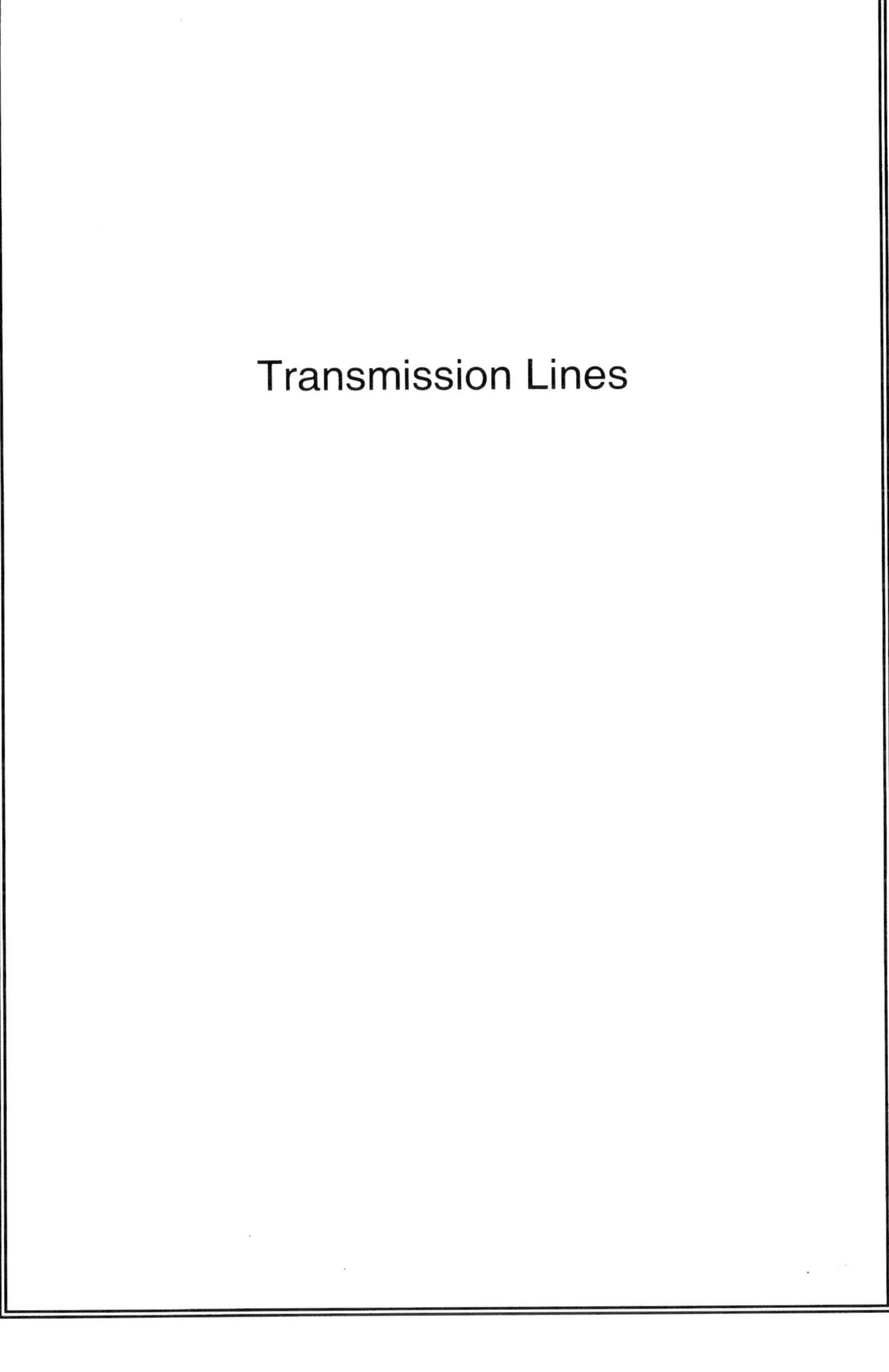

Transmission Lines

Distribution modelling – object-oriented implementation with Modelica and transmission lines

B JOHANSSON, P KRUS, and **J-O PALMBERG**
Fluid and Mechanical Engineering Systems, University of Linköping, Sweden

Abstract

Products are becoming increasingly complex with functionality and components from different engineering domains. Simulating these systems increases the demands on both modelling tools and methodology.

Modelica is a uniform object-oriented language for physical systems modelling. With functionality that gives the model a structure similar to the physical system, a straightforward modelling work is enabled.

The transmission line uses the delay in information transportation between components. The model becomes distributed implying a numerical time separation between components, resulting in robust and efficient simulation.

In this paper, it is illustrated how to combine Modelica with transmission lines for distributed simulation. It is also discussed what the advantages are compared to approaches that are traditionally used for the same type of problems.

1 INTRODUCTION

Simulation models of engineering components and sub systems are increasingly put together resulting in large and complex system models. This also results in a need for tools and methods to keep a good structure of the model and all information associated with the model. One approach is the object-oriented modelling technique, which means that the simulation models are built by objects with the same structure as the physical objects. The interaction between components is also comparable to physical connections between objects. To realise this in computer simulation models it is necessary to use methods and tools suited for this task to get a straightforward implementation. Two important modelling concepts are object-oriented modelling and the non-causal modelling technique.

Object-oriented languages are widely used in traditional software development as the complexity has increased rapidly in that area. In physical systems modelling, the most common languages are special design languages or traditional languages like C or Fortran. Using an object-oriented modelling language adds the possibility to build a hierarchic model structure and enables reuse of parts coded in an earlier stage or from a library of components.

The main advantage in using these constructs is to remove some of the necessary skill in computer programming and numerical analysis and add that skill to computer software. The designer should be able to focus on describing the model instead of the actual low-level implementation. There are various approaches to this type of modelling environment. The Modelica language typifies one approach. It is a non-causal and object-oriented language for physical modelling. Modelica is designed to be a general modelling language and the goal is to become a standard within modelling of physical systems.

Using a distributed modelling approach in combination with an object-oriented description, a good structure can be kept also during the simulation. This paper shows a solution of how to create a distributed model using the high-level modelling language Modelica and the Transmission Line Modelling technique, TLM.

2 ABSTRACTION LEVELS OF A SYSTEM MODEL

High-level model modelling languages could be viewed as a layer between the user interface and the low-level code that can be compiled and simulated, see Figure 1. The desired interface to the user is most of the times a graphical tool where the models can be created and manipulated. The simulation program then perform a translation of the graphical model to a low-level code that can be simulated, compiled and executed, typically C or Fortran. When new components are to be created, it is not always possible to design the model only using the graphical tool, so the code has to be written manually. Inserting a high-level modelling language simplifies this process. Low-level programming knowledge is then not required and focus can be set on physical properties. High-level descriptions that are separated from the actual implementation can also be used for model exchange. It is desired for exchange formats to be standardised describing the model and not the implementation for a specific simulation package.

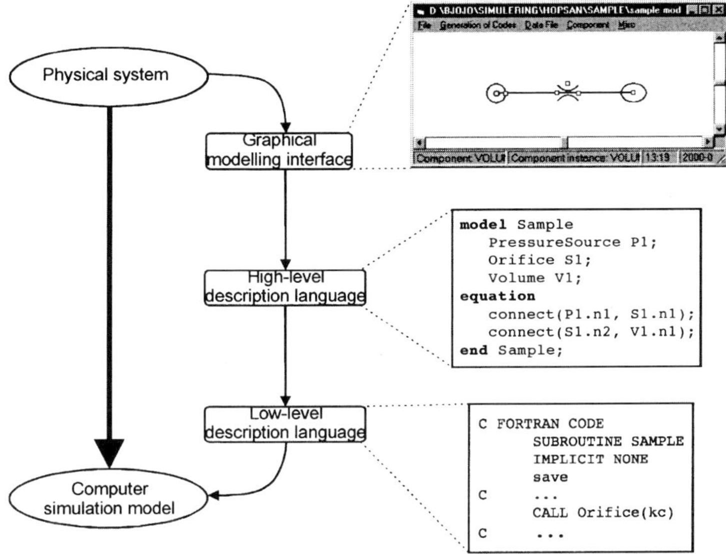

Figure 1: Abstraction levels of a simulation model

2.1 Common Modelling Principles

Mathematical models of physical systems can be made using a number of concepts. There are a few concepts that are more used than others. These are block diagrams, bond graphs and system decomposition or power port modelling.

2.1.1 Block Diagrams

Block diagrams are composed of causal input/output blocks linked together to create a system. This concept is for example used in the Matlab/Simulink simulation package [9]. This is a very useful method for example in modelling of control systems where it is interesting to follow specific signals through the system. When modelling physical systems, for example the hydraulic system showed in Figure 2, block diagrams is not so convenient.

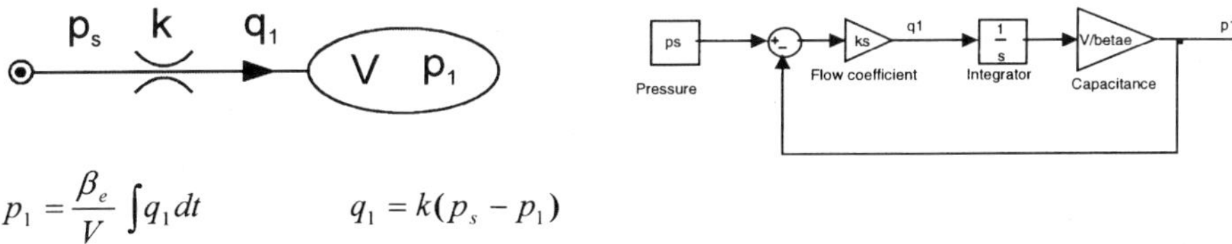

$$p_1 = \frac{\beta_e}{V} \int q_1 dt \qquad q_1 = k(p_s - p_1)$$

Figure 2: A simple hydraulic system **Figure 3: A block diagram of the system**

To model such systems using block diagrams, the equations describing the system must be expressed using statements where input and respective output must be chosen.

The block diagram in Figure 3 shows that the visual correspondence to the physical system in Figure 2 is not so clear.

2.1.2 Bond Graphs

The drawback of block diagrams is the need of an input/output representation of the system. Physical systems naturally do not have that behaviour. The major and common characteristic of all physical systems is the flow of energy between sub components. This energy does not flow in a specific direction, which means that it would be useful if the causality constraints could be left out in the early modelling phase. These considerations lead to the development of bond graphs [11]. The main idea is to model energy flow between components. Using bond graphs, the advantage is that the models often have good quality because the technique requires full understanding of the physical properties of the system. The basic idea is that models of physical objects can be described in terms of effort and flow variables. Examples of effort variables are electric voltage, force or hydraulic pressure. The corresponding flow variables are electric current, speed and hydraulic flow. In a connection between two physical sub components, these two types of variables can be identified. One example is the fluid in a hydraulic pipe where the flow and pressure are the variables describing the exchange of energy. In a mechanical joint velocity and force can be identified. For almost all types of physical systems, analogies can be made to describe exchange of energy between sub systems.

2.1.3 System Decomposition – Power Port Modelling

When modelling complex systems, decomposition into a number of smaller systems is a way to bring structure to the problem [6]. Decomposing the system could of course be made in a number of ways. The question is how the components should be defined and how the connections between the components should be described. A large number of physical systems are already built of components interfaced by distinct connections. For example in the hydraulics domain, hydraulic components are connected by hydraulic pipes. The electric domain is characterised by electric components connected by wires. When building models of the physical systems it is useful if the physical structure could be used in the model. Power port modelling is a concept, which focuses on these considerations. The system model is built by components representing the physical components that are connected by *connectors* representing the physical connections, see Figure 4. The connectors are also referred to as *cuts*, *nodes*, *ports* or *attachments points*.

Figure 4: System modelled using system decomposition and power ports

This modelling technique is non-causal because the signal flow between components can be bi-directional. The variables in the connectors are of two types, recognised from bond graphs modelling: effort (or across) and flow (or through) variables. The two types have different constraints in the connectors. When connecting components, the corresponding effort variables from the connected components are identical, whereas the sum of the corresponding flow variables is zero. This is well illustrated within an electric circuit. In a connection point, all voltage potentials from the connected components are equal and current is summed to zero (Kirchhoffs law). Analogies for this constraint can be found in other physical domains as well.

3 OBJECT-ORIENTED MODELLING

Object-oriented *modelling* languages are developed with the main ideas from the more general object-oriented *programming* languages. Examples of those are C++ and Ada95. The main idea behind object-oriented programming is to be able to build computer programs that have a similar structure as physical objects. The software development process will then be more straightforward especially in the development of large and complex systems. Other important features in object-oriented programming languages are the possibilities to build a hierarchic structure of the program code and inherit or reuse already coded parts.

3.1 Main Characteristics of Object-Oriented modelling languages

The object-oriented modelling languages are adapted for modelling of physical systems. Most paradigms of these languages are the same as for those for object-oriented programming languages. The main characteristics can be summarised as [1]:

- Encapsulation of knowledge: It should be possible to model an object in one place and in a compact fashion. It should be possible to create well-defined interfaces to the outside.
- Topological design capability: The modeller should be able to interconnect components in the same way as the equipment is connected in the real world. The ability to "plug" components without considerations of the directions of signal flow requires a non-causal equation declaration.
- Hierarchical modelling: It should be possible to group a collection of objects to a new component with the same interfaces as the consisting objects.
- Object instantiation: The modeller should have the possibility to code generic descriptions of object classes or templates, and instantiate complete objects from these classes.
- Class inheritance: Inheritance in this respect means the possibility to inherit properties from base classes. The advantage is that knowledge not need to be coded several times at different places in a model.

3.2 Modelica – An Example of an Object-Oriented Modelling Language

One example of an object-oriented modelling language that fulfils the characteristics described above is *Modelica* [3, 10]. It is a language developed in an international effort with contribution from both academia and industry. The ambition is to develop a language that becomes a standard within modelling of physical systems. Modelica is suited for modelling in several engineering domains, for example the mechanical, electrical and hydraulic domains. The language is object-oriented and uses a non-causal structure that makes it possible to use equations in a natural way in the modelling code. The language supports connections between sub components using nodes as in the power port modelling described earlier in this paper.

To give a short overview of Modelica and of object-oriented modelling a simple hydraulic system will be modelled. The model consists of a hydraulic pump transforming mechanical rotation to hydraulic flow and a motor performing the opposite transformation.

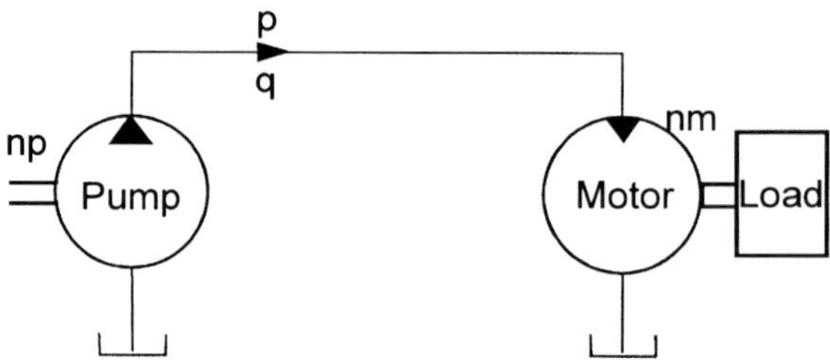

Figure 5: Hydraulic transmission

First, variable types can be defined to suit the variables in the actual systems. Additional information such as quantity, unit and description can be included in the type description.

```
type HydPressure = Real(quantity = "Pressure", unit = "Pa");
type HydFlow = Real(quantity = "Flow", unit="m3/s");
type Torque = Real(quantity = "Torque", unit="Nm");
```

To connect components coded in Modelica there is an object called *connector* defining the interface between components.

```
connector Port
  HydPressure p;
  flow HydFlow q;
end Port;
```

It is possible to define base components for reuse, a so-called *partial* model. It could for example be a simple hydraulic component only defining that it is built by two hydraulic ports and that the pressure drop is the pressure difference between the ports. As a model can inherit properties from objects already defined consisting variables itself, the notation for relating these variables are: `object.variable` or `object.subobject.variable`. This is a common syntax of defining the path to the declaration of the variable in object-oriented languages.

```
partial model TwoPort "Superclass of elements with two hydraulic ports"
  Port n1, n2;
  HydPressure pdif;
  HydFlow q;
equation
  pdif = n1.p - n2.p;
  0 = n1.q + n2.q;
  q = n1.q;
end TwoPort;
```

The partial and incomplete model could now be extended to a complete model of a component using the keyword *extends*. The complete model of the hydraulic pump will then be:

```
model HydraulicPump "Simple model of a hydraulic pump"
  extends TwoPort;
  parameter Real omega(unit="rad/s") "Rotational speed";
  parameter Real Dp(unit="m3/rad") "Pump displacement"
  Torqe M "The input torque on the pump shaft";
equation
  q = Dp*omega;
  M = Dp*pdif;
end HydraulicPump;
```

With a similar model of the hydraulic motor, a system model can be described. The motor is assumed to have a mechanical connector where the external load can be attached. The components are connected using a *connect* statement which generates the equations that defines the connection.

```
model SimpleHydraulicSystem
  HydraulicPump P1;
  HydraulicMotor M1;
  MechanicLoad L1;
  Tank T1,T2;
equation
  connect(T1.n1, P2.n1);
  connect(P1.n2, M1.n1);
  connect(M1.n2, T2.n1)
  connect(M1.n3, L1.n1)
end SimpleHydraulicSystem;
```

The code only describes the actual model, no information is given how to actually execute this simulation model. The natural implementation is to flatten the code and transform it to a description that can be numerically solved by the computer. It is though free for any software developer to implement support for Modelica in a desired way.

4 DISTRIBUTED MODELLING

Distributed modelling is a concept to numerically separate components in a model during the simulation. This is done introducing a time delay between the components. The concept will be described below and compared to the opposite solution, centralised modelling.

4.1 Distributed modelling vs. centralised modelling

The distributed modelling concept can be viewed as an opposite solution to the centralised modelling technique in how the system is solved. A brief description of the solving of a centralised model is that all equations describing the dynamic system are collected to one large system of equations. An integration algorithm is then applied to this large system and all state variables are solved for in each time step. This concept is very common and used in most simulation software. Advanced symbolic manipulation of the equations is also possible to achieve a set of differential and algebraic equations that can be solved efficiently. The drawback of the centralised simulation technique is that the whole system must be stated in a uniform way. Another drawback is that fast dynamics in one part of the system effect the step-size and simulation time of the whole system.

The opposite approach is as mentioned to use a distributed modelling technique. The main idea is to include the propagation of information through the system. In all physical systems, there is a delay in time when a signal propagates from one component to another. For example a pressure pulse that is caused in a hydraulic valve does not immediately effect the whole system. The delay is depending of a number of parameters, as distance and fluid type. This delay in information propagation is a valuable piece of extra information that can be used to separate the components numerically. Another criteria for a distributed model is that the components are individuals that are designed to execute and solve themselves. This means that a component solves its own equations and delivers the information through the nodes to the connected components. One approach to implement a distributed modelling architecture is the transmission line modelling concept [5], which will be described below.

4.2 Distributed modelling using transmission lines

The core component in Transmission Line Modelling, TLM, is the unit transmission line, UTL. This component can be viewed as carrier of energy with a time delay. Depending of the system that is to be modelled the medium passing through the transmission line is different, but the mathematical variables are flow and effort variables, the same way as in the Bond-graph modelling technique.

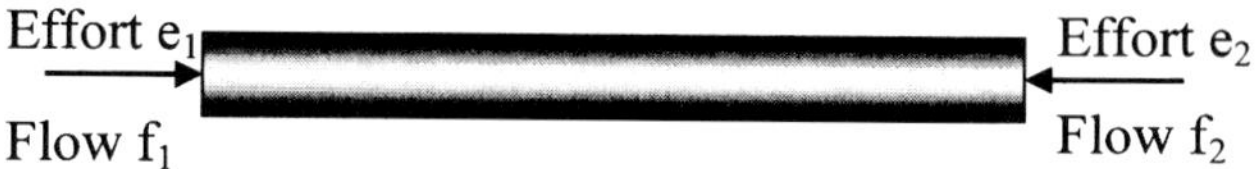

Figure 6: Unit Transmission Line

The equations describing the transmission line are

$$e_1(t+T) = c_1(t) + Z_c f_1(t+T) \tag{1}$$

$$e_2(t+T) = c_2(t) + Z_c f_2(t+T) \tag{2}$$

where Z_c is the characteristic impedance, c_1 and c_2 are the characteristics describing the wave from the connected component. The characteristics are of the same magnitude as effort and are defined as

$$c_2(t) = e_1(t) + Z_c f_1(t) \tag{3}$$

$$c_1(t) = e_2(t) + Z_c f_2(t) \tag{4}$$

At each component between the lines, the equations that needs to be solved are the following set of equations

$$\mathbf{f} = \mathbf{f}(\mathbf{e}) \tag{5}$$

$$\mathbf{e} = \mathbf{c} + \mathbf{Z_c f} \tag{6}$$

where **f** is a vector containing all the flow variables, **e** is the corresponding effort variables, **c** is the characteristics and $\mathbf{Z_c}$ is a diagonal matrix with the characteristic impedances in the diagonal.

4.3 Transmission line components in a system

In a system, the components modelled as TLM- components are divided into two groups of components.

- Q-components, calculating flow
- C-components, calculating characteristic impedance and the characteristic (the wave, effort variable)

During the simulation, the different types of components are executed alternately. First, the C-components are called, updating the characteristics and characteristic impedances. Then the Q-components are called calculating the state variables as flow and effort.

5 MODELICA COMPONENTS IN A DISTRIBUTED SYSTEM MODEL

As mentioned above, the Modelica language is a high-level object-oriented modelling language. The use of Modelica simplifies the modelling work in several ways thanks to the non-causal technique, the object-oriented structure and the straightforward syntax. The drawback of Modelica implementations available today is that the structure built in the models is not used in the simulation.

The TLM-concept keep the objects separated through the simulation with benefits as numerical efficiency, possibilities to trace cause effects from single components, parallel computation etc. This means that a solution of Modelica components that are used in a TLM-environment would have the potential to show a number of advantages compared to a traditional implementation using the centralised technique.

In this work, a tool for automatic translation of Modelica components has been created using the Mathematica package [12]. Mathematica is a symbolic math package, enabling high-level programming. Besides the mathematical processing, advanced text operations can be performed as well. The tool translates a Modelica component to a component written in Fortran that can be inserted in a HOPSAN [4] simulation model.

HOPSAN is a simulation package using the distributed modelling technique described in section 4. The models that are simulated in HOPSAN are normally written in the Fortran language, but components are stored in a library and system models are created using a graphical modelling tool. The following parts will demonstrate how a Modelica component is adapted to fit in a TLM environment.

5.1 Translating Modelica components

Distributed modelling using the TLM-methods requires models described in a specific way. The components should have extra variables for the wave propagation and a solver for independent solving of that component. The Modelica component could be stated directly in that way, still following the Modelica syntax. It is desired however that the model designer should not have to care about issues that are specific to the implementation. The main advantage with Modelica is that the model code should be separated from the implementation. This means that the Modelica component must be manipulated to fit into the TLM-environment.

Extending a previously created tool in the Mathematica package, called *Compgen* [7], the component file written in Modelica can be processed according to a sequence that can be described as follows:

Step 1: Read the Modelica component file.

Step 2: Find connectors, variables and parameters.

Step 3: Sort out the equations describing the component.

Step 4: Transform the equations to Mathematica notification.

Step 5: Define the type of component, Q or C-component, see section 4.3.

Step 6: Add the boundary equations.

Step 7: Call the *Compgen* algorithm, performing the translation to a HOPSAN component.

Step 8: The component can now be connected to HOPSAN components through the graphical interface

Through step 1-4, the text file containing the component is scanned by the Mathematica program. Using the Modelica syntax definition, the information is then sorted and grouped. The mathematical notification in Modelica and Mathematica is not the same, which is why a translation must be performed so that the equations can be analytically manipulated by Mathematica. As several components can be designed as either Q or C-components, it is for the user to select accordingly. Depending on the type different variables are updated during the execution.

The tool is only able to translate a subset of the Modelica language. The effort has been put into creating a tool that works for single components and the main purpose will be to facilitate exchange of components between tools and people. The major limitation is that inheritance is not handled, which means that a component must be fully described with the complete set of equations.

As showed in section 3.2, a system of components can be described in Modelica using the **connect** statement. This is not yet implemented in the tool presented here. It is though interesting to notice that the system description for the program generator DYNMOC, developed by Arne Jansson for the HOPSAN environment, has a very similar syntax as in Modelica, see Figure 7. The minor syntactical differences mean that it would be straightforward to create a translator also for the topological description so that complete Modelica systems can be simulated in HOPSAN.

<table>
<tr><th>Modelica topological description</th><th>HOPSAN topological description</th></tr>
<tr><td><pre>
model SimpleHydraulicSystem
  HydraulicPump P1;
  HydraulicMotor M1;
  MechanicLoad L1;
  Tank T1,T2;
equation
  connect(T1.n1, P2.n1);
  connect(P1.n2, M1.n1);
  connect(M1.n2, T2.n1)
  connect(M1.n3, L1.n1)
end SimpleHydraulicSystem;
</pre></td><td><pre>
component HydraulicPump P1
component HydraulicMotor M1
component MechanicLoad L1
component Tank T1
connect T1 n1 to P2 n1
connect P1 n2 to M1 n1
connect M1 n2 to T2 n1
connect M1 n3 to L1 n1
</pre></td></tr>
</table>

Figure 7: Topological descriptions in Modelica and HOPSAN showing a similar syntax

One problem is that in Modelica, the system description is necessarily not separated from the descriptions of the components. Connections are transformed to normal equations when the model is transformed to low-level code. HOPSAN uses a separate file to describe the connections on system level with the syntax shown in Figure 7. This means that to be able to transform a Modelica system to a distributed environment like HOPSAN, the code must be restricted. It must be stated what defines the system description, expressed either in a separate file, or separately between identifiers.

5.2 Example

Consider a simple system describing a hydraulic transmission. The system is modelled in the HOPSAN package using components from the component library. Using the model translator presented in this paper, a model of an electric motor written in the Modelica language can be automatically translated and connected to the hydraulic transmission.

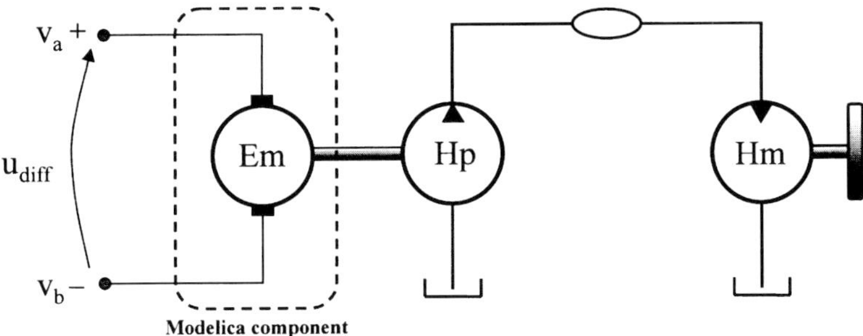

Figure 8: Hydraulic transmission in the HOPSAN environment

The hydraulic system consists of a pump and a motor with an intermediate volume. A mechanical load is attached to the hydraulic motor. To the hydraulic pump, a prime mover can be connected through a mechanical shaft, modelled as a stiff spring. This is where the Modelica model of an electric DC-motor is to be attached.

5.2.1 System of equations

The simplest model of an electric DC-motor can be viewed as an inductance, resistance and a converter describing relations between electrical and mechanical quantities. The following equations describe the component:

$$\frac{di_a}{dt} = \frac{(v_a - v_b) - R_a \cdot i_a - \omega_x \cdot \psi_m}{L_a} \tag{7}$$

$$J \cdot \frac{d\omega_x}{dt} = i_a \cdot \psi_m - b \cdot \omega_x - M_x \tag{8}$$

$$i_a = -i_b \tag{9}$$

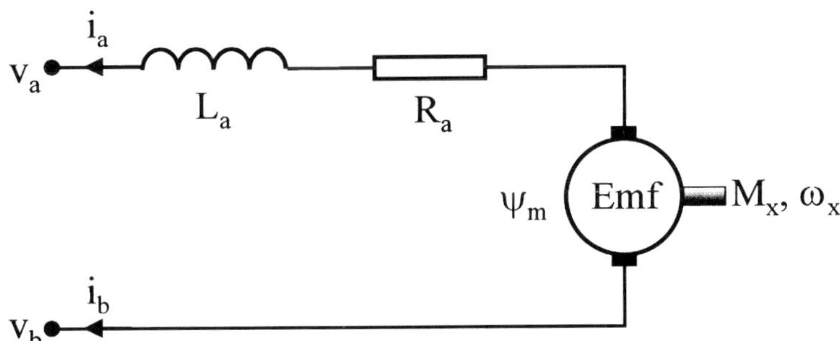

Figure 9: DC-motor circuit

If the electric motor is viewed as one component, there are three connections to surrounding components: two electrical nodes and one mechanical. For this component to fit in a TLM-environment there must be at least four variables in the connector to model the exchange of energy. These variables are flow, effort, the characteristic and the characteristic impedance. In the component, this is handled by adding one extra equation for each node describing the boundary conditions. The three boundary equations are:

$$v_a = c_a + Z_c \cdot i_a \tag{10}$$

$$v_b = c_b + Z_c \cdot i_b \tag{11}$$

$$M_x = c_x + Z_{cx} \cdot \omega_x \tag{12}$$

where c_a, c_b, and c_c are the waves that are calculated from the connected nodes. Z_c and Z_{cx} are characteristic impedance also delivered from the connected components.

5.2.2 Solving the system

The model must be equipped with a numerical solver to become an independent component in the TLM system. This can be done using any suitable solver. In this work, the Newton-Raphson iteration algorithm is used based on that it has shown very good robustness in combination with the TLM-method. A short description of the steps performed inserting the solver to the component will be given below. For a detailed description of the process, see [7].

The equations (7)-(9) describe the internal relations in the component and are those that we found from the Modelica base component. Call this set of equations F_a and the boundary equations (10)-(12) F_b. The total set of equations can then be expressed as:

$$F(F_a, F_b, t) = F(y, \frac{dy}{dt}, \frac{d^2y}{dt^2}, \ldots, \frac{d^ny}{dt^n}, t) = 0 \tag{13}$$

The following steps are:

- $F(F_a, F_b, t)$ is transformed into time discrete representation using bilinear transform. The result is called G
- The Jacobian J is evaluated from partial derivatives of G
- The equations are solved numerically using the Newton-Raphson iteration

$$y_{k+1}(t) = y_k(t) - J_k(t)^{-1} G(y_k(t)) \tag{14}$$

After performing these iterations, the component can be written to a Fortran file with the correct node interface and the solver. The procedure translating the DC-motor written in the Modelica language has now resulted in a component that now can be inserted into the HOPSAN environment and simulated. The system inserted in the graphical modelling tool *GDynmoc* [8] can be viewed in Figure 10.

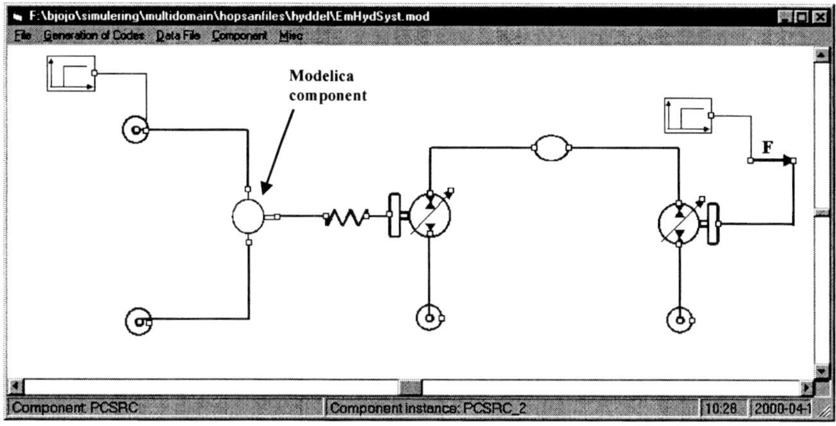

Figure 10. The simulated system

5.2.3 Simulation results

From the complete system model in the HOPSAN environment, some simulation results can be obtained.

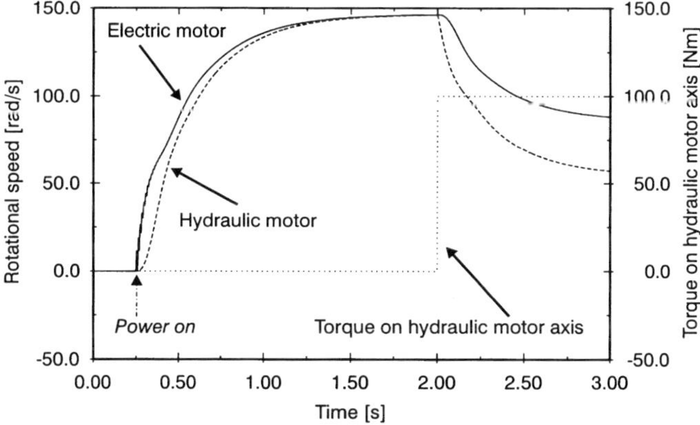

Figure 11: Results from the simulation

The simulation shows the speed of the hydraulic motor and the electric motor. When the power to the electrical motor is turned on, the speed of the electric motor increases followed by a similar behaviour of the hydraulic motor. The delay is due to the volume in the hydraulic system, which also affect the behaviour when a step in load is applied. The speed of the hydraulic motor decreases faster and more than the electric motor. The same model executed in HOPSAN has also been simulated in the Dymola environment, which has implemented support for Modelica [2]. The results show good accordance between the different simulation packages. The simulation of this system for a sequence of three seconds is performed in 0.4 seconds with a fixed time step of 1 ms.

6 DISCUSSION

Object oriented modelling shows great advantages in modelling of physical systems. With characteristics as encapsulation, object instantiation and node connections, the simulation models show a structure that is flexible, logical and with strong similarities to the physical system. Non-causal modelling removes the need of state-space models, reducing error-prone transformations. Instead, differential and algebraic equations from handbooks can be used directly.

The Modelica approach is a promising effort to a standard modelling language for dynamic systems. If a standard become widely accepted, it would mean a large step forward in simulation of multi-domain dynamic systems. As simulation packages are becoming more general and capable to perform simulation in different domains, the possibilities with exchanging models between different participants in a design group are significant.

The approach presented in this paper is an implementation of Modelica where component models are simulated in a distributed environment. The demonstrated translator is capable of translating a subset of Modelica with the major limitation that inheritance is not implemented. In a scenario where components are to be exchanged between members of a design group, inheritance is though considered to be of secondary importance. Although, the modelling effort can be reduced using inheritance, the code is more straightforward and easy to follow if inheritance is not used. The limitation is only of significance for the import function to HOPSAN and not in the opposite direction as there are no modifications to the Modelica syntax done to perform the translation.

The outlook using an object-oriented modelling language in a distributed environment is to handle large simulation models by partitioning the system not only through the modelling process but also through the simulation. A flexible, numerically robust and efficient environment can be achieved suited for large and complex systems. Using a standardised language as Modelica also adds advantages as increased integration between project members and supported knowledge exchange between engineering disciplines.

7 CONCLUSIONS

Using an object-oriented modelling language together with a distributed modelling technique, the objects in a system model can be kept partitioned as objects through both modelling and simulation.

The implementations of the Modelica language that are available today are based on the centralised modelling technique, showing drawbacks compared to distributed modelling when it comes to large and complex models. In this paper, it has been demonstrated how to use Modelica in a distributed environment. Component models are automatically translated from standard Modelica code to fit in the distributed environment.

The authors consider the presented combination of Modelica and transmission lines to be a powerful modelling environment for design of robust simulation models supporting model exchange and efficient simulation.

8 REFERENCES

[1] Cellier F. E., "Object-Oriented Modeling: Means for Dealing With System Complexity," presented at 15th Benelux Meeting on Systems and Control, The Netherlands, 1996.

[2] Elmquist E., Brück D., and Otter M., *Dymola - User's Manual*, Dynasim AB, 1999.

[3] Elmqvist H., Mattson S. E., and Otter M., "Modelica - An International Effort to Design an Object-Oriented Modeling Language," presented at Summer Computer Simulation Conference, SCSC'98, Nevada, USA, 1998.

[4] Jansson A. and Krus P., "HOPSAN a Simulation Package User's Guide", http://hydra.ikp.liu.se/hopsan.html, 1998.

[5] Johns P. B. and O'Brien M., "Use of transmission line modelling (t.l.m) method to solve nonlinear lumped networks," *The Radio Electron and Engineer*, vol. 50, pp. 59-70, 1980.

[6] Krus P., "Modelling and Simulation of Heterogeneous Engineering Systems", http://hydra.ikp.liu.se/~petkr/SMSOcourse/ModellingSimulation.pdf.

[7] Krus P., "An Automated Approach for Creating Components and Subsystems for Simulation of Distributed Systems," presented at 9th Bath International Fluid Power Workshop, Bath, UK, 1996.

[8] Larsson J., "GDynmoc User's guide", IKP-R-1088, Dept. of Mech. Eng., Div. of Fluid and Mech. Eng. Systems, Linköping University, Linköping, Sweden, 1999.

[9] Math Works Inc., *Simulink - A Program for Simulating Dynamic Systems*, Math Works Inc., 1992.

[10] Modelica Design Group, "Modelica A Unified Object-Oriented Language for Physical Systems Modeling, Language Specification", http://www.Modelica.org, 1999.

[11] Paynter H. M., *Analysis and design of engineering systems*. Cambridge, Mass, MIT Press, 1961.

[12] Wolfram S., *The MATHEMATICA Book*, 4 revised edition, Cambridge University Press, 1999.

Pneumatic pipes – experimental and simulation approach

E BIDEAUX and **S SCAVARDA**
Laboratoire d'Automatique Industrielle, INSA de Lyon, Villeurbanne, France

Abstract : In fluid power systems, junctions and pipes may have a predominant effect on dynamics. In this paper, as a part of the modelling process, we propose an experimental analysis of pneumatic pipes. The experimental approach will be related to the modelling problem. Various experimental results will be analysed and discussed: influence of the main parameters related to the technology, i.e. diameter, length, material, etc. The spatial distribution of the pressures will be observed and the results will be used to criticise the models based on finite difference or volume schemes.

NOTATION

e	Specific internal energy (J/kg)
P	Absolute pressure (Pa)
T	Temperature (K)
ρ	Specific mass (kg/m^3)
E_T	Total energy (J)
q_m	Mass flow rate (kg/s)
h	Specific enthalpy (J/kg)
m	Mass (kg)
u	Velocity of gas (m/s)
p	Momentum (kg.m/s)
T_{out}	External temperature (K)
Cp	Constant-pressure specific heat (J/kg/K)
f	Friction factor (kg/m^3)
K	Thermal exhange constant (J/K/m^2)
S	Pipe section (m^2)
d	Pipe diameter (m)
l	length of pipe (m)
x	Position along the pipe (m)
t	Time (s)
Sx	External exchange area (m^2)
Δx	Distance between two nodes or length of control volume (m)
i	Indices for the nodes
I	Indices for the volumes

1 - INTRODUCTION

The phenomena introduced by long pipes may have crucial incidences in some industrial application (7, 9). In the truck or railway industry, the brake system uses pneumatic energy and the dynamics of this system influence the safety and the lifetime of the vehicle. In such applications, the control signal to the brake cylinder is usually a pneumatic pressure carried by pipes. However electro-pneumatic systems allow better performances in driving the brake system, the time delay and the oscillatory phenomena induced by long pipe are still essential. The main goals of this experimental approach are an evaluation of the significance of these phenomena in the case of pneumatic brake system and an estimation of the detail level required by the simulation model.
In a basic approach, there are three conservation laws giving an expression of the general behaviour of gas flow in a pipe. These conservation laws are usually completed in CFD (10, 11) by the k-ε laws, which represent others energy forms (turbulent phenomenon). Although the k-ε scheme could be very useful in he case of gas, we will limit our work to the use of mass conservation, momentum conservation and energy conservation. These laws are commonly known as the following equations in a mono-dimensional case :

Mass conservation : $\frac{\partial \rho}{\partial t}+\frac{\partial \rho}{\partial x}=q_s$ where q_s is the flow source term (i)

Momentum conservation : $\frac{\partial \rho u}{\partial t}+\frac{\partial \rho u^2}{\partial x}+\frac{\partial P}{\partial x}=\frac{\partial \tau}{\partial x}$ where $\frac{\partial \tau}{\partial x}=-f\frac{u|u|}{2d}$ (ii)

Energy conservation : $\frac{d\rho e}{dt}=-P\frac{\partial u}{\partial x}+R+\Phi+\lambda\frac{\partial^2 T}{\partial x^2}-h_T(T-T_{out})$ (iii)

R is the chemical, radiation, phase change power
Φ is the power of the local viscosity forces
λ is the thermal conduction
h_T is the thermal exchange coefficient

2 – EXPERIMENTAL BENCH AND PARAMETERS ESTIMATION

Our first approach was to observe the static behaviour of the gas flow in pipes. The implemented bench (figure 1) consists of two pressure servo-valves controlling the upstream and the downstream pressure and the pipe to study fitted out with 5 miniature pressure sensors (sensing element diameter : 1.2 mm). The pressure sensors were flush-mounted in specific flanges in order to avoid the introduction of perturbation in the flow. A range of mass flowmeters enables the measure of the flows less than 2500 Nl/mn[1] (about 50 g/s).

[1] 1 Nl/mn (≈ 0.02 g/s) is the equivalent volume flow rate at normal conditions of pressure and temperature (1 atm, 288.16°K)

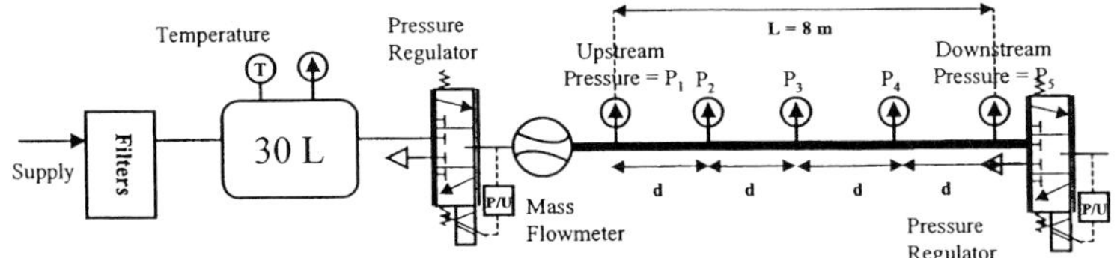

Figure 1 : Bench used for the pipe characterisation

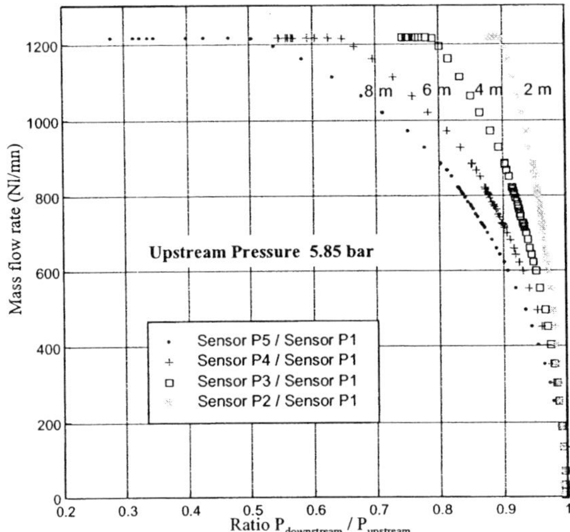

Figure 2 : 8 metres long pipe static characterisation

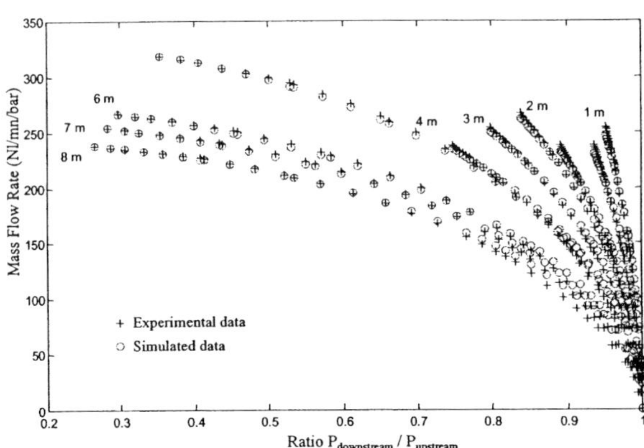

Figure 3 : Static characterisation of 8 metres long pipes using different length

A first set of trials has been realised with 6, 9 and 12 millimetres diameter polyamide tubes. The mass flow rate and the pressure distribution along the pipe were studied in order to obtain static characteristics of the pipe according to its length and its diameter. Different values of upstream and downstream pressure were applied and for each configuration, we plot the mass flow rate versus the ratio between the upstream pressure and the pressure at some position on the pipe.

In this stationary case, the pipe behaves as an orifice and shows sonic and subsonic regions. The deduced sonic conductance is related to the length and to the diameter of the pipe. As an example, figure 2 shows the mass flow rate versus the pressure ratio at different positions in a pipe of length 8 metres (9 mm diameter) and illustrates the sonic shock phenomenon. Figure 3 shows a pipe flow characteristic (9 mm diameter) for different lengths and at different position of the pressure sensor along the pipe.

In a second step, these measurements were used to evaluate distributed pressure drop and the friction coefficient. In stationary conditions, the momentum conservation law gives :

$$\Delta P = \tfrac{1}{2} fr(Re) \frac{\Delta x}{dS^2} q_m^{\,2} \qquad \text{(iv)}$$

From this relation, we can assume that the friction coefficient is :

$$2 \frac{dS^2}{\Delta x} \frac{\Delta P}{q_m^{\,2}} = fr(Re) \qquad \text{(v)}$$

The friction is a function of the Reynolds number in theory (3) and may be approximated by the Poiseuille and Blasius laws for Reynolds number less than 10^5 although it depends on the roughness of the pipe material.

$$\begin{cases} f = \dfrac{64}{Re} & \text{for} \quad Re < 2000 \quad \textit{(Poiseuille)} \\ f = 0.316 Re^{-1/4} & \text{for} \quad 4.10^3 < Re < 10^5 \quad \textit{(Blasius)} \end{cases} \qquad \text{(vi)}$$

$$\text{and } Re = \frac{\rho d \bar{u}}{\mu} = \frac{d}{S\mu} q_m \qquad \text{(vii)}$$

Comparing the theoretical to the evaluated values of the friction coefficient according to the Reynolds number, we observe a good correlation for Reynolds number greater than 10^4 (figure 4). Because of the low viscosity of gas, it is practically very difficult to observe small Reynolds numbers and in order to correlate the theory with trials, very accurate pressure sensors are requested to obtain pertinent data. This observation is nevertheless sufficient to validate the use of the theoretical law for the friction coefficient in the case of dynamic behaviour as the small Reynolds numbers only appear when the pneumatic system is around its steady state.

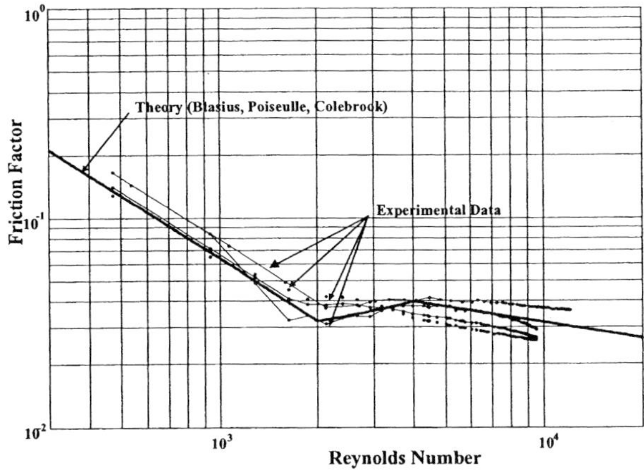

Figure 4 : Comparison in between theoretical and evaluated friction coefficient

The comparison in between the measured and the theoretical maximum mass flow rate (figure 5) shows that using the theoretical model of the friction coefficient we obtain rather good results for any studied lengths and diameters.

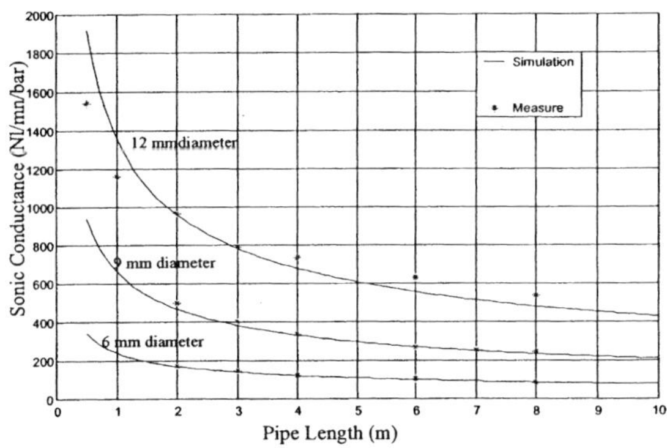

Figure 5 : Comparison of maximum mass flow rate versus length and diameter

3 – Simulation Model

In the case of gas flow, the choice of the right variables for the model are not definitively fixed, and, as the state equation for a perfect gas imposes three independent variables, many combinations are available (1). Our choice is guided by the fact that the considered pipes have a constant section and that the discretization scheme will impose a fixed length for each control volume. This means that the traditional specific mass ρ may advantageously be replaced by the gas mass and the velocity by mass flow. Knowing the specific mass, it is then easy to compute temperature from the total energy (viii), and then pressure and enthalpy mass flow (ix). In what follows, we will try to use mass, mass flow and total energy as independent variables, the others variables are given by the following equations in a control volume i :

$$\begin{cases} \rho = \dfrac{m}{S\Delta x} \\ u = \dfrac{q_m}{m}\Delta x = \dfrac{q_m}{S\rho} \\ T = \dfrac{1}{C_v}\left[\dfrac{E_T}{m} - \dfrac{u^2}{2}\right] \end{cases} \quad \text{and} \quad \begin{cases} P = \dfrac{mrT}{S\Delta x} \\ h = C_p T \end{cases} \qquad \text{(viii) and (ix)}$$

This choice of independent variables is also convenient for the user of the model because the initial conditions are relatively easy to obtain as they are mass, mass flow, and total energy distribution.

The implementation of a finite volume scheme is conceptually different from the finite difference scheme although it may provide similar models (6). This approach is usually closer to the physical system as it considers a volume of fluid instead of nodes. Each conservation law is integrated in this control volume and uses approximations of the quantities at the volume boundaries (8). To increase the performance of our model, we will use a discretization scheme with two sets of volume, as shown in figure 6 :

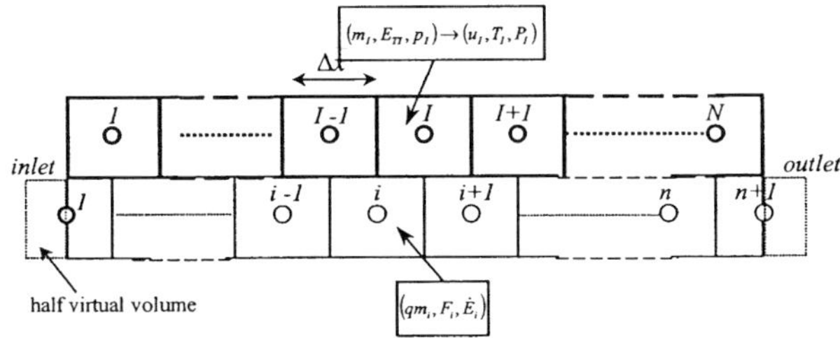

Figure 6 : Finite volume scheme.

The advantage of this interlaced discretization scheme is that it integrates the effort and flux variables separately. In the first profile, the total energy and the mass conservation are computed, giving mass and total energy and then temperature, specific mass and pressure, these are effort variables. In the second profile, we only consider the momentum conservation law, which gives mass flow and velocity, these are flux variables. This method is a good way of expressing the quantities going in and out a volume where effort variables are computed and also of expressing the tension between two volumes that leads to acceleration of the flow , i.e. the flux variables, mass flow and velocity.

At the limits of a volume, the quantities are approximated according to the following schemes :

$$\text{Central Scheme : } f_i = \frac{1}{2}(f_{I-1} + f_I) \qquad \text{(x)}$$

$$\text{Upwind Scheme : } f_i = f_{I-1} \qquad \text{(xi)}$$

The model obtained is as follows in the case of the central scheme :

$$\begin{cases} \dfrac{dm_I}{dt} = qm_i - qm_{i+1} \\ \dfrac{dE_{TI}}{dt} = \left(qm_i C_p T_i - qm_{i+1} C_p T_{i+1}\right) - \left(qm_i \dfrac{u_i^2}{2} - qm_{i+1}\dfrac{u_{i+1}^2}{2}\right) - KS_x\left(T_I - T_{out}\right) \\ \dfrac{dp_I}{dt} = S\left(\rho_i u_i^2 - \rho_{i+1}u_{i+1}^2\right) + S\left(P_i - P_{i+1}\right) - \dfrac{f}{2d} S\Delta x u_I \left|u_I\right| \end{cases} \tag{xii}$$

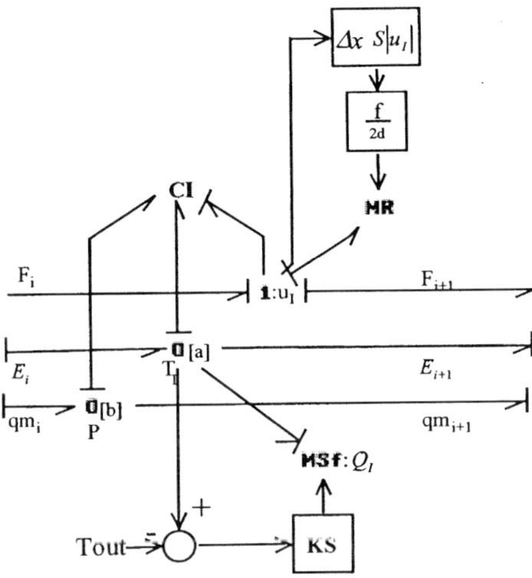

Figure 7 : Pseudo bond graph of the finite volume scheme.

This numerical scheme may be represented by a pseudo bond graph (4, 5), as shown on figure 7. This approach is detailed with some other bond graph approaches in (2).

4 – A SIMPLIFIED MODEL

The previous model is still difficult to handle and may present numerical instability. In most of the cases, it is possible to assume that kinetic energy is negligible compared to internal energy and this allows to simplify the energy conservation law. The total energy is equal to the internal energy and we obtain an algebraic relation in between internal energy and temperature. It leads to the following relation in the case of finite volume scheme :

$$\frac{d(me)_I}{dt} = \left(q_{mi-1} C_p T_{i-1} - q_{mi} C_p T_i\right) - KS_x\left(T_I - T_{out}\right) = C_v T_I \frac{dm_I}{dt} + m_I C_v \frac{dT_I}{dt} \tag{xiii}$$

This equation is equivalent to the equation of the temperature dynamic in a chamber with a constant volume. Consequently mass conservation and energy conversation laws may advantageously be replaced by the equivalent ordinary differential equations of the dynamics of a chamber using pressure and temperature as state variables.

It is also possible, if the spatial discretization is not too small, to consider that the convection-diffusion term of the momentum conservation law is negligible compared to pressure gradient during the dynamical phases.
On this basis, we developed a model which allow to consider a pipe as series of inertia with friction phenomenon and chambers and each part of the pipe may be represented by the pseudo bond graph of figure 8.

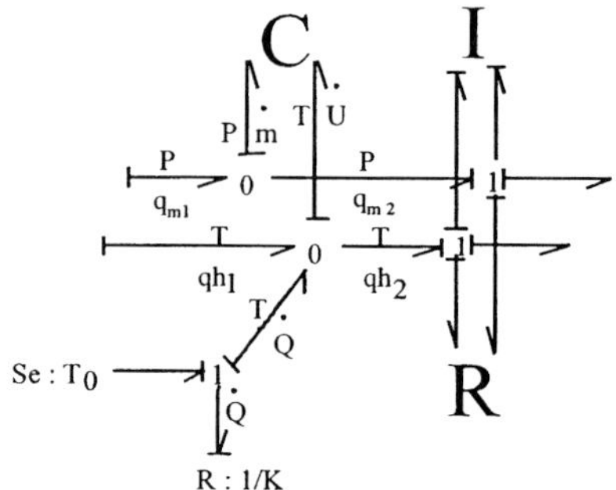

Figure 8 : Pseudo bond graph representation of a part of pipe with the simplified model

As it is usually difficult to characterise properly the heat exchange in between the gas and the pipe, we may then assume that the thermal behaviour is following a polytropic law. This last simplification allows to reduce the order of the model by one and to use only 2 state variables per part of pipe. In the case of a finite volume approach using a central scheme, the simplified model is finally :

$$\begin{cases} \dfrac{dq_{mi}}{dt} = \dfrac{S}{\Delta x}(P_i - P_{i+1}) - \dfrac{1}{2}\dfrac{fr(Re)}{dS} q_{mi}|q_{mi}| \\ \dfrac{dP_{i+1}}{dt} = \dfrac{rT_{i+1}}{S\Delta x}(q_{mi} - q_{mi+1}) + \dfrac{P_{i+1}}{T_{i+1}}\dfrac{dT_{i+1}}{dt} \\ \dfrac{dT_{i+1}}{dt} = \dfrac{(\gamma-1)T_{i+1}}{S\Delta x P_{i+1}}\left[(q_{mi-1}C_pT_{i-1} - q_{mi}C_pT_i) - KS_x(T_I - T_{out})\right] \end{cases} \qquad \text{(xiv)}$$

5 – REAL TRIALS AND SIMULATION RESULTS OF THE DYNAMIC BEHAVIOR

The proposed models have been implemented on AMESim (Advanced Modelling Environment for Simulation), a software dedicated to fluid power systems.
In the first approach, the static characterisations show that the proposed model allows a good representation of the stationary behaviour. As an example, figure 9 presents the comparison between the static characteristic obtain from simulation and measurements for an 8 m long pipe (diameter 9mm). These results also show the good adequacy of the fluid dynamic theory to the reality and provide the values of the parameters we need before starting the dynamical approach. The only parameter, which was not evaluated, is the thermal exchange coefficient, which is indeed relatively difficult to obtain as it may vary with the pressure, flow and temperature conditions.

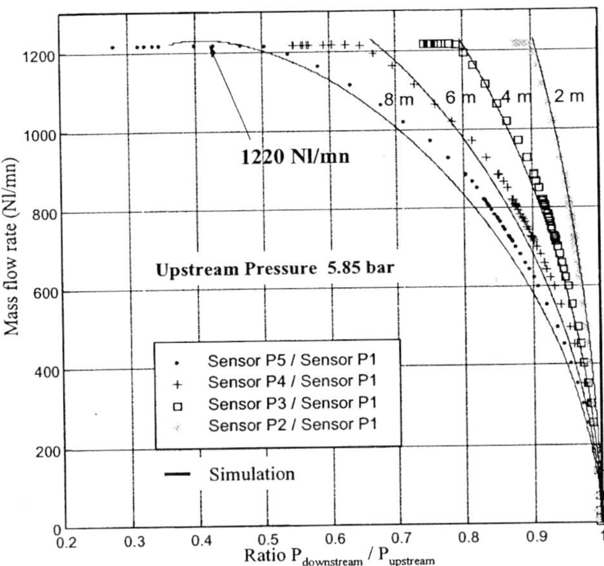

Figure 9 : Simulated and experimental static characteristic of 8 metres long pipe (9 mm diameter)

In the second approach, the bench (figure 10) was used to measure the dynamic behaviour of pipes. As in the previous paragraph, we studied different kind of pipes using various lengths and diameters For the validation of the model, 3 different volumes were added at the end of the pipe in order to observe the behaviour of the system according to the boundary condition. We also used various fittings for which we had known mass flow rate characteristics, to connect the pipe to the volume. The measured pressure input to the system is given by a pressure valve, which is driven by a pressure step. This circuit representing a very common configuration in the pneumatic brake systems, it seems to be a good test for the proposed model.

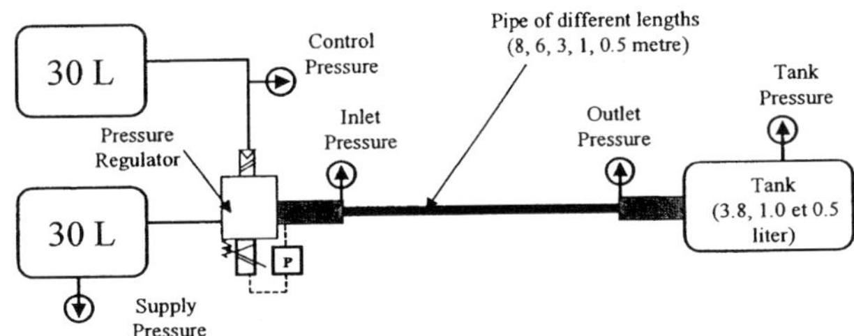

Figure 10 : Circuit used for the dynamic validation of the pneumatic pipe model

Here we will only present the results concerning a pipe with 9 mm diameter although we obtain similar results in others configurations. In all the following simulations, the parameters related to the thermal exchanges have been fixed and remain unchanged. The figures display the comparison in between the measured and simulated pressures at the end of the pipe and in the volume. Figure 11, 12, and 13 present the results when the capacity at the extremity of a 8 metres long pipe is changed from 3.8 l to 1.0 l and 0.5 litre. Figure 13, 14, 15, and 16 presents the results when the length of the pipe is changed from 8 m, 6 m to 3 m and 1 metre for a 0.5 litre capacity. The oscillatory phenomenon has to be related to the coupling in between the inertia of the pipe and the volume of the capacity.

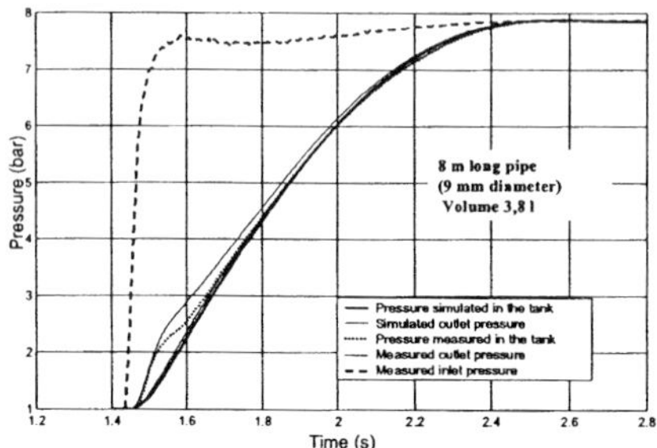

Figure 11 : Simulation vs. real trials

Figure 14 : Simulation vs. real trials

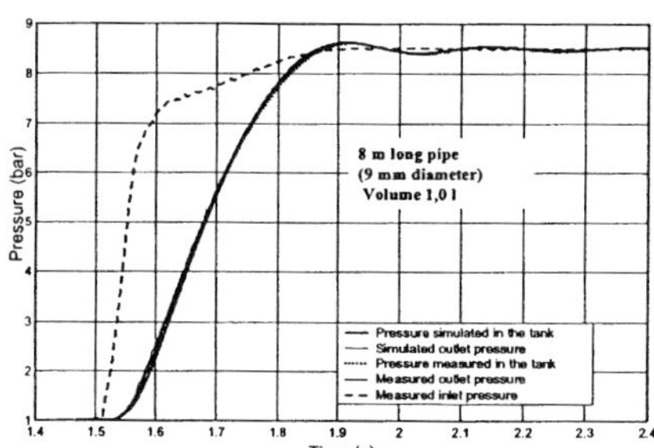

Figure 12 : Simulation vs. real trials

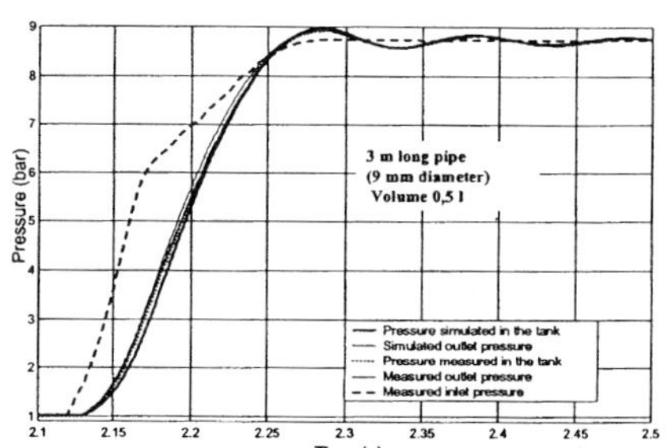

Figure 15 : Simulation vs. real trials

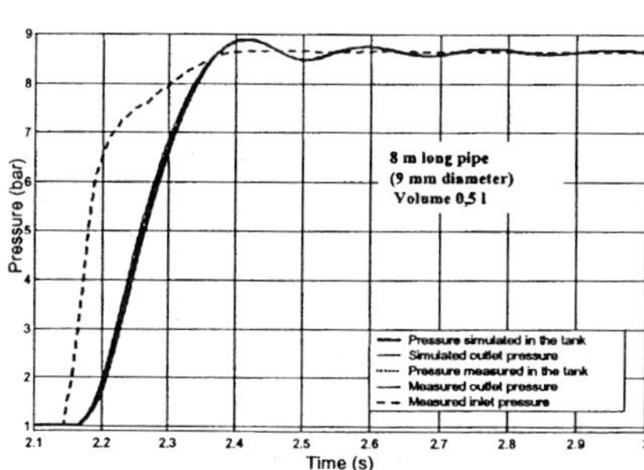

Figure 13 : Simulation vs. real trials

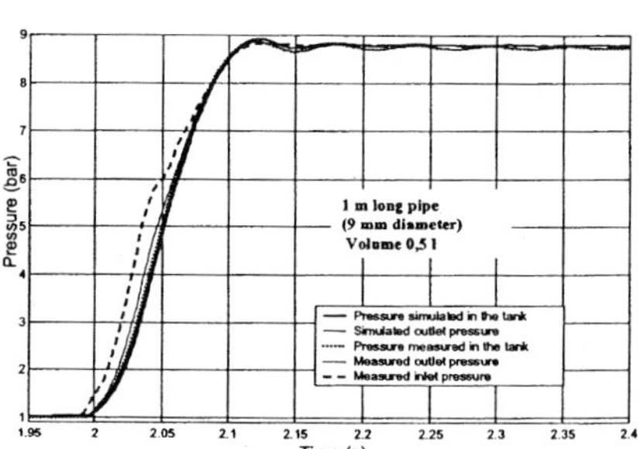

Figure 16 : Simulation vs. real trials

The results show then the good adequacy of the simulation to the real trials. Figure 17 and 18 illustrates the accuracy of the model in representing the time delay introduced by the inertia of the pipe and the coupling in between the pipe inertia and the capacity of the volume.

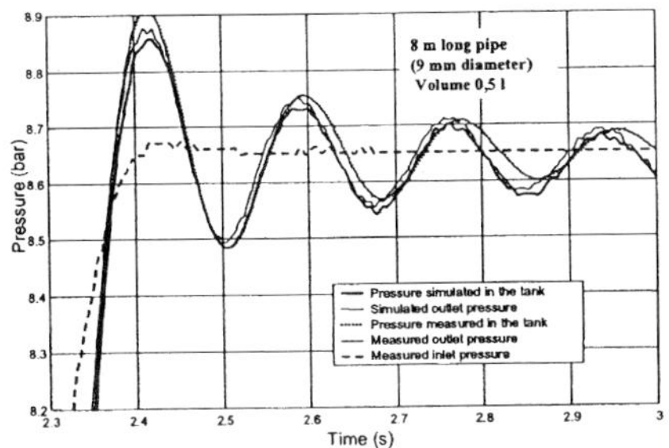

Figure 17 : Coupling representation

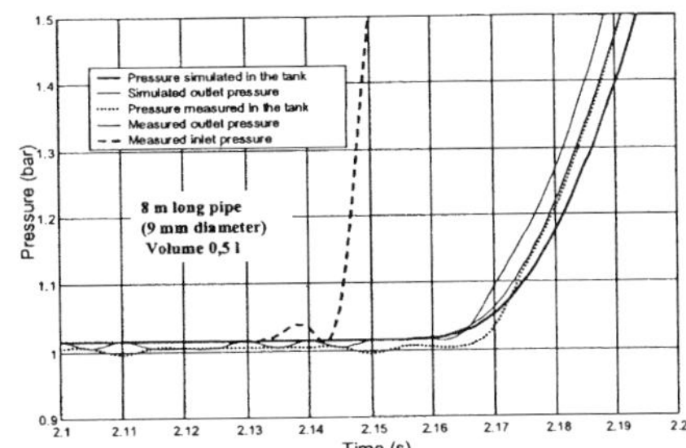

Figure 18 : Time delay representation

6 – FREQUENCY APPROACH

In a last approach, as the finite volume is only giving an approximation of the conservation laws (i, ii and iii), it is interesting to observe the conditions which are required by the model to obtain a correct modal representation. Each control volume is adding a new mode and has also an influence on the previous modes. For pneumatic pipes, the values of the modes are well-known in the case of specific boundary conditions and acoustic approximation ; they are given by formula xv (free-fixed boundary condition) and xvi (free-free boundary condition).

$$f_i = (2i-1)\frac{c}{4L} \qquad i \in \{1, \cdots, n\} \tag{xv}$$

$$f_i = i\frac{c}{2L} \qquad i \in \{1, \cdots, n\} \tag{xvi}$$

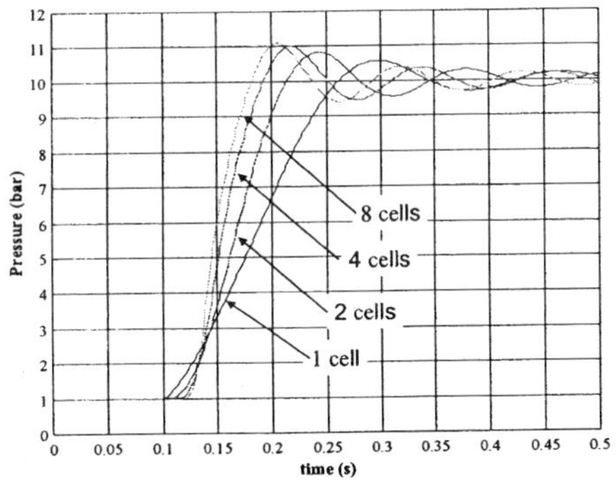

Figure 19 : Model simulation using 1 to 8 control volumes.

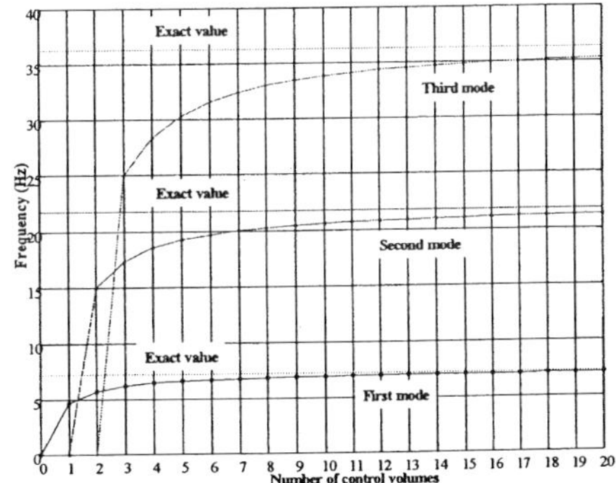

Figure 20 : Approximation of the first mode.

The figure 19 shows the pressure simulated in a 10 metres long pipe (diameter 8 mm) using 1, 2, 4, or 8 control volumes when the pipe is closed at one end and a pressure step is imposed at the other end. As shown on figure 20, the convergence of the mode value is a function of the number of cells used in the model. The error on the value of the first mode is less than 10 % if at last 5 control volumes are used and it is less than 5 % with 10 control volumes.
As this approach is not especially oriented to obtain a correct frequency representation of the pipe behaviour, the cost in term of the numerical model dimension may seem important. This means that some consideration is required of the robustness of the model in the frequency domain or the usable frequency range.

7 - CONCLUSIONS

The finite volume approach seems a good approach and gives good and straight forward results. The choice of user friendly variable is also crucial to allow a good understanding and analysis of the phenomenon. This first piece of work may provide interesting developments in the simulation of pneumatic circuits although it has to be extended to the study of longer pipes (for railway applications) and to the validation of the frequency accuracy with spectrum analysis. The experimental bench allowed to tune the model and to verify in different configurations the validity of the developments. The results obtained with the simulation show that the proposed model provides an accurate representation of the dynamic and static behaviour of pipes. The choice of user friendly variable is also crucial to allow a good understanding and analysis of the phenomenon.
Actually, the generalisation of the use of electro-pneumatic brake systems (truck and railway industry) is generating new problems related to the stability of the system. As the feedback loop is usually the pressure at the outlet of the power modulator (pressure regulator) and not the pressure in the actuator, the time delay and especially the modes of the pipe turn out to be determinant effects.

ACKNOWLEDGMENTS

This work has been carried out with the help a regional founding (FIT) in 1998 and in collaboration with the company RENAULT Véhicules Industriels in 1999.

REFERENCES

(1) BREEDVELD, P.C. *Physical systems theory in terms of bond graphs.* Ph.D. thesis, Twente Un. of Technology Enschede, The Netherlands, ISBN-n°90-9000599-4, 1984, 201p.

(2) BIDEAUX, E. and SCAVARDA, S., *Pneumatic Pipes Modelling with Bond Graph,* WMC'99 International Conference on Bond Graph Modeling, San Francisco, January 1999, pp. 297-302.

(3) CANDEL, S. *Mécanique des fluides.* Paris : Dunod Université, 1990, 451p.

KARNOPP, D.C., MARGOLIS, D.L., and ROSENBERG, R.C., *System Dynamics : A Unified Approach*, New York : John Wiley & sons, 1990.

(4) KARNOPP, D.C. *State variables and pseudo bond-graphs for compressible thermofluid systems*. Journal of Dynamic Systems Measurement and Control, 1979, Vol. 101, september, pp. 201-204.

(5) KARNOPP, D.C. *Modelling and simulation of adaptive vehicle air suspensions with pseudo bond graphs, CAMP and ACSL*. Proc. 11th IMACS World Congress, Oslo, Norway, 1985, Vol. 4, pp. 297-300.

(6) MOKSNES P.O., ENGJA, H., and MARINTEK, E.P., *Modelling of One-dimensional gas flow in an oil Plateform Riser using Bond Graph*, ICGBM'95, pp.261-266.

(7) LIN, X.F. *Contribution à la modélisation et à la commande d'un vérin pneumatique de grande longueur*. Thèse Sci. : Institut National des Sciences appliquées de Lyon, 1992, 193p.

(8) PATANKAR, S.V., *Numerical Heat Transfer and Fluid Flow*, Hemisphere Publishing Corporation, 1980.

(9) TOKASHIKI, L. R., FUJITA, T., KAGAWA, T., PAN, W. *Dynamic characteristics of pneumatic cylinders including pipes*. Preprints of 9th Bath International Fluid Power Workshop, Bath, September 1996, 14 p.

(10) TORO, E., Riemann solvers and numerical methods for fluid dynamics, Springer-Verlag, 1997.

(11) VERSTEEG, H.K., MALALASEKERA, W., *An introduction to computational fluid dynamics*, Longman Scientific & Technical (UK), Wiley (US), 1995.

Calculation of eigenvalues by locus method for stability analysis of oil hydraulic circuits

S WASHIO, Y NAKAMURA, and **K SYOJINO**
Department of Mechanical Engineering, Okayama University, Japan

ABSTRACT

If a mathematical model is provided for an oil hydraulic circuit, its stability can be discussed on the basis of the poles of the governing transfer function derived from the model. When the system contains long pipes where the changes of pressure and flow are propagated as waves, however, it is no easy task to find the exact values of those poles. In order to obtain a practical method to calculate the poles of such a complicated transfer function, the following mathematical theorem was utilized: when a complex variable z counter-clockwise makes a trajectory along an arbitrary closed loop on the complex plane, the complex function $F(z)$ counter-clockwise turns around the origin by $N_Z - N_P$ times, where N_Z and N_P represent the numbers of zeros and poles of $F(z)$ inside the loop, respectively. Taking a poppet valve connected to a pipe with a constant-pressure end as an example circuit, the suitability of a calculus developed on the basis of the theorem was tested. First, attention was paid to the trajectories the eigenvalues described on the complex plane as the pipe length was varied. Then how each of the model parameters affects the "unstable" eigenvalues was examined by calculations. As a result it has turned out that the viscosity needs to be taken into consideration for the stability analysis of a circuit including pipes. Moreover it has been recognized that various parameters can affect the stability of the eigenvalue in different ways of their own, which probably explains why self-induced vibration seems so unpredictable in actual oil hydraulic systems.

NOMENCLATURE

A: cross sectional area
c: velocity of sound in oil
x: poppet displacement
x_0: initial spring compression

C: closed loop
c_P : damping factor of poppet motion
d: diameter
$X_i(z), F_i(z)$: propagation operator
k_P : spring constant
l: pipe length
m_P : poppet mass
p: pressure
P: Laplace transform of p
q: flow rate
Q: Lapalce transform of q
s: Laplace operator
S: opening area of constriction
t: time
X: Laplace transform of x
z: complex variable

α : nonlinear loss coefficient
β : linear loss coefficient
θ : oil temperature
ν : kinetic viscosity of oil
ρ : density of oil
ϕ : half zenithal angle of poppet

SUBSCRIPT

T: small tube
D: poppet drain
P: poppet
L: pipe

1 INTRODUCTION

Engineers working on oil hydraulic systems are often annoyed by self-induced vibration which occurs unpredictably and uncontrollably. So it is natural that there have been many researches which tackled the problem with an intention to keep those vibrations under control [1-4]. If the vibration can be analyzed by a linear model, prediction and control of the self-induced vibration come down to a matter of finding eigenvalues for the characteristic equation. For the actual self-induced vibrations in oil hydraulic systems, however, the eigenvalue analysis does not necessarily succeed. Probably there are two reasons for that; firstly, the mathematical system models conventionally used for that purpose are not accurate enough for the purpose. Secondly, a pipe, which is usually included in the oil hydraulic circuit to be analyzed and behaves as a distributed-parameter component, makes it difficult to find the zeroes for the characteristic equation.

The present paper intends to propose a practical measure to solve the latter problem. The "Locus Method" formerly developed by one of the present authors [5] is introduced as a tool to find the eigenvalues for the complicated characteristic equations. The suitability of the method is examined with an example circuit composed of a poppet valve, a tank and a connecting pipe in-between. In order to make the mathematical model of the example circuit as accurate as possible, the actual poppet valve for which the present authors have previously measured the characteristics in detail [6,7] is assumed as a model valve. The effects of various model parameters, especially the pipe length and the viscosity, on its instability are also discussed through calculations.

2 CALCULUS OF EIGENVALUES

2.1 Principle

An ordinary way to analyze mechanical instability is formulating the characteristic equation for the vibration system and finding its roots. When it comes to such a system as an oil hydraulic device connected to a pipe, the characteristic equation can become complicated, including hyperbolic functions if the pipe is treated as a distributed-parameter model, or even Bessel functions if viscosity is taken into consideration. For the purpose of solving that kind of difficult equation, the traditional calculi developed for algebraic equations [8] are completely useless but there are two methods available; one is "Eigenvally Method" proposed by Scarton and Rouleau [9] and the other is "Locus Method" formerly advocated by one of the present authors [5]. The latter will be employed in the present study.

If a complex function $F(z)$ is regular and single-valued on and inside a closed loop C on the complex (z) plane except at a finite number of isolated poles, the following "argument principle" is derived from the residue theorem [10].

$$\frac{1}{2\pi j}\int_C \frac{F'(z)}{F(z)}dz = N_Z - N_P \tag{1}$$

where N_Z, N_P are the numbers of zeroes and poles of $F(z)$ within C, respectively, $F'(z)$ is the derivative of $F(z)$ and j denotes the imaginary unit $\sqrt{-1}$.

Since the integration in Eq.(1) gives log $F(z)$, the left-hand side represents how many times and in which direction the locus of the complex vector $F(z)$ turns around the origin when z makes a counterclockwise trajectory along C. Consequently an idea comes out from Eq.(1) that observing this locus can reveal the number of zeroes (or poles) of $F(z)$ within C, if the number of poles (or zeroes) within C is known beforehand. The Locus Method is based upon this idea.

2.2 Demonstration by example

The concrete procedure of the Locus Method is demonstrated by taking the 2nd order spherical Bessel function of 2nd kind

$$n_2(z) = (-1)^3\sqrt{\frac{\pi}{2z}}J_{-5/2}(z) = -\frac{1}{z^3}\left\{(3-z^2)\cos z + 3z\sin z\right\} \tag{2}$$

as an example. It is known that the spherical Bessel function has a finite number of complex roots and an infinite number of real ones [11]. In order to avoid overflow in computing cos z and sin z on the complex plane, $F(z)$ is rearranged as follows;

$$F(z) = z^3 n_2(z)/\cos z = z^2 - 3z\tan z - 3 \tag{3}$$

A circle of radius R with its center at the origin, that is $z = R\cdot e^{j\theta}$, is chosen as the closed loop C, to begin with. If $F(z)$ is an even function like Eq.(3), the vector $F(z)$ traces the same trajectory twice when z goes round the circle. So it is enough to let z go just half round from $\theta = -\pi/2$ to $\theta = \pi/2$ for Eq.(3), as is shown in **Fig.1**.

Several examples of actual loci obtained for different values of R are shown in **Fig.2**. All the loci are drawn in such a way as the maximum $|F(z)|$ has the same length and the arrow indicates the direction of change as z moves along the circle. Since the poles of $F(z)$ are identical with the roots of cos z, which are $z_k = \pm(2k+1)\pi/2$ ($k = 0,1,2,\cdots$), observing how many times and in which direction the loci in **Fig.2** go round the origin reveals the number of zeroes within the circle $z = R\cdot e^{j\theta}$. The number of zeroes thus discovered is shown in each figure with the value of R and the number of poles inside the circle. It should be noticed that the actual numbers are double those listed.

The above procedure leads to discovering a required number of concentric circles that partition the complex plane into ring-shaped domains containing 1 or 2 zeroes. The

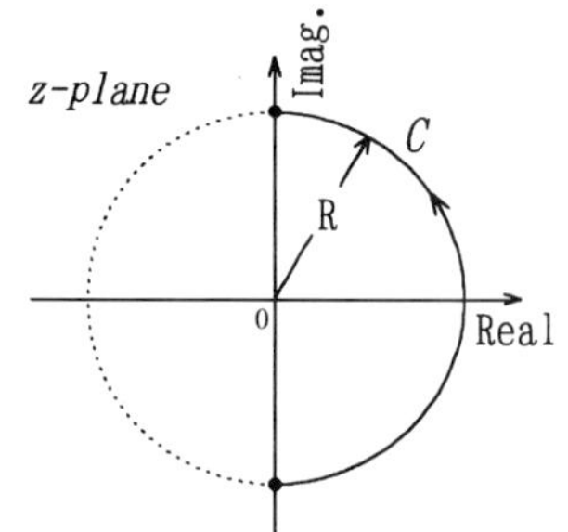

Fig.1 Half circle initially selected as closed loop *C*

Fig.3 Fan-shaped region bordered by two radii and two circles

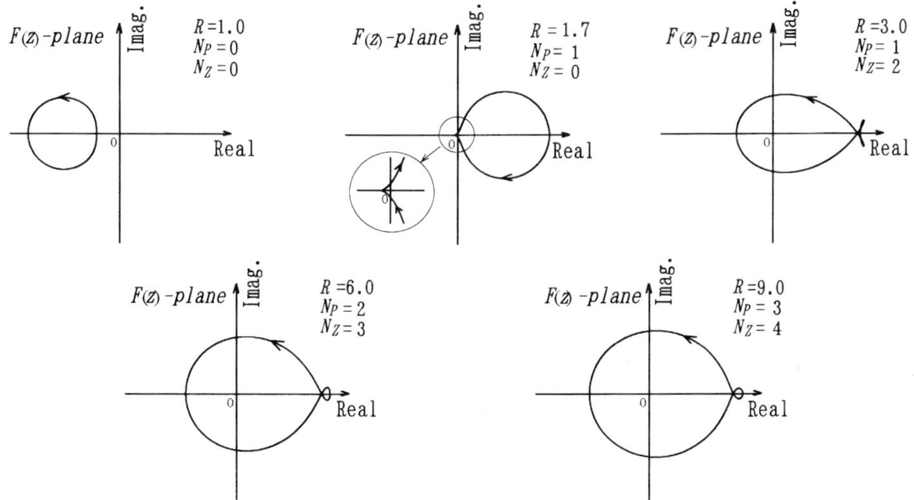

Fig.2 Loci of Eq.(3) when *z* moves half round the circle

procedure is repeated by decreasing the outer radius and increasing the inner one of the adjacent circles until a new pair of concentric circles having a small gap and containing the zeroes between them are found. For Eq.(3), the radii R_1, R_2 of these pairs of circles are obtained as follows;

$R_1=1.9 \quad R_2=2.1 \quad N_Z=2$

$R_1=3.8 \quad R_2=4.1 \quad N_Z=1$

$R_1=7.3 \quad R_2=7.6 \quad N_Z=1$

Next, as is illustrated in **Fig.3**, the ring domains between the above-obtained pairs of circles are partitioned by the radii into many fan-shaped sub-domains, to which contours the Locus Method is applied again. **Figure 4** shows two typical loci appearing in the application of the method; the locus on the left side goes round the origin counterclockwise, indicating there exists one zero in the examined sub-domain. The right-side one, which corresponds to the adjacent sub-domain, does not have the origin in its inside, showing that this sub-domain does not contain zeroes. After all the fan-shaped domains are examined, those containing just one zero can be singled out.

Each domain which has turned out to contain one zero is divided by many radii and arcs forming a lattice, and $|F(z)|$ is computed at every lattice point. Then at the lattice point having the minimum $|F(z)|$ is chosen as the first approximation of the zero $z = z_0$. Starting from this first approximation, the Newton method

$$z_2 = z_1 - \frac{F(z_1)}{F'(z_1)} \tag{4}$$

is usually applied to improving accuracy of the zero value.

In case the Newton method is not applicable or $F'(z)$ is too complicated to compute, the "spotting" method is introduced. First, consider a circle of a radius $\Delta R = (R_2 - R_1)/2$ with its center at the same z_0 as the first approximation. Then divide its circumference into many equal arcs (100 e.g.) and find out one division point z_1 where $|F(z)|$ takes the minimum value. If $|F(z_0)| < |F(z_1)|$, put $\Delta R \rightarrow \Delta R/2$ and repeat the procedure. If $|F(z_0)| > |F(z_1)|$, shift the center of the circle from z_0 to z_1 and repeat the same procedure, until ΔR which is halved at every step satisfies the condition $\Delta R/|z_i| < \varepsilon$ ($\varepsilon = 10^{-6}$ e.g.). Then z_i will give the zero as accurately as expected. Although the spotting method requires more time than the Newton method, it can surely get to the target point within the accuracy of computing $F(z)$.

In practice, the calculation of the loci of $F(z)$, which were exhibited above for the convenience of explanation, are not actually required. Checking a variation of $F(z)$ argument would be enough to find how many times and in which direction the loci go round the origin. One of the present authors previously published a FORTRAN program of the Locus Method, which gives the designated number of roots for any calculable complex function [12]. Applying this program to the 2nd kind spherical Bessel function, Eq.(3) yields the initial and final values shown in **Table 1**. The roots above the 3rd have extremely small imaginary parts compared to the real ones.

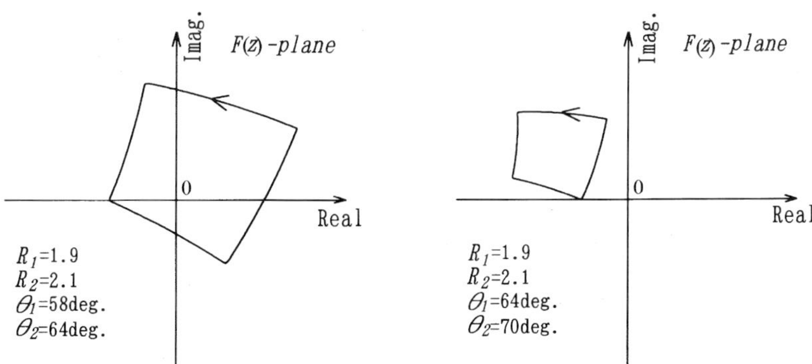

Fig.4 Loci of Eq.(3) when *z* goes around the boundary of fan-shaped sub-domain

Table 1 Results of root computation for 2nd kind spherical Bessel function of 2nd order

No.	Initial $Z_0 = X_0$	$+ jY_0$	$F(Z_0)$	Final $Z = X$	$+ jY$	$F(Z)$
1	9.3735180E-01	1.7022710E+00	1.21E-03	9.3819520E-01	1.7028030E+00	6.43E-08
2	9.3769130E-01	-1.7020830E+00	1.07E-03	9.3819520E-01	-1.7028030E+00	6.30E-08
3	3.9588860E+00	-2.7148160E-03	3.93E-02	3.9595280E+00	-3.8745650E-09	6.95E-07
4	7.4522580E+00	-5.1104050E-03	2.77E-01	7.4516100E+00	-6.0905450E-09	8.34E-07
5	1.0705970E+01	-7.3416470E-03	1.36E+00	1.0715650E+01	6.0936150E-10	3.62E-05
6	1.3918480E+01	-9.5446330E-03	1.92E+00	1.3921690E+01	-6.9361490E-09	3.10E-05
7	1.7109140E+01	-7.0117326E-02	3.76E+00	1.7103360E+01	-4.1909730E-09	2.89E-05
8	2.0269770E+01	-1.3900050E-02	5.71E+00	2.0272370E+01	9.2745720E-09	9.96E-05

Table 2 Dimensions of pilot valve and pipe

d_T	1.5×10^{-3} m	l_T	8.0×10^{-3} m
d_P	4.7×10^{-3} m	ϕ	0.349 rad
d_L	9.2×10^{-3} m	k_P	20.3×10^{3}N/m
m_P	4.7×10^{-3} kg	x_0	0.7×10^{-3} m

They really have null imaginary parts, which can be proved by computing real roots of Eq.(3) by some traditional method.

3 STABILITY ANALYSIS BY EIGENVALUES

3.1 Circuit to be analyzed

With a combination of a pipe and a poppet valve illustrated in **Fig.5** taken as an example circuit the stability problem is discussed in terms of eigenvalues. Since it is desirable to have as accurate mathematical models as possible for each component, the pilot module of the piston-type relief valve which the present authors have previously investigated in detail [6,7] is chosen as the poppet valve. Dimensions of the pilot module, which actually comprises a poppet valve and a small tube in series, as well as those of the pipe are shown in **Table 2**.

(1) Small tube The dynamic model for the small tube constriction is given by the following equation [7];

$$\frac{p_T - p_P}{\rho} = \frac{1}{S_T^{\,2}}\left(\alpha_T q_T^{\,2} + \beta_T \nu l_T q_T\right) + \frac{l_T}{S_T}\frac{dq_T}{dt} \quad (\alpha_T = 0.496,\ \beta_T = 8\pi,\ S_T = \pi d_T^{\,2}/4) \tag{5}$$

The first term on the right hand side shows the pressure loss, while the second term represents a kinetic differential pressure caused by the fluid motion in the tube.

(2) Poppet valve The dynamic behavior of the poppet valve can be described by the three equations below [7]. The first one is the relationship between the differential pressure across, the flow rate through and the poppet displacement of the valve;

$$\frac{p_P - p_D}{\rho} = \frac{1}{S_P(x)^2}\left(\alpha_P q^2 + \beta_P \nu \pi d_P q\right) \tag{6}$$

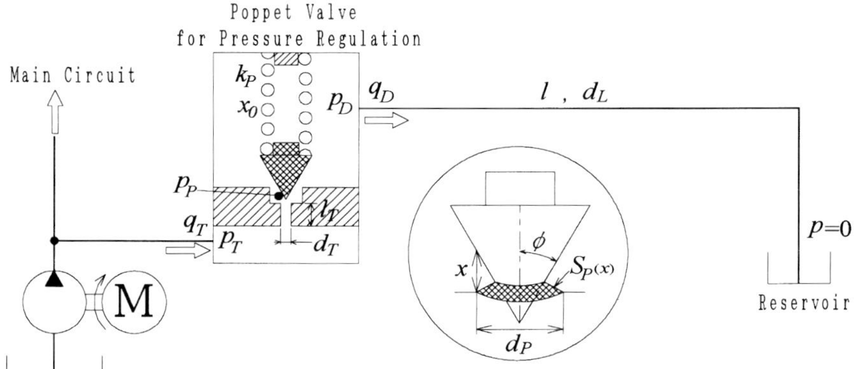

Fig.5 Example circuit for stability judgement by eigenvalues

where $\quad S_P(x) = \pi x \sin\phi_P\left(d_P - \dfrac{1}{2}x\sin 2\phi_P\right)$ (7)

The second one expresses the poppet motion;

$$m_P\frac{d^2x}{dt} + c_P\frac{dx}{dt} + k_P(x + x_0) = A_P(p_P - p_D) \tag{8}$$

And the third one shows that the flow rate into the poppet valve is not equal to the net one through the constriction when the poppet moves;

$$q_T = q_P + A_P\frac{dx}{dt} \tag{9}$$

Furthermore the flow rate into the small tube is equal to that out of the poppet valve;

$$q_T = q_D \tag{10}$$

(3) Pipe The pipe model gives the relationship between the pressure $p_D(t)$ and the flow rate $q_D(t)$ at the downstream port of the poppet valve. Solving the wave equation for the pipe with the boundary condition of constant pressure at the tank yields the following relationship in the Laplace domain;

$$P_D(s)\cosh sX_i(z)T_l - Z_iX_i(z)Q_D(s)\sinh sX_i(z)T_l = 0 \tag{11}$$

where $\quad T_l = l/c\,,\ \ Z_i = 4\rho c/\pi d_L^2$ (12)

and the propagation operator $X_i(z)$ is defined as follows;

$X_i(z) = 1$ (13)₁

(neglecting viscosity in wave transmission, called "inviscid case" hereafter)

$X_i(z) = F_i(z) = \left\{\dfrac{I_0(z)}{I_2(z)}\right\}^{\frac{1}{2}} \qquad (z = \sqrt{\chi_0 s}\,,\ \ \chi_0 = d_L^2/4\nu\,)$ (13)₂

(considering viscosity in wave transmission, called "viscous case" hereafter)

In Eq.(13)₂ $I_0(z)$ and $I_2(z)$ are 0-th and 2-nd order modified Bessel functions of the first kind.

3.2 Characteristic equation for stability judgement

With respect to small changes of each variable around its average (designated by $\bar{\ }$), Eqs.(5) and (6) are linearized and Eqs.(7) to (9) are rearranged. Taking the Laplace transformation of the results yields the following equations;

$$P_T - P_P = (b_1 + sb_2)Q_T \tag{14}$$

$$P_P - P_D = b_3 Q_P - b_4 X \tag{15}$$

$$(m_P s^2 + c_P s + k_P)X = A_P(P_P - P_D) \tag{16}$$

$$Q_T = Q_P + sA_P X \tag{17}$$

$$Q_T = Q_D \tag{18}$$

where

$$b_1 = \frac{\rho}{S_T{}^2}(2\alpha_T \bar{q}_T + \beta_T \nu l_T), \qquad b_2 = \frac{\rho l_T}{S_T} \tag{19}$$

$$b_3 = \frac{\rho}{S_P(\bar{x})^2}(2\alpha_P \bar{q}_P + \beta_P \nu \pi d_P), \quad b_4 = \frac{2\rho S_P{}'(\bar{x})}{S_P(\bar{x})^3}\left(\alpha_P \bar{q}_P{}^2 + \beta_P \nu \pi d_P \bar{q}_P\right) \tag{20}$$

From all the equations (14) to (18) as well as (11), the transfer function $G(s)$ between the upstream pressure P_T and the poppet displacement X is derived as follows;

$$G(s) = \frac{X(s)}{P_T(s)} = \frac{b_3 A_P H_C}{B_1(s)H_S + B_2(s)H_C} \tag{21}$$

where

$$H_S = Z_i X_i(z)\sinh sX_i(z)T_l, \quad H_C = \cosh sX_i(z)T_l \tag{22}$$

$$B_1(s) = m_P s^2 + c'_P s + k'_P,\ B_2(s) = (b_1 + b_2 s)(m_P s^2 + c'_P s + k'_P) + b_3(m_P s^2 + c_P s + k_P) \tag{23}$$

$$c'_P = c_P + b_3 A_P^2, \quad k'_P = k_P + b_4 A_P \tag{24}$$

Eventually the characteristic equation turns out to be

$$F(s) = B_1(s)H_S + B_2(s)H_C = 0 \tag{25}$$

3.3 Calculation of eigenvalues

The stability of the oil hydraulic circuit in **Fig.5** is judged by the eigenvalues of the characteristic equation (25). Speaking from the traditional viewpoint, however, it can be distressing trying to find zeros for an equation including hyperbolic functions like Eq.(25), let alone Bessel functions in addition. When the Locus Method was applied to that purpose, however, it turned out to be effective enough. If the zeroes of Eq.(25) are computed by the Locus Method, it becomes a requisite to know the poles beforehand. Actually they can be easily found for the viscous case, because they are equal to the zeroes of $I_2(z)$ which are commonly known [11].

The first concern is how the pipe length effects the eigenvalues. The principal parameters for the calculation are arbitrarily chosen as follows;

$\bar{q}_T = \bar{q}_P = 38\text{cm}^3/\text{s}, \quad \bar{x} = 0.278\text{mm}, \quad \theta = 25^\circ\text{C}$

Moreover the working fluid is assumed to be a VG46 mineral oil, whose kinetic viscosity ν is related to the temperature θ by the following empirical equation;

$$\nu = \{\exp(\exp(22.031 - 3.6\log(\theta + 273.15))) - 0.6\} \times 10^{-6} \quad \text{m}^2/\text{s} \tag{26}$$

The results of the calculation are plotted in **Fig.6 (a)** and **(b)** for the viscous and inviscid cases, respectively, as trajectories of the eigenvalues on the complex plane when the pipe length l is increased from 0. The imaginary axes are graduated in Hz, which means the imaginary parts of eigenvalues are divided by 2π in the plotting.

When l=0 to begin with, it follows that $H_S = 0$ and $H_C = 1$, and accordingly the characteristic equation (25) becomes

$$B_2(s) = 0 \tag{25$_1$}$$

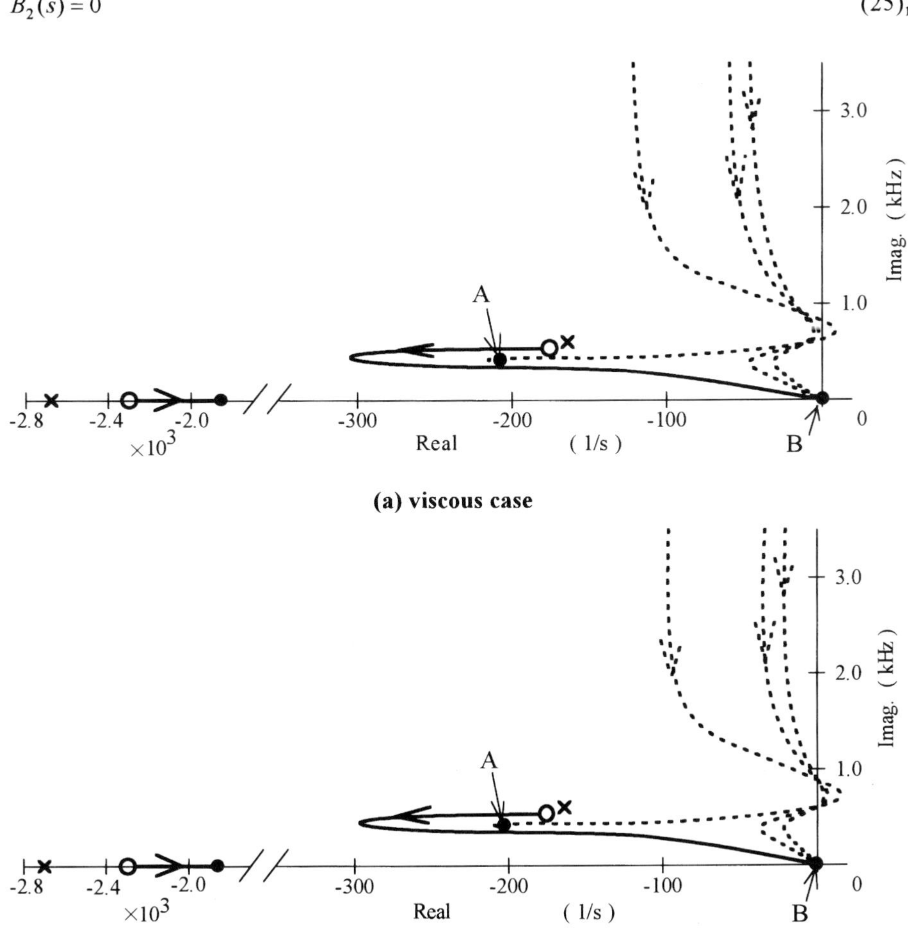

(a) viscous case

(b) inviscid case

Fig.6 Eigenvalue trajectories as a pipe length is increased from 0

Since Eq. $(25)_1$ is an algebraic equation of the third degree, it has one real root and a pair of complex ones, which are indicated by circles (○) in **Fig.6**. These three eigenvalues, which are what Shirai et al called the "valve mode" [4], are supposed to be the starting points of the trajectories.

As the pipe length l increases, these three eigenvalues start to move; the trajectories and directions of their movements are shown by the solid curves and arrows, respectively. When $l \neq 0$, there exist an infinite number of other eigenvalues which Shirai et al named the "pipeline mode" [4]. These new eigenvalues emerge at infinity and come down along the imaginary axis with an increasing l. Some of these trajectories are also shown by the dotted curves in **Fig.6**.

When l is small, all the eigenvalues stay exclusively on the left-half complex plane, which indicates that the poppet valve is stable. As l increases, one of the pipeline-mode eigenvalues goes into the right-half plane for the viscous case, or three of them do for the inviscid case. These eigenvalues are called "unstable" hereafter. The behaviors of the unstable eigenvalues are shown in more detail in **Fig.7** for the viscous and inviscid cases, respectively. The numbers attached to the points on the trajectories indicate pipe lengths in m. Obviously it causes substantial difference to the unstable eigenvalues whether the fluid viscosity is taken into account in the wave equation or not. Therefore the viscosity should not be neglected in the wave transmission model for the pipe.

When the pipe becomes longer, each eigenvalue approaches the specific ultimate point. These points are indicated by solid circles (●) in **Fig.6**. For the inviscid case they turn out to be the origin and the other three points given by the roots (one real and two complex) of the following algebraic equation;

$$B_2(s) - Z_i B_1(s) = 0 \qquad (27)_1$$

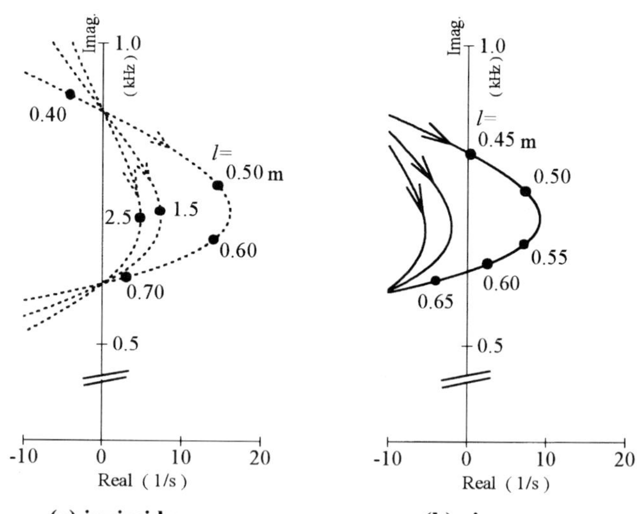

(a) inviscid case **(b) viscous case**

Fig.7 Unstable eigenvalues with and without considering viscosity

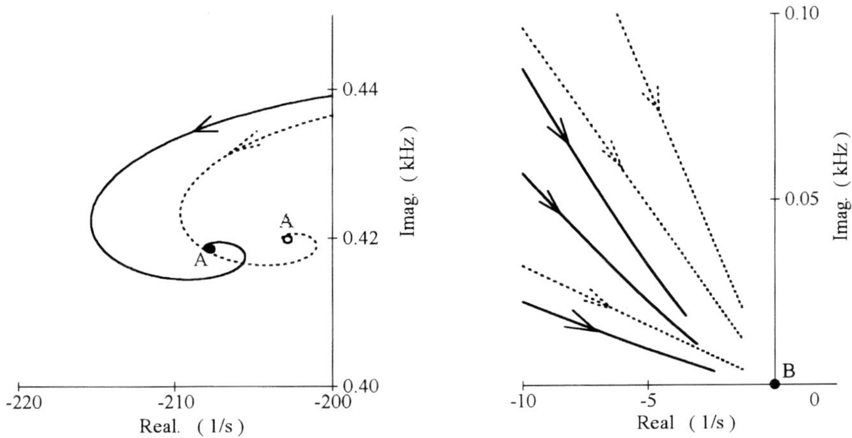

(a) Ultimate point A **(b) Ultimate point B**

Fig.8 Eigenvalue behaviors in the vicinity of ultimate points

For the viscous case the ultimate points are given by the origin and the roots of the following equation;

$$B_2(s) - Z_i F_i(z) B_1(s) = 0 \tag{27$_2$}$$

On the other hand, if impedance matching is realized at the pipe end and the reflection can be neglected in the wave transmission, Eq.(11) must be replaced by

$$P_D(s) - Z_i X_i(z) Q_D(s) = 0 \tag{11$_1$}$$

and accordingly the characteristic equation becomes

$$B_2(s) + Z_i X_i(z) B_1(s) = 0 \tag{25$_1$}$$

The eigenvalues of this characteristic equation $(25)_1$, which are plotted by the crosses (×) in **Fig.6**, do not agree with the ultimate points given by Eq.$(27)_1$ or Eq.$(27)_2$. These seemingly conflicting results imply that an infinitely long pipe is not theoretically the same as a pipe with no reflection. In that sense it should be mentioned that the ultimate points given by Eqs.$(27)_1$,$(27)_2$ are just theoretical.

Moreover the eigenvalue trajectories enlarged near the ultimate points A and B are shown in **Fig.8**; the solid and dotted lines indicate viscous and inviscid cases, respectively. A pair of valve-mode eigenvalues and all the pipeline-mode eigenvalues except one monotonously approach the point B, whereas one pipeline-mode eigenvalue which can be unstable approaches the point A describing a spiral trajectory.

3.4 Effects of parameters on instability

The effects of model parameters on the unstable eigenvalues are examined next. The parameters affecting the eigenvalues are divided into two categories, that is, working parameters and structural ones. The former depend on the actual working condition of the circuit and include the average flow rate $\overline{q}_T = \overline{q}_P$, the average poppet displacement and the oil temperature θ, for example. The latter are determined from the structure of the circuit and are given in **Table 2**.

The variations of the unstable eigenvalue as one specific working parameter is increased or decreased without changing the other parameters are illustrated in **Fig.9(a)~(c)**. The average flow rate $\overline{q}_P$, the average poppet displacement $\overline{x}$ and the oil temperature θ are chosen as the changing parameter. The circles indicate the initial point where the eigenvalue is located at the beginning. The initial values of the working and structural parameters are tabulated in **Table 3**.

According as the working parameter is increased or decreased, the eigenvalue moves on the solid line in the "Inc" direction or on the dotted one in the "Dec" direction. The numbers attached at the ends of those lines show the values of the changing parameter at the points. These results reveal the following; when it comes

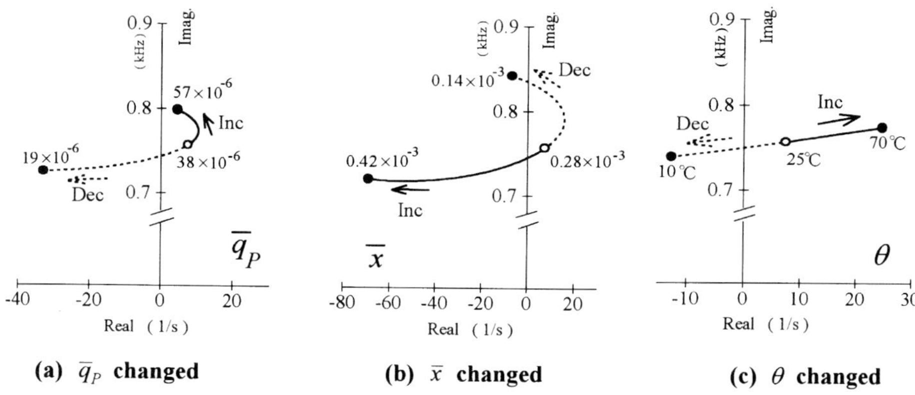

(a) $\overline{q}_P$ changed **(b) $\overline{x}$ changed** **(c) θ changed**

Fig.9 Variations of unstable eigenvalue with one working parameter changed

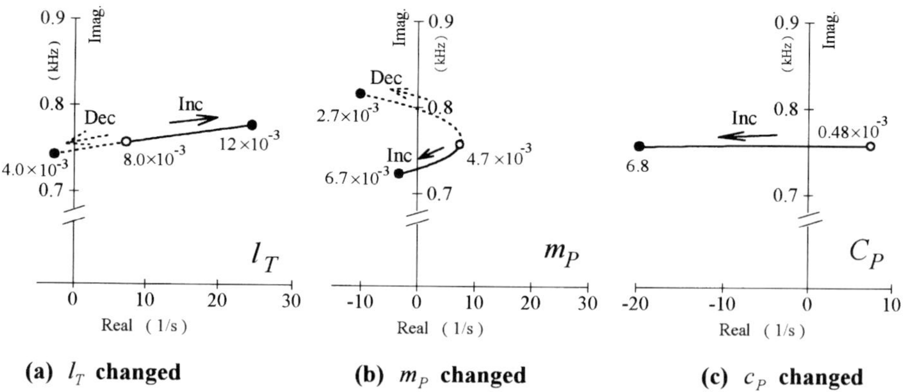

(a) l_T changed **(b) m_P changed** **(c) c_P changed**

Fig.10 Transitions of unstable eigenvalue with one structural parameter changed

Table 3 Initial parameter values

m_P	4.7×10^{-3} kg	l_T	8.0×10^{-3} m
x_0	0.7×10^{-3} m	θ	25.0℃
$\bar{x}$	0.28×10^{-3} m	l	0.5 m
c_P	0.48×10^{-3} kg/s	q_P	38.0×10^{-6} m³/s

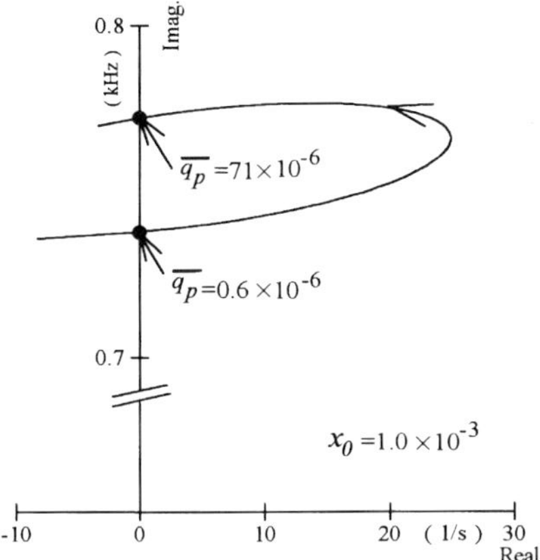

Fig.11 Instability in normal valve operation starting from cracking

to the average poppet flow rate and displacement, they can cause instability only in the limited ranges of their values. On the other hand an increase of the oil temperature acts to shift the eigenvalue toward the unstable side and vice versa.

The effects of the structural parameters, such as the length of the small tube l_T, the poppet mass m_P and the damping factor for the poppet motion c_P, have been examined too. The results are illustrated in **Fig.10(a)～(c)**; an increase of the damping factor and an decrease of the tube length result in shifting the eigenvalue toward the stable side. With regard to the poppet mass, however, not only the increase but also the decrease of mass eventually works to move the eigenvalue into the left-half plane.

As is demonstrated in these figures, the working and structural parameters affect the stability in their own complicated ways. This result probably explains why self-induced vibration looks so capricious in actual systems.

3.5 Instability in the override process

When the poppet valve is used to control pressure, it starts to operate when the line (upstream) pressure reaches the set value (the cracking pressure). The subsequent process where the poppet displacement increases with an increasing flow rate is called "override", which is appropriate for experimentally examining the validity of above-discussed instability analysis.

The trajectory of an unstable eigenvalue in the override process is shown in **Fig. 11**. The initial compression of the poppet spring x_0 is chosen to be 1.0 mm, which corresponds to the cracking pressure of 941 kPa. The values of other parameters are the same as the foregoing calculation. After the cracking the eigenvalue enters the right-half plane at a small flow rate of 0.6 cm^3/s and keeps staying there until the flow rate becomes 71 cm^3/s.

4 CONCLUSIONS

The stability problem of a poppet valve connected to a pipe was numerically evaluated on the basis of eigenvalues. The Locus Method was introduced as a tool to compute the eigenvalues of a complicated characteristic equation derived from the fundamental mathematical models including a distributed-parameter model for the pipe. The suitability of the Locus Method was fully supported through numerical demonstrations. It was also confirmed that there are various working and structural parameters which have substantial influence on the occurrence of self-induced vibration.

REFERENCES

(1) **Funk, J.E.** *Poppet Valve Stability*, Trans. ASME, J. Basic Eng., Ser.D, Vol.86, No.2 (1964) 207.
(2) **Hayashi, S.** *Instability of Poppet Valve Circuit*, Trans. JSME(B),Vol.57, No.541(1991) 2867.(in Japanese)
(3) **Hayashi, S.**, **Yamamoto, H.** and **Iimura, I.** *Influence of Drain Orifice on Poppet Valve Stability*, JHPS J., Vol.24, No.2(1993) 283. (in Japanese)
(4) **Shirai, A.**, **Hayashi, S.**, **Hayase, T.** and **Wang, W.** *Eigenvalue Analysis of Direct-acting Poppet Valve Circuit*, Proc. 4th JHPS International Symposium on Fluid Power (1999), 303.
(5) **Washio, S.** and **Konishi, T.** *Locus Method and Mode Calculation for Fluid Vibration in a Double Pipe*, Trans. JSME(B), Vol.52, No.478 (1986) 2345. (in Japanese)
(6) **Washio, S.**, **Nakamura, Y.** and **Yu, Y.** *Static Characteristics of a Piston-type Pilot Relief Valve*, Proc. Inst. Mech. Engrs., Pt.C, Vol.213 (1999) 231.
(7) **Nakamura, Y.**, **Washio, S.** and **Yu, Y.** *Modeling of Dynamic Behaviors of a Poppet Valve*, Proc. 4th JHPS International Symposium on Fluid Power, Tokyo'99, 315.
(8) **Muller, D.E.** Math. Tables Other Aides Comp.,10(1956),208.
(9) **Scarton, H.A.** and **Rouleau, W.T.** *Axisymmetric waves in compressible Newtonian liquids contained in rigid tubes: steady-periodic mode shapes and dispersion by the method of eigenvalleys*, J. Fluid Mech., Vol.58-Pt3 (1973),595.
(10) **Komatsu, Y.** *Theory of Functions*, Asakura Mathematics Lectures 11 (1964), 156, Asakura. (in Japanese)
(11) **Moriguchi, S.**, **Udagawa, Y.** and **Hitotsumatsu, S.** *Handbook of mathematical formula 3* (1965), 151, Iwanami. (in Japanese)
(12) **Washio, S.** and **Konishi, T.** *Locus Method and Its Programming to Calculate Complex Roots of Functions*, Reports of Research and Development, Okayama University Computer Center, Vol.2, No.3/4 (1986) (in Japanese)

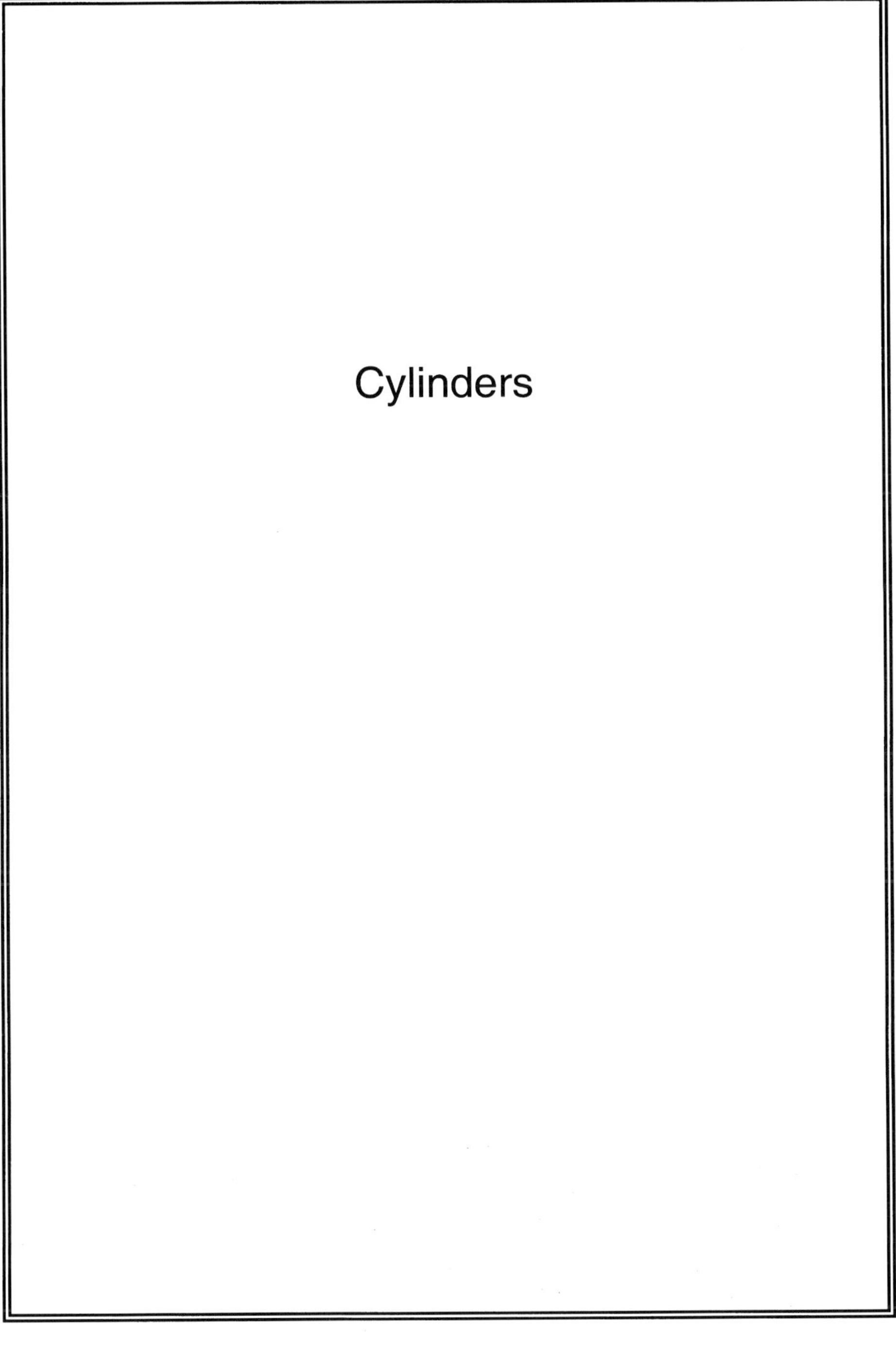
Cylinders

Actuator cushion performance simulation and test results

T LIE and **P J CHAPPLE**
Faculty of Mechanical Engineering, Norwegian University of Science and Technology, Norway
D G TILLEY
Department of Mechanical Engineering, University of Bath, UK

Abstract

This paper describes the simulation of cushioning systems for linear actuators. Comparisons are made with results obtained from experimental tests carried out on an actuator having a piston diameter 50 *mm* that was used to move an 890 *kg* inertial load mass at velocities up to 0.55 *m/s*. The cushion employed had a shaped plug, or spear, to create a variable restriction in the flow outlet from the actuator. The simulation results agreed satisfactorily with those from the tests with the allowance for likely non concentric positioning of the cushion spear in its bore as a result of manufacturing tolerances. The problem of obtaining a representative value for the fluid viscosity is also highlighted as this is subject to heating from fluid friction and from the metal components.

The work then shows how the simulation can be used for the design of cushions taking into consideration the uncontrolled parameters that are encountered in their application. This is particularly valuable as in most cases the testing of actuator cushioning is not feasible particularly if the range of known parameter variations were to be considered.

1 Introduction

Actuator cushioning is an important feature for absorbing the kinetic energy of moving loads at the end of the stroke. Cushioning systems for linear actuators have been in use for many years but their design for a given application or a range of applications creates problems because of uncontrolled parameters that are difficult to quantify. Traditionally cushioning has been carried out by the use of an adjustable restrictor placed in the outlet flow when this has been blocked by a plug attached to the actuator piston or rod. This method suffers from a number of problems including:

- the variation of the cushion pressure with the initial actuator velocity.
- the generation of excessive cushion pressures that can arise from incorrect adjustment of the restrictor. The appropriate setting of the restrictor requires transient measurement of the cylinder pressure a facility that is often not available to the user.
- the cushion pressure is at a maximum at the onset of cushioning which decays through the remainder of the stroke.

Alternative methods use shaped plugs, or spears, to create a variable restriction that can maintain the cushion pressure at a higher mean value thus increasing its energy absorbing capacity. For its general application there is the problem of determining a cushion geometry that will not create excessive cushion pressures with variations of:

- the load mass
- the initial actuator velocity at the commencement of cushioning
- the fluid viscosity.

Simulation techniques enable a sensitivity analysis to be carried out for designing a cushioning system for a specific application. The performance of a standard actuator range can also be evaluated using these methods where the manufacturer needs to determine the worst combination of parameters in order to ensure that the pressure does not exceed safe limits for the particular actuator. These necessitate the consideration of the effect of manufacturing tolerances on small clearances and the concentricity of assembled components. A major uncontrolled parameter is the fluid viscosity that, for a given fluid, will be a function of the inlet temperature and the amount of heat dissipation in the cushion.

A comparison of the simulated performance of some of the available cushion systems was reported by Chapple [1] but this topic has generally attracted little attention by way of published material.

The cushioning of a 50mm diameter actuator has been tested against an inertial load of 890 kg at velocities up to 0.6 m/s. This paper describes a basis for the simulation of a tapered cushion system together with a performance evaluation for the test application the results from which are compared with those obtained from the experimental work.

2 Tapered cushion system

There is a range of available geometrical shapes of cushion spears to give a desired retardation of the actuator that are aimed to provide as high a mean pressure as possible and low sensitivity to changes in fluid viscosity. Typical shapes include stepped diameters, tapers and a curved surface to the spear.

It must be realised that the changes in shape are small being of the order of 0.025 to 0.05 mm which makes it impossible for them to be seen with the naked eye and also to manufacture within close tolerances.

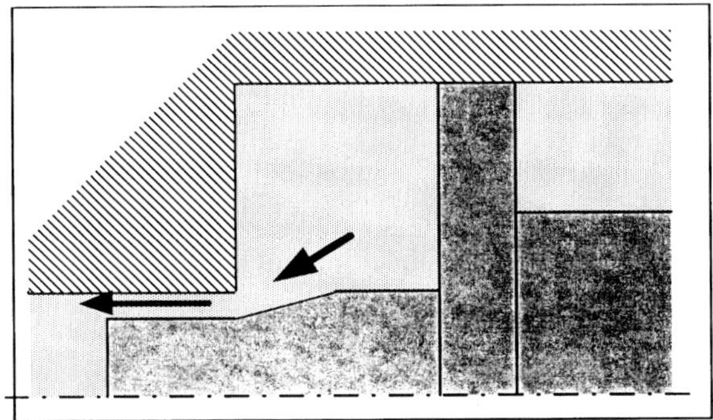

Figure 1 Linear actuator tapered cushion spear for retraction into the head

The paper is concerned with a tapered cushion as shown in Figure 1 that has two stages to its shape variation with displacement into the cushion. It had been found from simulation studies that a double taper could provide flexibility in the design to meet the required performance and the mathematical analysis is based on this approach. However, for the purposes of the tests it was found appropriate to make the first section parallel as shown in Figure 1.

3 Modelling the cushion process

The flow through the cushion space for a tapered spear can be analysed by using the laminar flow equation for concentric components with stationary boundaries:

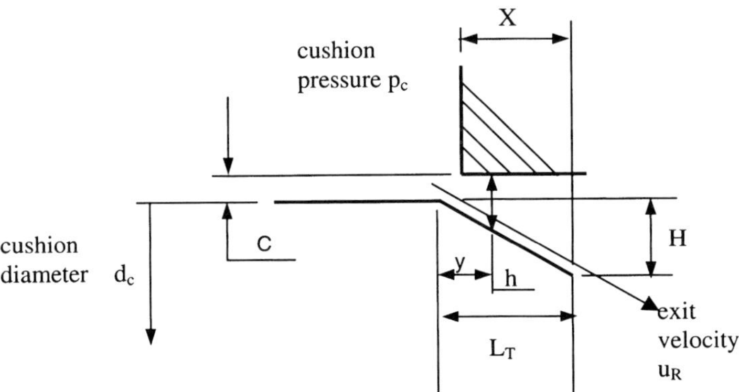

Figure 2 Cushion taper details

The slope of the taper is of the order of $0.2^0$and the two dimensional laminar flow analysis is applied so that the flow, q, through an element in the flow path with a clearance, h, is given by:

$$q = -\frac{h^3 \pi d_c}{12\mu}\frac{dp}{dy} \tag{1}$$

Defining the slope of the taper by B which is equal to H/L_T gives $h_l = H - By$

Thus: $h = h_l + c$ in Figure 2

$$\text{so} \int_{p_f}^{0} dp = -\frac{12\mu q}{\pi d_c}\int_0^x \frac{dy}{(c+h_1)^3} \tag{2}$$

Where p_f is the pressure caused by the viscous flow in the flowpath.
Therefore:

$$(c+h_1)^3 = (c+H-By)^3 = (a-By)^3 = B^3(b-y)^3$$

$$\text{where } a = c+H \text{ and } b = \frac{a}{B}$$

The integral in equation (2) gives:

$$\therefore \frac{1}{B^3}\int_0^x \frac{dy}{(b-y)^3} = \frac{1}{B^3}\left[\frac{(b-y)^{-2}}{2}\right]_0^x = \frac{1}{2B^3}\left[\frac{1}{(b-x)^2} - \frac{1}{b^2}\right]$$

$$= \frac{1}{2B^3}\frac{(2bx - x^2)}{b^2(b-x)^2} = \frac{1}{2B^3}\frac{x(2\frac{a}{B} - x)}{\frac{a^2}{B^2}(\frac{a}{B} - x)^2} = \frac{1}{2B^3}\frac{\frac{x}{B}(2a - Bx)}{\frac{a^2}{B^4}(a-Bx)^2} = \frac{x(2a-Bx)}{2a^2(a-Bx)^2}$$

Thus, for a given value of x, the pressure due to viscous friction in the flow path, p_f, is given by:

$$p_f = \frac{6\mu q}{\pi d_C}\frac{x(2(c+H) - Bx)}{(c+H)^2(c+H-Bx)^2} \tag{3}$$

When $x > L_T$, there will be an additional viscous friction term because the flow will have to pass through the radial clearance c. The pressure will then be given by:

$$p_f = \frac{6\mu q}{\pi d_C}\frac{L_T(2c+H)}{(c+H)^2 c^2} + \frac{12\mu q(x - L_T)}{\pi d_C c^3} \tag{4}$$

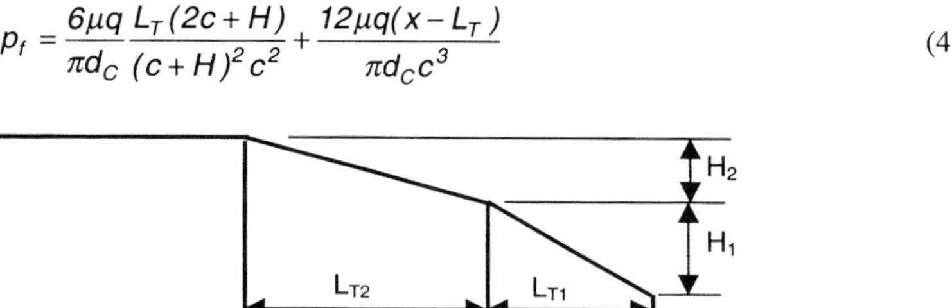

Figure 3 Double taper details

For a cushion spear having two tapered sections as shown in Figure 3 the analysis yields:

$0<x<L_{T1}$

$$p_{f1} = \frac{6\mu q}{\pi d_C} \frac{x(2(c+H_1+H_2)-B_1 x)}{(c+H_1+H_2)^2 (c+H_1+H_2-B_1 x)^2} \tag{5}$$

$x>L_{T1}$

$$p_{f1} = \frac{6\mu q}{\pi d_C} \frac{L_{T1}(2(c+H_2)+H_1)}{(c+H_1+H_2)^2 (c+H_2)^2} \tag{6}$$

$$p_{f2} = \frac{6\mu q}{\pi d_C} \frac{(x-L_{T1})(2(c+H_2)-B_2(x-L_{T1}))}{(c+H_2)^2 (c+H_2-B_2(x-L_{T1}))^2} \tag{7}$$

$x>(L_{T2}+L_{T1})$

$$p_{f1} = \frac{6\mu q}{\pi d_C} \frac{L_{T1}(2(c+H_2)+H_1)}{(c+H_1+H_2)^2 (c+H_2)^2} \quad \text{as equation (6)}$$

$$p_{f2} = \frac{6\mu q L_{T2}}{\pi d_C}\left[\frac{2c+H_2}{c^2(c+H_2)^2}\right] + \frac{12\mu q(x-L_{T2}-L_{T1})}{\pi d_C c^3} \tag{8}$$

The laminar flow pressure loss for any value of the spear position, x, can thus be obtained using the appropriate equations (5) to (8) but the effect of the entry and flow losses must also be considered. Koivula et al [2] examined this for their work on fine annular clearances using a pressure loss coefficient of 1.5.

At the commencement of cushioning the mean cushion flow velocity can be of the order of 150m/s with a Reynolds number. of around 1000. For this the dynamic pressure loss can be quite significant although it decays rapidly as the actuator velocity reduces and the insertion of the spear into the bore increases. The cushion pressure, p_c, will therefore be the sum of the viscous pressure losses and that due to the dynamic pressure loss.

4 Cushion dynamics and simulation

The dynamic system is assumed to consist of an inertial load mass, m, coulomb friction, F_C, and velocity dependent friction using the coefficient, C_f, which give:

$$m\frac{du}{dt} = -p_C A_C - F_C - C_F u \tag{9}$$

Equation (9) was incorporated into the Vissim simulation software together with the laminar flow equations and a dynamic head loss factor based on the exit fluid velocity. The effect of fluid compressibility was found to have a negligible effect on the system performance and for simplicity this was excluded from the simulation.

5 Test system

The test rig used for the experimental work has an inertial mass of 890kg mounted horizontally on rollers with a stroke length of 0.6m and was moved by an actuator of 50mm cylinder diameter with a rod diameter of 36mm. The actuator test cushion was fitted to the retract direction and enabled the cylinder pressure on the piston side to be recorded together with actuator position and the inlet pressure to the actuator annulus.

The cushion spear shape was parallel for the first stage and tapered for the second stage. Referring to Figure 2 the cushion dimensions were:

Taper heights $H_1 = 0$ $H_2 = 0.025mm$
Clearance $c = 0.08$ and $0.11mm$ (made possible by changing the head cap).

Taper lengths $L_{T1} = 10mm$ $L_{T2} = 8mm$

Cushion spear diameter $d_c = 25mm$.

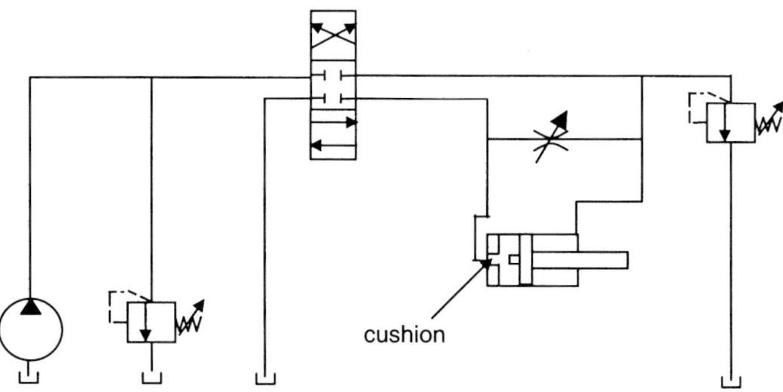

Figure 4 Hydraulic circuit for the test system

The test hydraulic circuit, shown in Figure 4, had provision for bypassing oil from the pump into the piston side return line with the actuator fully retracted. Thus, by adjusting the restrictor valve, the temperature of the end cap and oil in the return line could be set to values between 20 and 40^0C. Having set the temperature at the desired value the actuator was extended so filling the cylinder with heated oil and then retracted for the cushion performance test.

The pump flow of *36 L/min* can create a maximum theoretical retract velocity of 0.64 *m/s*. Because of the length of pipes in the return circuit and their diameters, the corresponding actuator outlet flow of *74.5 L/min* generates a piston pressure of around *50 bar*.

Consequently, during the retraction stroke, an annulus pressure of around *110 bar* was required, this being set by the relief valve attached to the annulus inlet port, as shown in Figure 4.

Different retract velocities could therefore be set by adjusting the annulus relief valve. However, the disadvantage with this arrangement is that as cushioning proceeds and the annulus flow reduces, its pressure increases due to the override of the relief valve. Ideally it would be desired to maintain this pressure constant in order to avoid having to include the supply circuit in the simulation. However, the Sun cartridge valve used had an override pressure of around *10 bar* for most of the operating pressures conditions which had a minimal effect on the simulated results.

6 Analysis of the test and simulation results

6.1 Test results

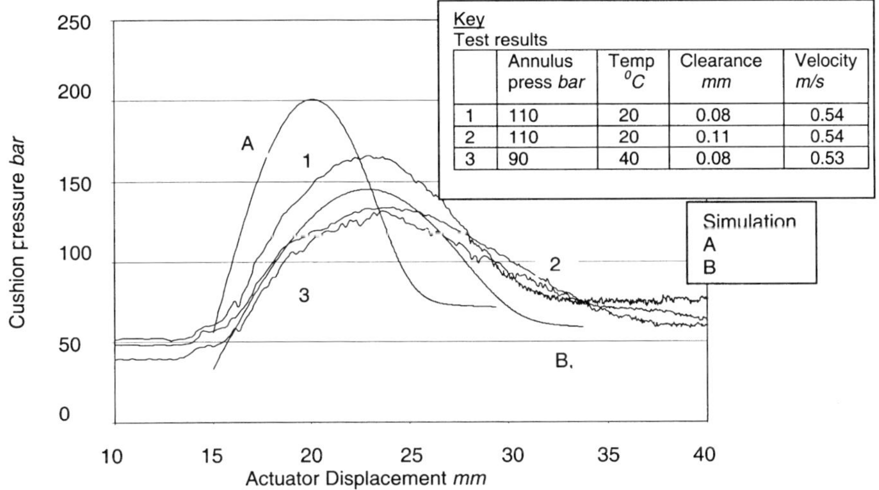

Figure 5 Test and simulation results

The assessment of cushion performance is primarily concerned with the cushion pressure and the actuator displacement into the cushion after the cushioning has commenced. The three test results shown in Figure 5 are for inlet temperatures of 20 and 40^{0}C for the smallest clearance head cap and 20^{0}C for the larger clearance at velocities in the region of 0.53 *m/s*.

As previously described, the initial piston pressure is created by the flow losses through the pipes, fittings and directional control valve. The variation in this pressure is due to the small differences in the velocity and fluid viscosity.

The measured actuator movements into the cushion for the tests in Figure 5 are shown in Figure 6 where it can be seen that both increased clearance and reduced viscosity both result in increasing the actuator displacement because of the reduced resistance in the cushion.

When the cushioning transient has ceased, as can be seen in Figure 5, the final piston pressure has increased because the cushion area is less than that of the piston.

At the end of the cushioning action the final actuator velocity is determined by the leakage through the clearance that gradually reduces with the increasing length of the leakage path until the piston contacts the cylinder cap.

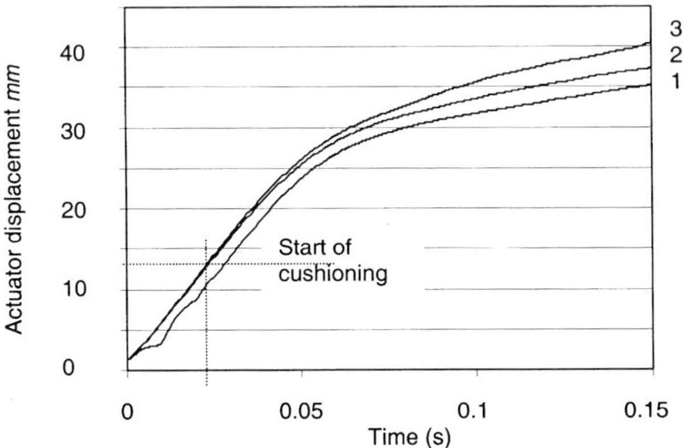

Figure 6 Actuator displacement during cushioning

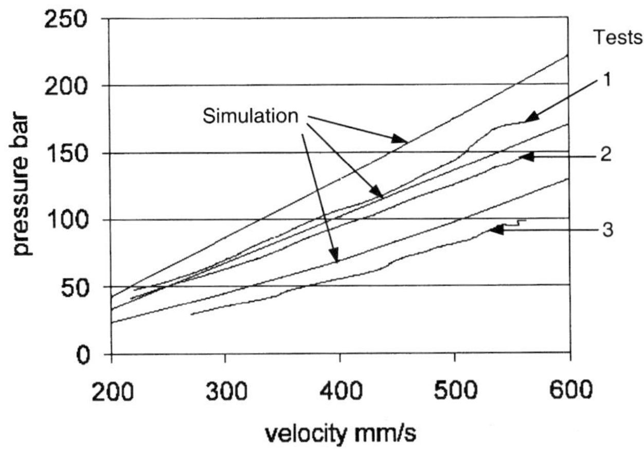

Figure 7 Variation in the peak cushion pressures with initial velocity

The variation in the peak pressures in Figure7, taken from all the of test results, show how these are related to the initial velocity for a given clearance and inlet fluid viscosity. This indicates that the pressure changes by a factor of approximately 2.5:1 for velocities in the range 0.25 and 0.5 *m/s*. Referring to Figure 7 the pressures are seen to vary in the range of 25 to 45 *bar* for a change in the inlet viscosity from 80 and 32 cSt (graphs 2 and 3) and upto 30 *bar* for a change in the clearance from 0.08 to 0.11*mm* (graphs 1 and 2).

Data in this form provides useful information in the design and application of cushion systems for the estimation of the peak pressure.

6.2 Simulation results

The application of equation (1) to cylinders having a small clearance, shows that the effect of eccentricity of the inner cylinder on the relationship between the flow and pressure can be expressed by $\{ 1 + 1.5(\frac{e}{c})^2 \}$, for an eccentricity, *e*, in the clearance space, *c* [2]. In assembled actuators it is likely that the eccentricity ratio will lie in the range 0.3 to 0.6 but this value can never be verified and may, in some instances, lie outside this range.

Eccentricity in the simulation is accommodated by using a flow factor, q_f. In the simulation results of Figure 5 and Figure 7 the value of the flow factor has been assumed to be 1.5 that corresponds to an eccentricity of the cushion spear in the bore of approximately 0.05 *mm*. The value of the fluid viscosity is assumed to be that which corresponds to the initial inlet temperature of the fluid in the cylinder and the pipes at the commencement of the retraction stroke.

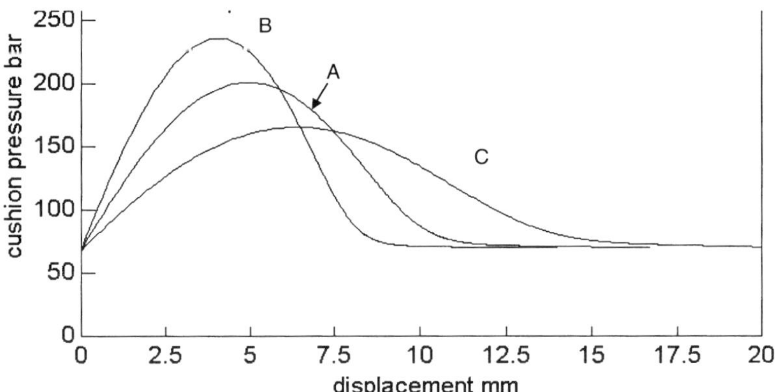

Figure 8 Simulation with reduced viscosity

The simulation result, A, in Figure 5 and also shown in Figure 7, is for 20^0C inlet temperature with the smallest clearance of 0.08 *mm*. Koivula [2] showed that for steady flow through concentric annular paths, using the viscosity that corresponded to the theoretical temperature generated in the flow path obtained theoretical flows that matched the measured values. Graph C in Figure 8 shows the effect of reducing the fluid viscosity to that corresponding to an increase in fluid temperature of 10^0C. This is seen to give an improved result when compared to that of test 1 in Figure 5. For concentric components (i.e. $q_f = 1$) the maximum pressure is increased by 35 *bar* which is shown in Figure 8 graph B.

The flow model incorporates a dynamic head loss equivalent to that applying to a restrictor flow coefficient of 1. This causes the initial pressure to rise as the cushion spear just enters the bore by an amount that will vary as the initial flow squared. It is in this region that compressibility has its major effect but the simulation studies do not indicate a significant change in the pressure rise rate with changes in the Bulk modulus.

CFD studies have shown general agreement with the flow characteristics of equations (3) to (8) but, at the time of writing, have not been used to examine the flow when cushioning starts. As part of an ongoing study this aspect will be investigated together with work on establishing the effect of eccentricity of the tapered cushion spear in the bore on the flow using the Star CD three-dimensional CFD software.

7 Cushion design

The results highlight the problems that are confronted in the design of cushioning systems for an application because of uncontrolled parameters that include:

- Fluid viscosity
- Manufacturing tolerances
- Eccentricity and alignment of the components in the assembled actuator

These are in conflict because of the range of peak pressures and cushion displacements that are obtained. For the actuator details already used, the performance can be simulated for a number of conditions to determine its envelope which will be a function of the maximum velocity and the level of assisting pressure.

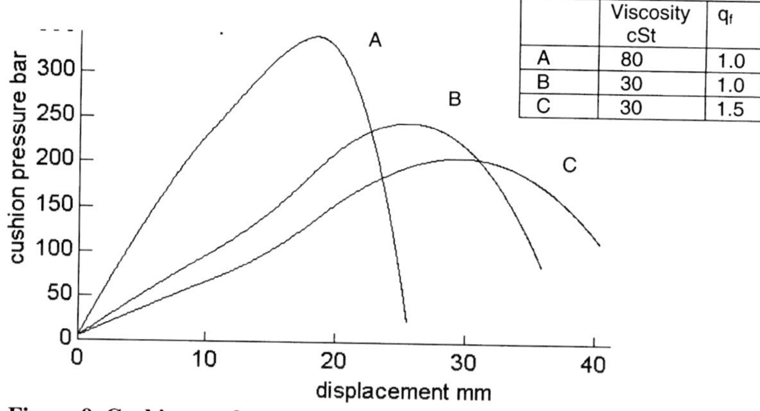

Figure 9 Cushion performance for an application

Taking an inertial mass of 40000 *kg* at a velocity of 0.2 *m/s* the results from the simulation are as shown in Figure 9 for two viscosities and eccentricities denoted by the flow factor q_f.

The maximum pressure from curve A of 340 *bar* is a normal maximum value of fatigue pressure for a standard cylinder. However, considering the possibility of eccentricity and lower viscosity in curve C reduces the maximum pressure to 184 *bar* but with an increase in the cushion distance of 15 *mm*. For many applications the fluid viscosity range could well be in excess of the values considered in this study and the simulation provides a valuable tool to consider all the likely parameter variations.

The ideal cushion system would be one in which the cushion pressure is maintained at a constant level. This provides a constant deceleration so that the actuator displacement X_m into the cushion for zero final velocity is given by:

$$X_m = \frac{mU_m^2}{2p_C A_C}$$

Thus for the example this ideal displacement would be:

Pressure p_C (bar)	Displacement X_m (mm)
200	27
300	18

The level of the assisting pressure can also be uncertain in an application but the simulation can be used to assess the performance for constant values of this pressure. Considering other conditions e.g. a velocity of 0.5 *m/s*, an inertial mass of 4800 *kg*, 0.08 *mm* clearance, a q_f factor of 1.5 and 80 *cSt* viscosity gives the simulated cushion performance shown in Figure 10, A, for pressure and velocity with cushion displacement.

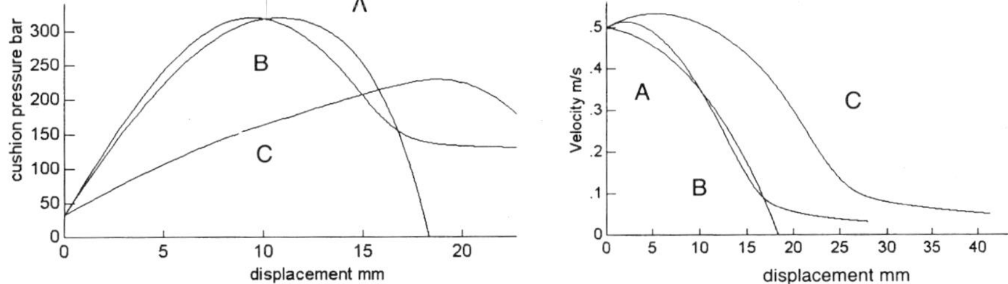

Figure 10 Cushion pressure and velocity for 0.5 *m/s* initial velocity

For the conditions of 200 *bar* assisting pressure the final velocity is created by the assisting force at a level that depends on the viscosity as shown by graphs B and C in Figure 10. For an available maximum cushion length of, say, 25*mm* the cushioning effect is probably adequate. The energy absorbed for the three cases considered are as follows:

Velocity *m/s*	Mass *kg*	Assisting pressure *bar*	Energy *J*
0.2	40000	0	800
0.5	4800	0	600
0.5	2100	200	263

Alternative cushion designs having different spear surface configurations and diameters can be assessed using simulation techniques which are of particular benefit to the designer especially as the testing of actuators with large inertial masses is not usually feasible. The additional advantage is that the design can be assessed for its sensitivity to the uncontrolled parameters outlined at the beginning of this section.

8 Conclusions

The use of simulation techniques has been shown to provide satisfactory performance estimation for the cushioning of the linear actuator cushioning system when compared to the results of the tests. The simulation also provides an indication of the problems involved in designing a cushion for a given application because of the number of uncontrolled factors that are encountered.

In particular the establishment of the value to be used for the fluid viscosity creates a problem because the temperature of the metal components can have an overriding effect on the fluid as it flows through the cushion. Clearly this depends on the environment in which the actuator is situated, the duty cycle and the availability of cooling for the fluid.

Cushion systems are frequently used in applications in which the performance of the system is not clearly specified at the design stage so the manufacturer needs to be able to understand the effect of some of the problems that can be expected. Some flexibility can be provided by the use of an adjustable restrictor that bypasses the cushion but this has the attendant problem of user abuse.

9 Acknowledgements

This work has been carried out in association with Helipebs Controls Ltd and the authors wish to thank them for their assistance with the project and the provision of the actuator for the test work. Thanks are also due to the technical staff at the University of Bath for their help in the laboratory test work.

10 References

[1] Chapple, P.J.

Using simulation techniques in the design of actuator cushioning.
Drives and Controls Conference, Telford, UK; March 1999

[2] Koivula,T., Ellman, A. and Vilenius, M.

The effect of oil type on flow and cavitation properties and annular clearances.
Bath Workshop on Power Transmission and Motion Control (PTMC99);
Bath, UK.; September 1999

Suppression of pressure transients in on/off control of low-pressure water hydraulic cylinder

M LINJAMA, K T KOSKINEN, and **M VILENIUS**
Institute of Hydraulics and Automation, Tampere University of Technology, Finland

ABSTRACT

The paper analyses and tests experimentally two methods to suppress pressure transients in an on/off controlled water hydraulic cylinder. The on/off position control is realised with inexpensive two-way solenoid valves and the basic three-state control is used as a starting point. In the first method, termed as the modified three-state control, the control algorithm is modified such that two velocities are achieved with the same valves as in the three-state control. In the second method two damping volumes are added into the system via small orifices. The effect of the size of the orifice and damping volume on the damping factor of the system is analysed by using a linearised model. Results show that both methods significantly reduce the pressure transients. The drawback of the modified three-state control is that it is more sensitive to changes in the load force than the three-state control. The use of the damping volumes is effective in reducing the pressure transients but the position error increases slightly.

KEYWORDS: on/off position control, low-pressure water hydraulics, solenoid valves.

1 INTRODUCTION

Low-pressure water hydraulics (LPWH) is a new fluid power technology between traditional oil hydraulics, high-pressure water hydraulics and pneumatics. The pressure level is typically between 10 and 40 bar and the aim is to combine good properties of each technology: low price level of pneumatics, good controllability, stiffness and efficiency of oil hydraulics as well as environmental safety of water hydraulics. The basic lines of the LPWH technology are presented in [1] and some initial test results are given in [2].

Typical applications of the LPWH technology are low and medium power systems where an extreme power density is not required and where pneumatic systems do not offer sufficient force and/or controllability. Because the technology is new, only a few components exist that are optimised for low-pressure water hydraulics. High-pressure water hydraulic components are all too expensive and pneumatic or oil hydraulic components are not directly suitable for water usage. Therefore, component development will play an important role in the development of the LPWH technology. The current situation is such that only on/off type valves are available in reasonable prices.

In this study the on/off position control of a LPWH hydraulic cylinder is studied. Inexpensive two-way solenoid valves are used and the properties of these valves have been analysed in [3]. The basic control schemes have also been tested experimentally in [4] and in this study the emphasis is in the analytical analysis and in the suppression of harmful pressure transients. Pressure peaks are a big problem in LPWH systems because of the low pressure levels.

2 CHARACTERISTICS OF DIRECTLY OPERATED TWO-WAY SOLENOID VALVES

The static and dynamic properties of directly operated two-way solenoid valves are analysed in detail in [3] and only the main points are discussed here. The operation principle of a normally closed type valve is shown in Figure 1. The valve has only one moving part driven by the solenoid force. Usually a rubber seal is used and therefore the valve has no leakage. The main drawback of these valves is that the flow capacity is limited. When the orifice size d increases, the closing pressure force also increases, which causes the maximum operation pressure to decrease. The highest operation pressure is achieved with a normally closed type valve driven by an AC coil. The typical flow capacity of this kind of valve is 2.5 l/min at a pressure differential of 1 bar, and the maximum operation pressure is 30 bar. With reasonable pressure differences the maximum flow through a single valve is less than 10 l/min. The price level of these valves is low (15-40 €) and therefore a parallel connection can be used when more flow is required.

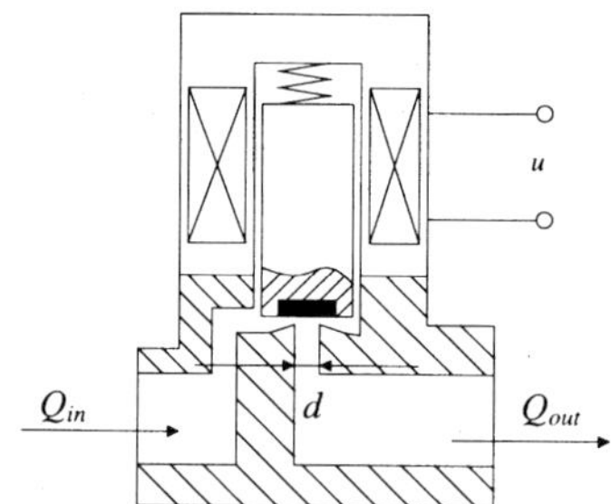

Figure 1. Operation principle of a normally closed two-way solenoid valve.

The opening and closing delays and their variations are essential when on/off valves are used in position control. The effect of a delay can be compensated if the velocity is known but the variation of the delay affects directly the position accuracy. The opening and closing delays were measured in [3] and the results were: opening delay 20–32 ms, closing delay 3–12 ms. The variation in the delays is mainly caused by the AC current. The pressure level or flow has only a minor effect on the delays. The closing delay is more important in the position control because it determines the position accuracy. The durability of this kind of valves was also tested in [3] and the durability was about ten million openings and closings when the supply pressure was 16 bar. The durability increases when the orifice size decreases and decreases when the pressure increases.

3 ANALYSIS OF ON/OFF CONTROLLED CYLINDER

3.1 Introduction

This study concentrates on the on/off control of an asymmetric cylinder with four two-way solenoid valves. The system is shown schematically in Figure 2. The aim is to achieve the combination of good position accuracy, fast movement and small pressure peaks. Most often the stopping points are determined with adjustable switches. If programmable stopping points are needed, a continuous position measurement must be used. In this case it is also possible to estimate the piston velocity and utilise this in the control. This possibility is not considered in this paper although the piston position is méasured.

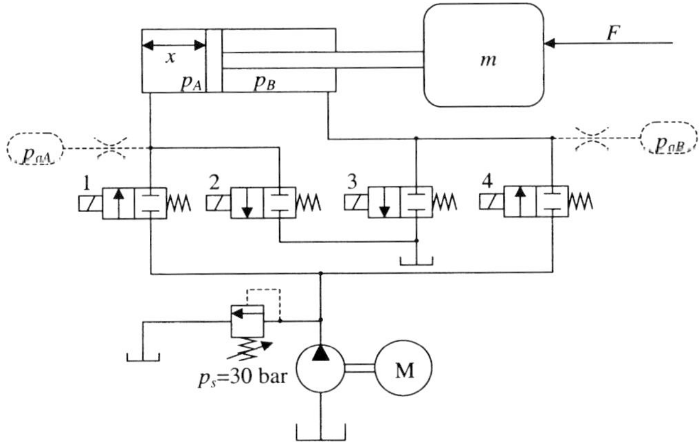

Figure 2. Hydraulic circuit of the on/off controlled cylinder

The position accuracy depends on the variation of valve closing delays, the velocity of the piston and the stiffness of the actuator. In general, the stiffness of an LPWH cylinder is very high because a typical supply pressure is 30 bar compared with 150 bar in high-pressure water hydraulic systems and 210 bar in oil hydraulic systems. If the same force is required from the LWPH cylinder, the area of piston will be at least five times higher. Because the stiffness is roughly proportional to the piston area, the stiffness of LPWH cylinders is much higher than in other fluid power systems. Usually the stiffness is so high that the dynamic effects due to compressibility of the fluid has only a minor effect on the position response and the system behaves mainly like an integrator. This in turn results in the position accuracy being almost independent of piston position or load mass. This situation differs dramatically from pneumatic systems.

Assume now that the cylinder behaves like an ideal integrator. The steady-state velocity of the piston depends on the sizes of valves, supply pressure and force acting on the cylinder. The closing delay of the valve is approximated by

$$\tau_d = \tau_{d0} + \text{rand}(-\Delta\tau_d, \Delta\tau_d) \tag{1}$$

where τ_d is the delay, τ_{d0} is the average delay and rand($-\Delta\tau_d$, $\Delta\tau_d$) is a random number between $-\Delta\tau_d$ and $\Delta\tau_d$. Similarly, the velocity of the piston just before stopping is

$$v = v_0 + \text{rand}(-\Delta v, \Delta v) \tag{2}$$

The uncertainty in the velocity is caused by variations in the supply pressure, friction forces and other external forces. In order to compensate for the valve delays, the valves must be closed a little bit before the stopping point. The required advance is

$$s = v_0\,\tau_{d0} \tag{3}$$

and with this advance the position error remains within following limits

$$-v_0\,\Delta\tau_d - \tau_0\Delta v - \Delta v\Delta\tau \;\le e \le\; v_0\Delta\tau_d + \tau_0\Delta v - \Delta v\Delta\tau \tag{4}$$

One important topic that must be considered in the valve sizing is the cavitation choking effect, which means that the flow through the valve does not increase when the downstream pressure decreases below a certain limit. The flow through a two-way valve can be approximated as follows

$$Q = \begin{cases} K_v\sqrt{p_u - p_d} & , \; if \; p_d / p_u \ge b \\ K_v\sqrt{(1-b)p_u} & , \; if \; p_d / p_u < b \end{cases} \tag{5}$$

The value of b is between 0 and 0.5 depending on the detailed construction of the valve. In tested valves the value was between 0.2 and 0.5.

3.2 Three-state control

The three-state control is the simplest possible on/off control scheme. Its control algorithm is

$$u_1 = u_3 = \begin{cases} 1 & , \; if \; e > s_{pos} \\ 0 & , \; if \; e \le s_{pos} \end{cases} \qquad u_2 = u_4 = \begin{cases} 1 & , \; if \; e < -s_{neg} \\ 0 & , \; if \; e \ge -s_{neg} \end{cases} \tag{6}$$

where u_1, u_2, u_3 and u_4 are the valve control signals as shown in Figure 2. The advances s_{pos} and s_{neg} are required to compensate for the valve closing delays of valves and the initial values can be calculated according to Equation 3. In the three-state control the steady-state velocities are

$$v_{pos} = \frac{K_{v1}}{A_B}\sqrt{\frac{\gamma p_S - F/A_B}{\gamma^3 + \left(K_{v1}/K^*_{v3}\right)^2}} \quad , \quad v_{neg} = \frac{K^*_{v2}}{A_B}\sqrt{\frac{p_S + F/A_B}{\gamma^3 + \left(K^*_{v2}/K_{v4}\right)^2}} \tag{7}$$

where $\gamma = A_A/A_B$, K_{vi} is the flow coefficient of valve i and $K^*_{vi} = K_{vi}\sqrt{1-b}$. In Equation 7 it is assumed that the steady state pressures at the cylinder chambers are bigger than $b*p_S$ and that the tank pressure is zero.

3.3 Modified three-state control

The main drawback in the three-state control is that the maximum velocity is limited. If the velocity increases, the position error and pressure peaks also increase. Even with infinitely fast valves the velocity cannot be high because of high pressure peaks. The basic approach to increase the maximum velocity is to use several velocity levels i.e. to use high velocity when the error is large and to switch to a smaller velocity near the stopping point. The most common way to achieve this is to add additional valves. For example in the five-state control four additional valves are required. This approach is effective but of course the price increases. Here an alternative solution is used in which two velocity levels are achieved with the same valves as in the basic three-state control. The control method is called here a modified three-state control. The basic idea is that when the error is smaller than the approach distance s_a, the valve is opened from the pressure side to the tank. This decreases the velocity and therefore pressure peaks and position error also decreases. The valve control signals as a function of position error are shown in Figure 3. The analytical expressions for the steady-state velocities in the approach phase are very complicated and will not be presented here.

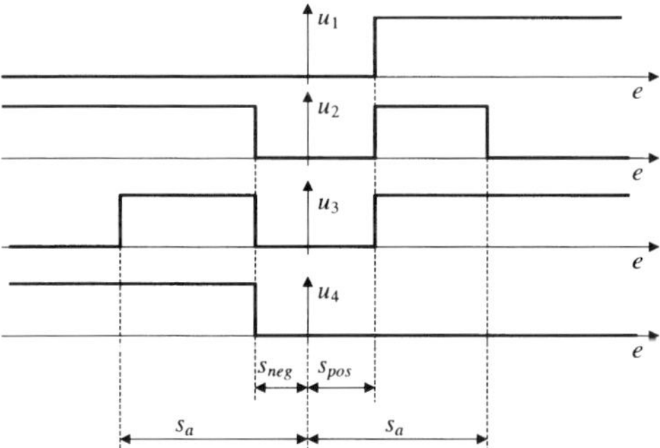

Figure 3. Valve control signals in the modified three-state control.

The external force has quite a strong effect on the steady-state velocity in the approach phase and therefore the modified three-state control can be used only in the cases where the variations in the load force are small or when the velocity is measured or estimated. If the external force is too big, the motion stops in the approach phase. The forces where the motion stops are given by

$$F_{min} = -\frac{K_{v4}^2 p_S A_B}{K_{v4}^2 + K_{v3}^{*\,2}} \quad , \quad F_{max} = \frac{K_{v1}^2 p_S A_A}{K_{v1}^2 + K_{v2}^{*\,2}} \tag{8}$$

These are the extreme limits where the modified three-state can be used.

3.4 Three-state control with damping volumes

An important drawback of the basic three-state control is the pressure peaks that occur in the stopping phase. The pressure peaks are not a problem with the modified three-state control but it requires fairly constant load force or velocity estimation in order to achieve sufficient position accuracy. Therefore, an alternative way to decrease the pressure transients in the three-state control is studied. The idea is to add small damping volumes (or accumulators) into the system via small orifices as shown by broken lines in Figure 2.

3.4.1 Linearised model

The aim of the damping volumes is to increase damping in the stopping phase such that the pressure peaks reduce. Therefore, a linearised model is derived in the situation when all valves are closed. Further, the friction forces are neglected and the volume and effective bulk modulus of the damping volumes are assumed constant. By linearising the orifice flows the following state-space model is achieved

$$\begin{bmatrix} \ddot{x} \\ \dot{x} \\ \dot{p}_A \\ \dot{p}_B \\ \dot{p}_{aA} \\ \dot{p}_{aB} \end{bmatrix} = \begin{bmatrix} 0 & 0 & A_A/m & -A_B/m & 0 & 0 \\ 1 & 0 & 0 & 0 & 0 & 0 \\ -BA_A/V_A & 0 & -BK_{LA}/V_A & 0 & BK_{LA}/V_A & 0 \\ BA_B/V_B & 0 & 0 & -BK_{LB}/V_B & 0 & BK_{LB}/V_B \\ 0 & 0 & B_aK_{LA}/V_{aA} & 0 & -B_aK_{LA}/V_{aA} & 0 \\ 0 & 0 & 0 & B_aK_{LB}/V_{aB} & 0 & -B_aK_{LB}/V_{aB} \end{bmatrix} \begin{bmatrix} \dot{x} \\ x \\ p_A \\ p_B \\ p_{aA} \\ p_{aB} \end{bmatrix} \tag{9}$$

where B is the effective bulk modulus of the cylinder chamber volumes V_A and V_B , B_a is the effective bulk modulus of damping volumes V_{aA} and V_{aB} , and K_{LA} and K_{LB} are the flow coefficients of the laminar orifices. Other symbols are defined in Figure 2. All other parameters can be evaluated quite accurately but the flow coefficients of the laminar orifices are difficult to obtain because the flow is in practice turbulent. Here the coefficients are determined such that if the pressure difference over the orifice is sinusoidal with a given amplitude Δp, the dissipated energy is the same in the turbulent and laminar orifice i.e.

$$p = \Delta p \sin(t) \quad , \quad Q_{turbulent} = K_v \operatorname{sgn}(p)\sqrt{|p|} \quad , \quad Q_{laminar} = K_L p$$

$$W_{turbulent} = \int_0^\pi p Q_{turbulent}\, dt = \int_0^\pi \Delta p \sin(t) K_v \sqrt{\Delta p \sin(t)}\, dt$$

$$W_{laminar} = \int_0^\pi p Q_{laminar}\, dt = \int_0^\pi K_L \Delta p^2 \sin^2(t) dt \tag{10}$$

$$W_{turbulent} = W_{laminar} \quad \Rightarrow \quad K_L = \frac{2K_v}{\pi\sqrt{\Delta p}} \int_0^\pi \sin^{\frac{3}{2}} t\, dt$$

This problem does not have an analytical solution but it is straightforward to get numerical values.

3.4.2 Stiffness reduction due to damping volumes

The high stiffness is advantageous in the on/off position control because it reduces the position error caused by the compressibility of fluid. The steady state stiffness of a hydraulic cylinder without damping volumes is

$$K_{h1} = \frac{A_A^2 B}{V_A} + \frac{A_B^2 B}{V_B} \tag{11}$$

and when damping volumes are added, the steady state stiffness becomes

$$K_{h2} = \frac{A_A^2 B}{V_A + \dfrac{B}{B_a} V_{aA}} + \frac{A_B^2 B}{V_B + \dfrac{B}{B_a} V_{aB}} \tag{12}$$

which is clearly smaller than K_{h1}. The ratio B/B_a is large and therefore the damping volumes must be very small in order to retain sufficient stiffness. However, because the stiffness of LPWH cylinders is very high, even a large decrease in the stiffness has probably only a small effect on the position error. Even if the stiffness decreases by 50 percent, it is still higher than in an oil hydraulic cylinder with the same force.

3.4.3 Realisation of damping volumes

It is essential that damping volumes can be realised without any significant increase in the price level of the system. If damping volumes are too expensive, it is more economical to use other solutions to decrease the pressure peaks, such as five-state control. Normal water hydraulic accumulators are rather expensive and it is also difficult to find sufficiently small accumulators. Therefore, an alternative solution is developed in which small pieces of flexible hose are used as damping volumes. The hose is plugged at one end and the other end is connected to the system via the orifice. In order to avoid long hoses, the bulk modulus of the hose must be as low as possible. Also the price of the hose and fittings must be low, which prevents the use of stainless steel fittings.

4 ANALYSIS OF THE EXAMPLE SYSTEM

4.1 Introduction

The example system is shown in Figure 4 and its hydraulic circuit is similar to Figure 2. Only the valves and their sub-plates are true LPWH components and other parts are realised with high-pressure water hydraulic components. The pressurised water is produced with a small Danfoss Nessie® Power Pack (PPH 6.3) and the cylinder is from SSH Stainless Ltd. (AQUA 70 series). The piping is made with stainless steel pipes and fittings. Valves are manufactured by Cosmotecnica and they are sub-plate type with aluminium bodies. The load mass is adjustable between 30 and 200 kg and its position is measured by a linear potentiometer. The other measurements are the chamber and supply pressures. The main parameters of the system are shown in the Table 1. Smaller valves are used in the piston rod side to compensate for the asymmetry of the cylinder.

Figure 4. The example system with damping volumes.

Table 1. Main parameters of the example system.

Cylinder size [mm]	32/16×500
Coulomb friction force [N]	≈ 600
Valve flow at Δp = 1 bar, valves 1 and 2 [l/min]	1.1
Valve flow at Δp = 1 bar, valves 3 and 4 [l/min]	0.76
Supply pressure [bar]	30
Load mass [kg]	200

4.2 Three-state control

The calculated steady-state velocities of the example system with three-state control are shown in Figure 5. When friction forces are considered, the steady-state velocities are 72 mm/s when extending and 58 mm/s when retracting. Therefore, the velocity still depends on the direction of the movement although the situation is better than in the case of equally sized valves. It is seen that the effect of the external force on the steady-state velocity is about 20 mm/(s kN). The effect on the position error is according to Equation 4 about 0.16 mm/kN. Therefore, the position error is quite insensitive for variation in the load force.

4.3 Modified three-state control

The steady-state velocities of the example system with modified three-state control in the approach phase are shown in Figure 5. When friction forces are considered, the steady-state velocities are 37 mm/s when extending and 26 mm/s when retracting indicating that the velocity drops about 50 percent when the error becomes smaller than the approach distance. It is seen that the force has greater effect on the velocity than in the three-state control and the useable force range is also much smaller. The effect of the force on the velocity is about 30 mm/(s kN) resulting in a position error 0.24 mm/kN.

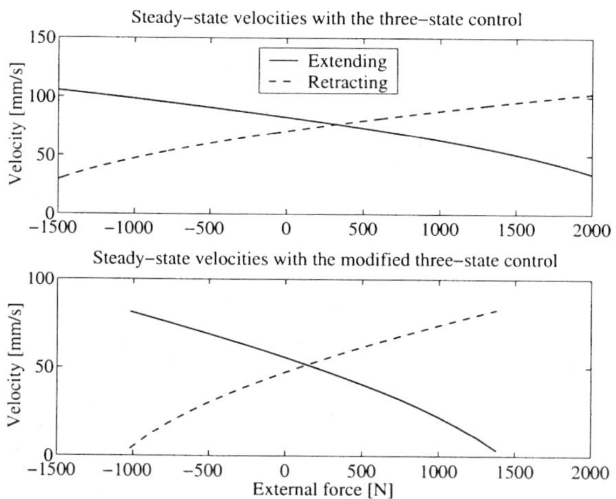

Figure 5. Calculated steady-state velocities of the example system with three-state and modified three-state control.

4.4 Three-state control with damping volumes

The damping volumes and orifices are assumed equal at both sides of the cylinder in order to simplify the analysis (i.e. $K_{LA} = K_{LB} = K_L$ and $V_{aA} = V_{aB} = V_a$). The analysis is made at the position $x = 0.25$ m. The damping volumes are selected such that the steady-state stiffness reduces by 50 percent which gives $B_a/V_a = 4.5\times10^{12}$ Pa/m^3. Achievable damping is studied by using the linearised model of Equation 9. The model has one complex pole pair corresponding to oscillatory motion of the load mass and the other poles correspond to pressure dynamics. Figure 6 shows the damping factor and natural frequency of the complex pole pair as a function of orifice diameter. It is seen that the damping factor has a maximum value with a certain orifice diameters (in this case around 0.7 mm) and the maximum damping factor is 0.21. When the orifice size decreases, the natural frequency of the system approaches the natural frequency without damping volumes and when the orifice size increases, the natural frequency approaches the natural frequency without orifices. The optimal damping is achieved when the natural frequency is the average of these two values.

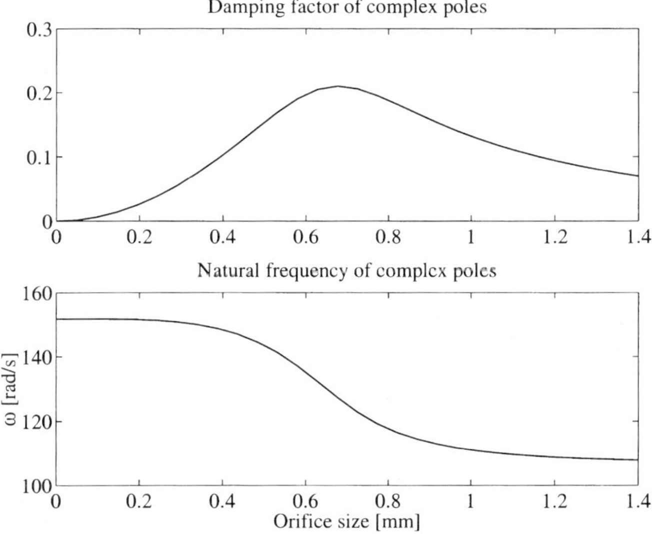

Figure 6. Damping factor and natural frequencies of the complex poles of the linearised model when the orifice size is varied. The orifice linearisation is made with $\Delta p = 4$ bar and the ratio B_a/V_a is 4.5×10^{12} Pa/m^3.

The achieved damping factor is 0.21, which is not too high. However, because the friction forces are neglected, the damping is larger in the real system. The damping can be increased if larger damping volumes are allowed. Figure 7 shows the damping factor and natural frequency of the system when the orifice size is kept at 0.7 mm and the ratio B_a/V_a is varied. It is seen that the damping factor increases rapidly when the ratio B_a/V_a decreases from the nominal value of 4.5×10^{12} Pa/m^3. However, with fixed orifices, the damping does not increase after a certain point (in this case when $B_a/V_a < 2\times10^{12}$ Pa/m^3).

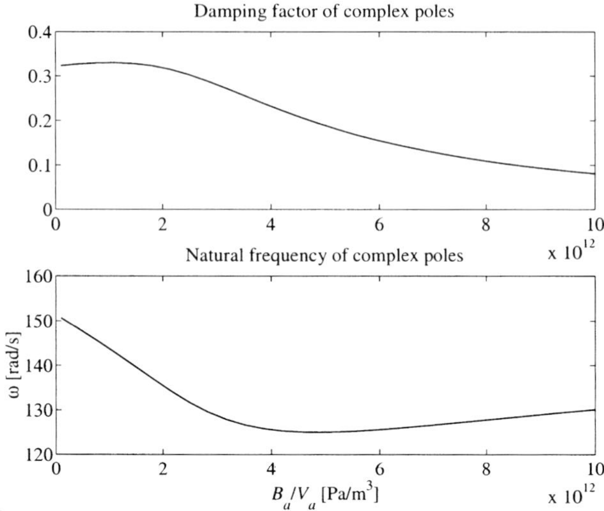

Figure 7. Damping factor and natural frequencies of the complex poles of the linearised model when the size of damping volumes is varied. The orifice linearisation is made with Δp = 4 bar and the orifice size is 0.7 mm.

5 EXPERIMENTAL RESULTS

5.1 Introduction

Certain parameters must be considered in the realisation of the control schemes. The sample time must be small enough in order to achieve sufficient time resolution. The sample time is selected to be 2 ms, which can be considered as an additional delay in Equation 1. The position measurement is noisy and needs filtering so that false openings/closings due to measurement noise can be avoided. The filtering is made by software with a second order filter (natural frequency 200 rad/s and damping factor 0.7). The sampling and filtering requirements resulted in the advance being slightly larger than calculated value. The position error varies randomly and a lot of responses must be obtained to find out the range where the error lies with a high probability. This is why all the responses are repeated 50 times. The highest pressure peaks occur when the motion stops near the end of the cylinder. Because the velocity is bigger and because the braking is made with a smaller piston area for the extending movement, the worst situation occurs near the maximum stroke. Only this situation is studied here and the responses in other positions can be found in [4].

5.2 Results with three-state control

The three-state control has been tested experimentally in [4] and results can be summarised as follows:

- The position error stays within ±0.3 mm despite of variations in the load mass or piston position
- The full velocity is achieved after 1.5 mm movement and the position error is also independent of stroke length if it is at least 2 mm

Figure 8 shows 50 position responses from 0.47 m to 0.49 m with the three-state control and corresponding distribution of the position error. The used advance is 1.2 mm while the calculated is according to Equation 3 only 0.6 mm. The maximum pressure peaks are 43 bar, which is really high when compare to the 30 bar supply pressure.

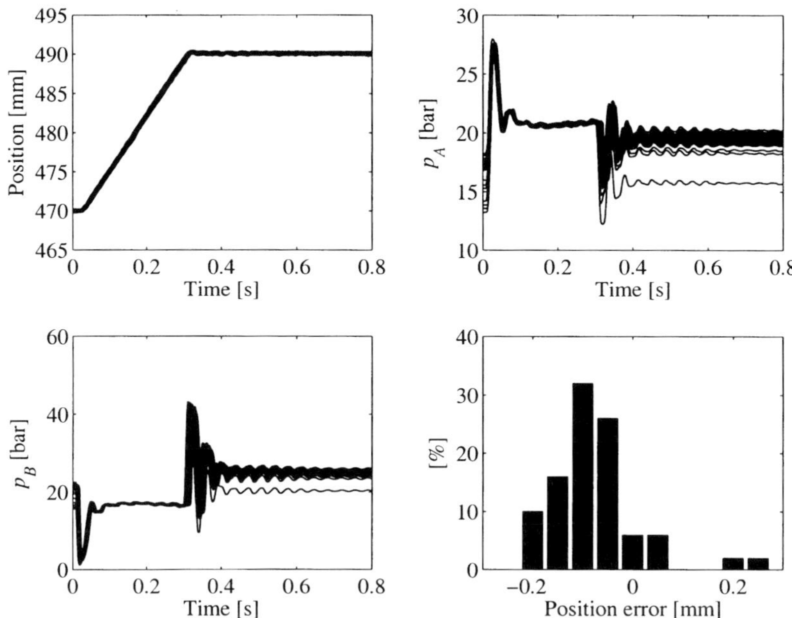

Figure 8. Fifty position responses from 0.47 m to 0.49 m with the three-state control. The advance is 1.2 mm.

5.3 Results with modified three-state control

The modified three-state control has also been tested experimentally in [4] and the achieved position accuracy was ±0.25 mm. As in the three-state control case, the position error is also independent of the load mass, piston position and stroke length if the stroke is at least 2 mm. Figure 9 shows 50 position responses from 0.47 m to 0.49 m with the modified three-state control and corresponding distribution of the position error. The used advance is 0.6 mm and the approach distance s_a is 5 mm. Again equation 3 predicts too small an advance of 0.3 mm. It is seen that the pressure peaks are not a problem with the modified three-state control.

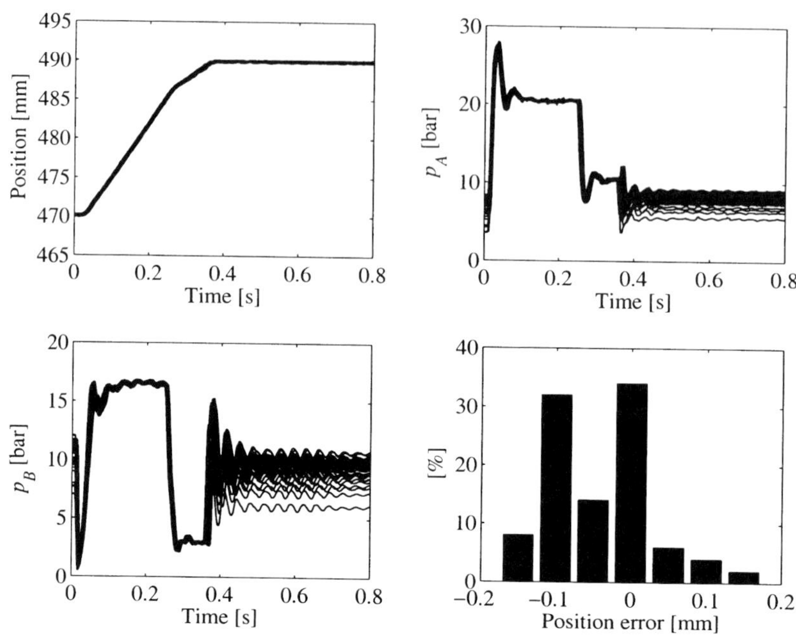

Figure 9. Fifty position responses from 0.47 m to 0.49 m with the modified three-state control. The advance is 0.6 mm and the approach distance x_a is 5 mm.

5.4 Results with three-state control and damping volumes

The damping volumes are formed from Vega Manuli hoses. The hose is textile braided rubber hose and its pressure rating is suitable for LPWH systems. The bulk modulus of the hose is about 150 MPa at the pressure level 2 MPa and therefore the length of hose remains reasonable. The inner diameter of the hose is 13 mm and the demanded B_a/V_a ratio is achieved with a 220 mm long hose. The fittings used are Parker Push-Lok brass fittings that offer a cost-effective solution. The estimated price for one damping volume is 20 € without the orifice. Figure 10 shows the responses with the three-state control and damping volumes. The maximum pressure peaks are now 32 bar and the pressure oscillations have disappeared. The variation in the position error is slightly more than without damping volumes and the advance has been increased to 1.5 mm.

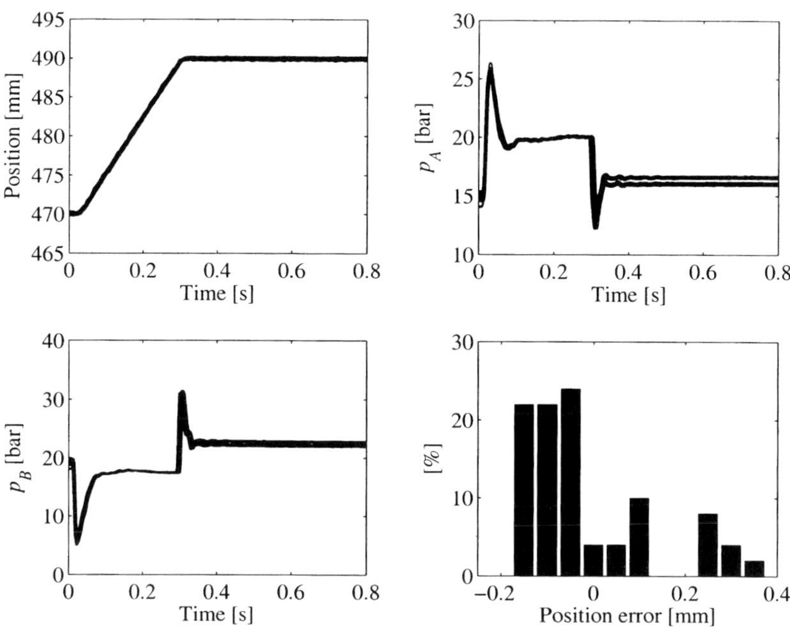

Figure 10. Fifty position responses from 0.47 m to 0.49 m with three-state control and damping volumes. The advance is 1.5 mm and the parameters of the damping volumes are: B_a = 130 MPa, V_a= 29 cm^3, $d_{orifice}$ = 0.7 mm.

6 CONCLUSIONS

The on/off position control of a low-pressure water hydraulic cylinder has been analysed and tested experimentally. The basic three-state control offers good accuracy but the pressure peaks exceed the supply pressure. Two alternative ways for reducing the pressure peaks are developed. In the modified three-state control, two velocity levels are achieved with the same valves than in the three-state control. Its position accuracy is better than in the three-state control if the load force is constant but it is more sensitive to changes in the load force. In the example system, all the pressure transients stay well below the supply pressure when the modified three-state control is used. The three-state control with damping volumes offers the same insensitivity to changes in load force as the basic three-state control but the pressure peaks are much smaller. In the example system, the maximum pressure peaks are at the same level as the supply pressure. A drawback of this solution is that the position error slightly increases. The cost increase is minor, because the damping volumes can be made from lengths of flexible hose.

The results are encouraging and future research will focus on the further development of the control schemes. The estimation of velocity and the utilisation of it in the control is one interesting option because it would reduce the sensitivity of the position accuracy to the load force. Very small movements are also an unsolved problem.

REFERENCES

1 **Kunttu P., Koskinen, K.T.** and **Vilenius, M.** Low pressure water hydraulics – state of the art, *The Sixth Scandinavian International Conference on Fluid Power*, Tampere, Finland, 1999, pp. 67–75 (Tampere University of Technology).

2 **Aaltonen, J., Koskinen, K.T., Vilenius, M.** and **Kunttu, P.** Experiences on the low pressure water hydraulic systems, *The Fourth JHPS International Symposium on Fluid Power*, Tokyo, Japan, 1999, pp. 357–363 (The Japan Hydraulics and Pneumatics Society).

3 **Linjama, M., Tammisto, J., Koskinen, K.T.** and **Vilenius, M.** Two-way solenoid valves in low-pressure water hydraulics, to be published in *International Mechanical Engineering Congress & Exposition*, Orlando, Florida, USA, 2000.

4 **Linjama, M., Koskinen, K.T.** and **Vilenius. M.** On/off position control of low-pressure water hydraulic cylinder using low-cost valves, to be published in *FLUCOME*2000*, Sixth Triennial International Symposium on Fluid Control, Measurement and Visualization*, Sherbrooke, Canada, 2000.

Components and Systems

Modelling of a common rail fuel injection system

K HUHTALA, S TIKKANEN, and **M VILENIUS**
Institute of Hydraulics and Automation, Tampere University of Technology, Finland

ABSTRACT

This research work deals with the modelling of diesel fuel injection system. The fuel injection system is a Common Rail (CR). The application is the hydraulic free piston engine (HFPE) where the studied injection is used. The HFPE is very sensitive for the timing of the fuel injection and control of the amount of the fuel during the injection. The pressure oscillations might disturb the behaviour of the fuel injection event if the fuel injection system is not properly designed. The characteristics of the fuel such as bulk modulus, viscosity and density differ clearly between high (over 1000 bar) or normal atmosphere pressure. The results are different if the fluid properties are taken into account only as a constant value at normal atmosphere pressure. This paper contains the special features that should be taken into account when modelling the electro hydraulic fuel injection system compared to modelling of normal hydraulic systems. In addition this research work studies the behaviour of the Common Rail in different conditions by means of a verified simulation model of the system. Special interest has been focused to the volumes, injection time and fluid properties.

1 INTRODUCTION

The Common Rail (CR) fuel injection is one noteworthy alternative to carry out the fuel injection of the diesel engine. It has become a widely used injection system in diesel engine passenger cars. The delivery of the fuel and the injection is separated in the CR systems. The injection pressure is delivered independently into speed of engine and amount of injected fuel. The main difference between Common Rail and traditional fuel injection systems is that high pressure is always available in the common rail. The amount of injected fuel is controlled by means of an electrically controlled solenoid valve. The adjustable parameters such as injection timing, duration and pressure are decoupled from the engine speed and load. These parameters can be selected almost freely to achieve proper engine performance, noise and emissions.

2 SPECIAL FEATURES IN MODELLING OF FUEL INJECTION SYSTEMS

There are some special differences comparing in modelling of fuel injection systems than normal fluid power systems. Although the studied Common Rail fuel injection system is an electrically controlled hydraulic system, one very important feature is the fluid properties. Normally the fluid is handled as a constant parameter in hydraulic systems. The difference is not so clear in pressure range of 0...400 bar. Nowadays in diesel engines the fuel injection pressure is between 0...1500 bar or even higher and then the fuel characteristics have a more important role than before. There is illustrated in figure 1 one example of the fuel properties as a function of pressure. The viscosity varies in a large scale as a function of pressure. The value of the bulk modulus is about 2 times higher at pressure p = 1400 bar than at pressure p = 0 bar. It is worth-while to notice that one of the most important fact that effects to the value of the sonic speed is the amount of air in the fluid, not pressure.

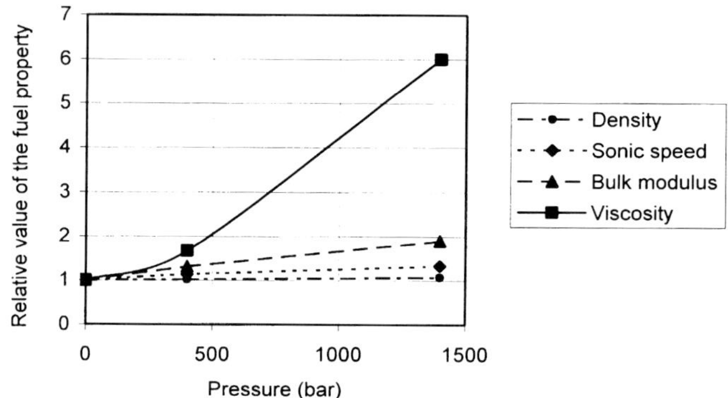

Figure 1. Example of the fuel properties as a function of pressure.

It is also noted that the values of fuel properties shown in figure 1 are only one example from one particular fluid. This example only shows that the fluid is as important a component as the others in fuel injection systems.

Another typical feature is the stiffness of the system. That is partly consequence of high pressure and partly small volumes used in fuel injection systems. The high stiffness may lead in some cases to numerical problems during the simulation.

The injection duration is very short. One talks of injection times between 0.5...10 ms. The pre-injection times are even shorter. For this reason it is easy to understand that the dynamic behaviour of the fuel injection system is very important. On the other hand the diameters of the fuel pipes are small and there exists oscillations. The modelling of the fuel pipes by means of volumes does not give correct answers. One has to use fluid transmission line models.

3 STUDIED COMMON RAIL SYSTEM

The studied Common Rail system is shown in figure 2. There are two injectors in the fuel injection circuit. The common rail is not, as usual, one pipe where the injectors are directly connected. In this case after the common rail there is an extra volume (fluid transmission line) just before both injectors. These volumes are designed to guarantee similar injection conditions to both injectors and prevent them from disturbing each other operation.

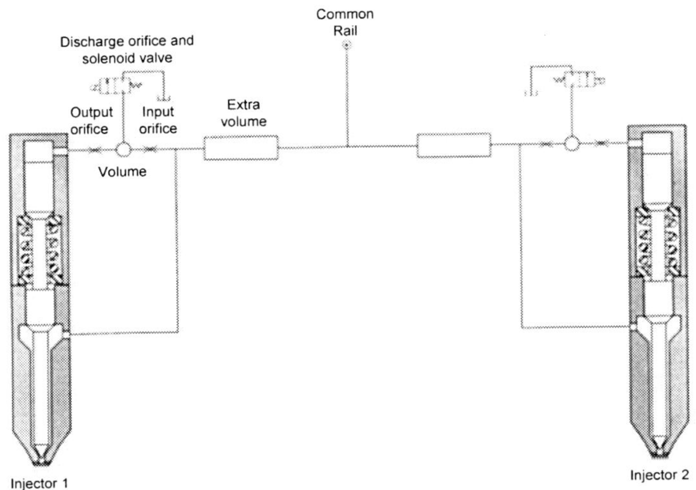

Figure 2. Studied common rail system.

The schematic principle and basis to the modelling work of the injection system is illustrated in figure 3. After the common rail the system separates into two identical injector parts. The connection line from volume 1 to the control valve and nozzle lift volume has been divided into several fluid transmission lines (pipes). The fluid transmission line model is based on the paper published in reference /1/. Pipes 1, 3 and 4 are the necessary connections that are required by the system. Pipe 2 is the extra volume to attenuate possible oscillations in the system. The tube fittings are modelled as orifices between the pipes. The principles of the control valve and injector are explained in references /2/ and /3/. The modelling work of the injector is also based on these papers.

In this kind of system the input and discharge orifices of the control valve (see figure 3) define the operation characteristics of the injector. By means of varying these two orifices the injection rate shape and also the speed of the injector can be changed.

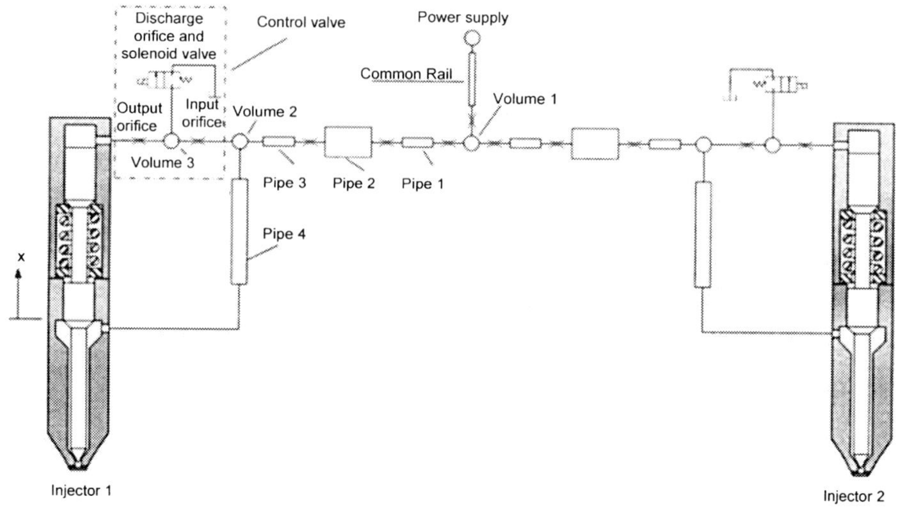

Figure 3. Schematic diagram of the simulation model.

4 VERIFICATION OF THE COMMON RAIL MODEL

The studied Common Rail system was measured on a test bench. The measured variables were the pressures p_1 and p_2 at volumes 1 and 2 and the nozzle lift x. The injection pressure is in many cases very difficult to measure without extra and difficult installations to injector. The verification of the simulation model is undertaken with measurements of pressure p_2 and nozzle lift x. In these measurements and simulation the injection time was 0.8 ms. The total range of the injection time is between 0.3 ...1.2 ms in this injection system.

The measured and simulated results of pressure p_2 are shown in figure 4. The simulated results are sufficiently accurate. There exists a pressure oscillation with frequency around 1 kHz. Small errors in damping are noted.

The measured and simulated results of nozzle lift x are illustrated in figure 5. Also, in this case, the correspondence of measured and simulated results is acceptable.

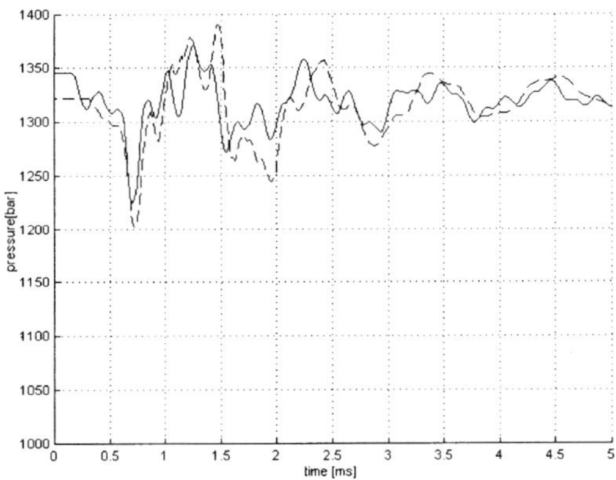

Figure 4. Simulated and measured results of pressure p_2 (-- measured and __ simulated).

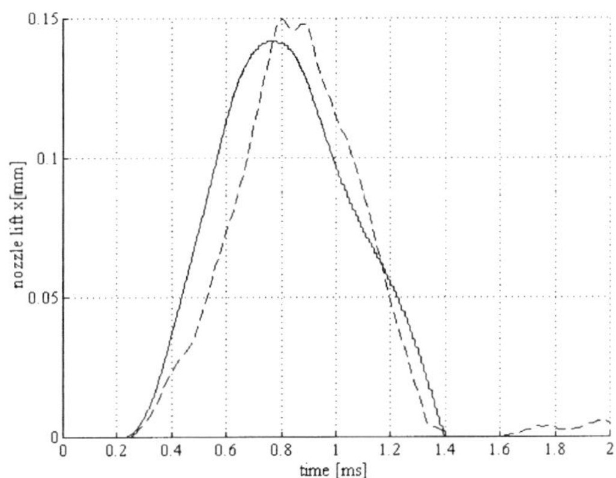

Figure 5. Simulated and measured results of nozzle lift x (-- measured and __ simulated).

5 SIMULATION RESULTS

5.1 Effect of fluid properties to characteristics of CR

The normal assumption in hydraulic calculations is that the fluid properties are constant in the operation range. A common value for bulk modulus is B = 1500 MPa in hydraulic systems. In this study the other constant fuel properties which corresponds to bulk modulus are (case 1):

- density $\rho = 850\ kg/m^3$ and
- viscosity $\nu = 7$ cSt.

On the other hand these properties change as a function pressure and temperature. In this study the pressure effect is taken into account. The properties of the used fluid are at pressure level p = 1350 bar in (case 2):

- density $\rho = 930$ kg/m^{3,}
- bulk modulus B = 2800 MPa and
- viscosity $\nu = 50$ cSt.

The results for simulations with these two different fluid properties are shown in figure 6. The solid line represents the simulation in which the pressure effect has been taken into account (case 2) and the dotted lines represent the case 1.

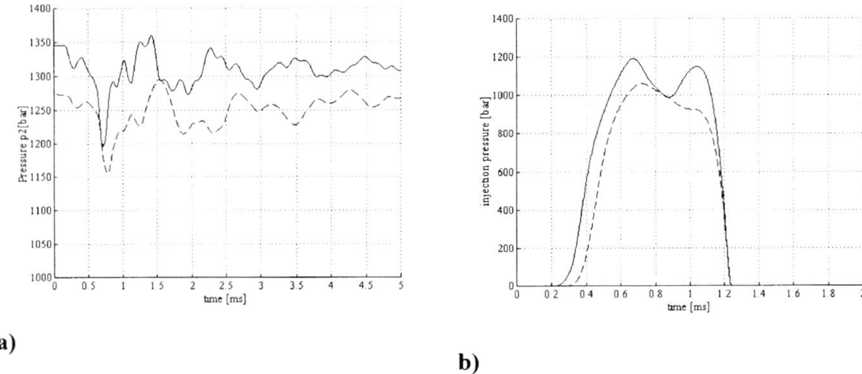

a) b)

Figure 6. Effect of the fluid properties to pressure p_2 (a) and injection pressure (b).

From figures 6 a) and b) it can be seen that the damping of the system is different in these two cases. This means that the real fuel injection system is stiffer from normal hydraulic systems. That means also that pressure oscillations have to be taken into account in designing fuel injection system. Figure 6 b) shows the injection pressure in both cases. In case 2 the delay between command signal ($t_{command} = 0$) is 0.2 ms and in case 1 the delay is 0.3 ms. It shows the difference characteristics of the fuel injection system between these two cases.

The rate shape of injection is also different (figure 6 b). In case 1 the injection is front-loaded and in case 2 there are two peaks in the rate shape at the start and end of injection. These two cases are different if the results are inspected in terms of emissions.

5.2 Effect of extra volume to characteristics of CR

The extra volume is represented in figures 2 and 3. The function of this volume is to attenuate oscillations of the injection system. The initial size of this volume is 50 cm^3 which is about 1000 times bigger than the maximum injected quantity per stroke. The size of the volume was decreased first to 25 cm^3. The third case was that when the extra volume was ignored. The simulation results of these three cases are represented in figure 7. The studied variable is pressure p_2.

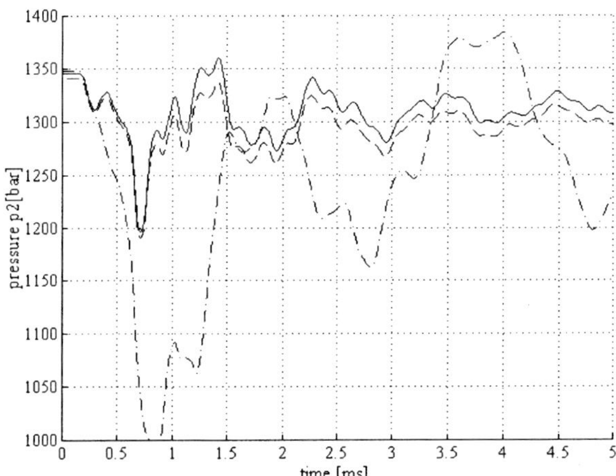

Figure 7. The simulated results of the effect of the extra volume change to pressure p_2 (solid line V=50 cm^3, dashed line V=25 cm^3 and dashdot line V=0).

The extra volume effectively damps the pressure oscillations whose frequency is around 500 Hz (see figure 7). For the system where the extra volume doesn't exist the amplitude of the pressure oscillations at frequency 500 Hz is large (dashdot line in figure 7). If the system has an extra volume the oscillation at this frequency is damped (solid and dashed lines in figure 7). The oscillation frequency of the system with the extra volume is about 1 kHz. According to these results the size of this volume could be half of the initial size.

The effect of the extra volume on nozzle lift and injection quantity is minor (see figure 8). The simulation results of both size of extra volume (V = 50 and 25 cm^3) are almost equal. If there is no extra volume then the results differ from each other but not significantly.

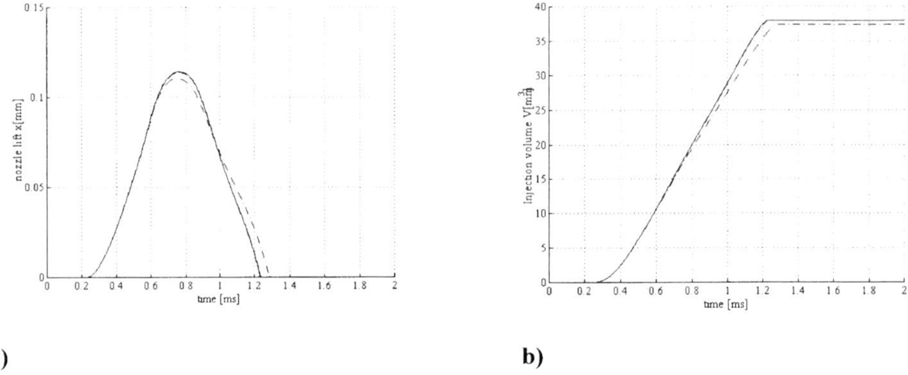

a) **b)**

Figure 8. The simulated results of the effect of the extra volume change to nozzle lift x (a) and to injection quantity (b) (solid line V=50 cm^3, dashed line V=25 cm^3 and dashdot line V=0).

The effect of the extra volume size to injection pressure is shown in figure 9. The same trend continues in this case. The injection pressure is different when the extra volume is missing. The rate shape of injection is different (figure 9) if the extra volume is missing. In this case the injection is front-loaded and in both of cases when the extra volume exists there are two peaks in the rate shape in the beginning and in the end of injection.

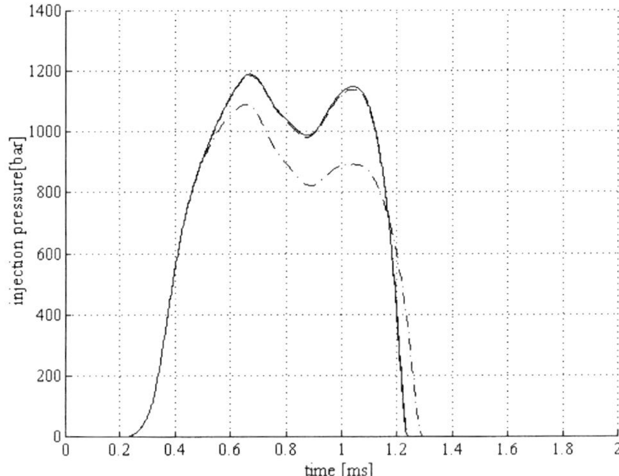

Figure 9. The simulated results of the effect of the extra volume change to injection pressure (solid line V=50 cm^3, dashed line V=25 cm^3 and dashdot line V=0).

5.2 Effect of injection time to characteristics of CR

The behaviour of the pressure p_2 varies as a function of injection time (figure 10). The extra volume size is V = 50 cm^3. The result is that with the shorter injection times the pressure oscillation amplitude is bigger. The injection times which have been studied are following :

- t = 1.2 ms (dotted line),
- t = 0.8 ms (solid line),
- t = 0.5 ms (dashdot line) and
- t = 0.3 ms (dashed line).

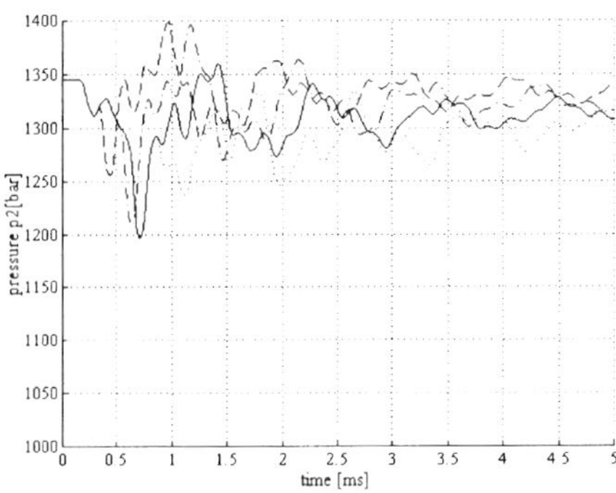

Figure 10. The effect of injection time to pressure p_2 (dotted t=1.2 ms, solid t=0.8 ms, dashdot t=0.5 ms and dashed t=0.3 ms).

The injection quantity as a function of the injection time is represented in figure 11 (a). The simulated results have been merged in figure 11 (b) where the injection quantity is represented as a function of injection time. The relation between injection time and quantity is not linear.

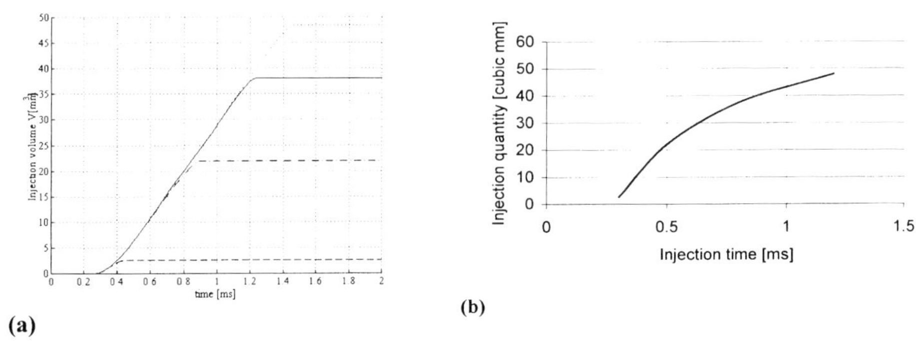

Figure 11. The effect of injection time to injection quantity (dotted t=1.2 ms, solid t=0.8 ms, dashdot t=0.5 ms and dashed t=0.3 ms).

The simulated results of the injection pressure shape at different injection times are shown in figure 12 (a). The simulated results of the injection pressure as a function of injection time is shown in figure 12 (b). The injection pressure shape (rate shape) is related to the injection time. The peak injection pressure decreases sharply when the injection time is shorter than 0.5 ms. The injection pressure drop at short injection times also explains the decrease of the injection volume in this range (see figure 11).

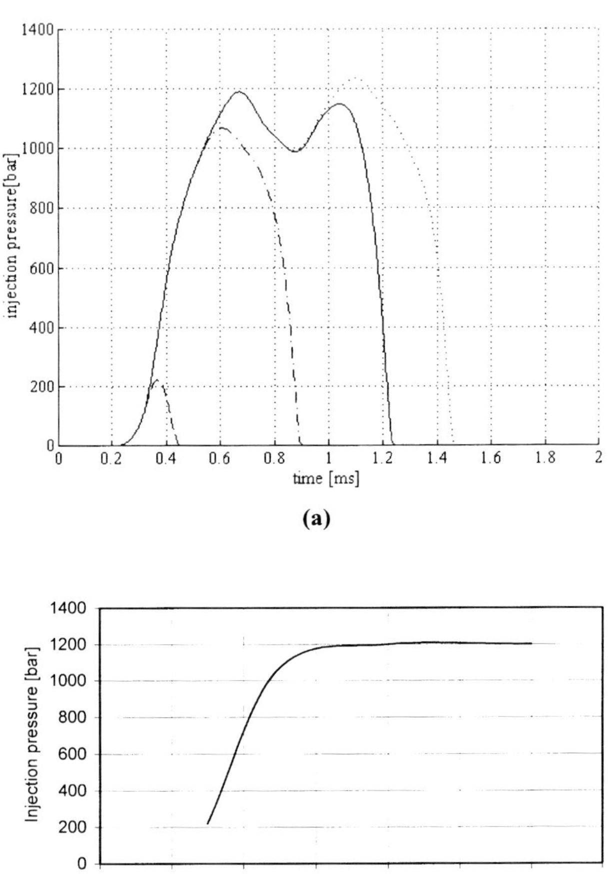

(b)

Figure 12. The effect of the injection time to the injection pressure (dotted t=1.2 ms, solid t=0.8 ms, dashdot t=0.5 ms and dashed t=0.3 ms).

6 CONCLUSIONS

The modelling of a Common Rail system was represented in this paper. The correspondence of simulated and measured results is good. The effect of fluid, extra volume and injection time was studied in the paper. The Common Rail fuel injection system is very stiff because of high pressure and as a consequence of that the fuel is compressed more and the value of bulk modulus is 2-3 times higher than in traditional hydraulic systems. The extra volume attenuates effectively the oscillations of the system. The relation between this volume and maximum injection quantity is about 1000. The size of this volume could be reduced. The effect of injection time is essential in short injection times (below 0.5 ms). The injection pressure decreases in short injection time range (< 0.5 ms). That leads to non-linear behaviour of injection volume as a function of injection time.

REFERENCES

/1/ Piche, R., Ellman, A., Vilenius, M., Integration of numerically stiff fluid power circuit models, 7th Bath International Fluid Power Workshop, Sept. 22-23, 1994, Bath, UK.

/2/ Ganser, M., Fröhlich, T., Betriebseigenschaften eines Common Rail Injektors fur Pkw-Motoren mit ausgerichteten Steuerbohrungen. Internationale Konferenz fur Common Rail Einspritzsysteme – Gegenwart und Zukuntftspotential, Nov. 7 1997. Zurich, Switzerland.

/3/ Ganser, M., Common Rail Injector with Injection Rate Control, SAE paper 981927.

Experimental verification of pulsation dampers and their simplified theory

M IJAS and **T VIRVALO**
Institute of Hydraulic and Automation, Tampere University of Technology, Finland

ABSTRACT

Hydraulic actuators can generate huge pressure peaks and vibrations in mobile hydraulic systems. That causes mechanical failures especially in hydraulic pipelines. This paper concentrates on the quietening of the low frequencies (10-100Hz), which are caused by the hydraulic actuator. The work has been done mostly in the laboratory by studying how different kinds of dampers operate in practice. The laboratory test equipment consisted of a pump, hose to be examined and a servo valve, which imitated a load The studied dampers are an accumulator, a Helmholz-resonator, a T-pipe and a hydraulic motor. A simplified theory, how the parameters of the dampers can be determined, has been presented.

1. NOMENCLATURE

a	Speed of sound	[m/s]
A_3	Area of connecting pipe	[m^2]
d_A	Internal diameter of pipe	[m]
D_{cyl}	Diameter of cylinder	[m]
f	Resonant frequency	[Hz]
J_{motor}	Moment of inertia of hydr. motor	[kg*m^2]
L_A	Length of line	[m]
L_T	Length of T-line	[m]
L_3	Length of connecting pipe	[m]
n	1, 3, 5...	
p_A	Working pressure	[bar]

p_{delta}	Ripple of pressure (peak to peak)	[bar]
p_0	Gas pre-charge pressure	[bar]
p_1	Minimum operating pressure	[bar]
p_2	Maximum operating pressure	[bar]
S	Stroke of the cylinder	[m]
Q	Oil flow	[m^3/s]
V_G	Gas volume at working pressure	[m^3]
V_0	Gas volume at pressure p_0	[m^3]
V_1	Gas volume at pressure p_1	[m^3]
V_2	Gas volume at pressure p_2	[m^3]
V_4	Volume of chamber of H-filter	[m^3]
ΔV	Available oil volume from the accumulator	[m^3]
κ	polytropic exponent	1.4
ρ	fluid density	[kg/m^3]
ω_A	angular velocity	[rad/s]

2. INTRODUCTION

The increase of pressure level has been a trend in mobile hydraulics. Then the hydraulic power can be transfered with lower pressure losses. At the same time reliability has become more and more important. Environmental protection assumes that hydraulic oil does not flow to the land. In that case the hose breakage of mobile machines should be strongly reduced. One way to reduce hose problems is to reduce the pulsations of the pressure. A lot of studies of dampers are available, but they usually concentrate on reducing the pressure ripple caused by a pump and then frequencies are quite high. However, in the mobile machines the most dangerous pressure vibrations may come from actuators (fig 1). In these cases frequencies are considerably lower, for example 50Hz. The most difficult situation is when a hose contains many vibrations at different frequencies. This is a very typical situation. In the mobile machines the operation point varies typically for example due to different loads. All this means that the determination of the damper is quite difficult.

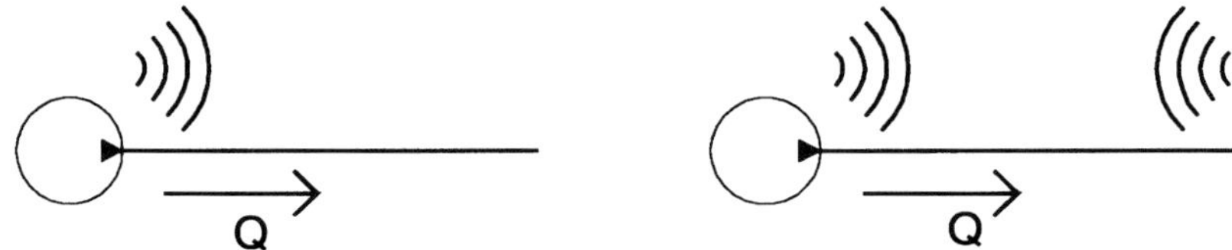

Figure 1. On left there is a situation in which the vibration caused by the pump and on right is a more normal situation.

The hydraulic dampers can be divided into two groups regarding to their operation principles [4]:

1. Absorption type dampers (for example accumulator)
2. Reflection type dampers (H-filter, T-filter, ECA-attenuator)

The best known damper is obviously an accumulator. Its operating principle is to absorb the vibrating energy. The damping characteristic is quite qood at low frequencies (below 300Hz) [4]. Actually, the operation in the frequency range higher than 100Hz is only possible with specially designed accumulators [1].

The operation of reflection type dampers is based on the reflecting of the pressure wave back. Helmholtz-resonator (H-filter) and T-pipe (T-filter) are branching line type dampers. It is characteristic to them that the damping is good, but frequency range is poor [3], [5]. It is possible that they operate even as an amplifier at some frequency.

An expansion chamber attenuator (ECA) consists of a volume connected to a pipe. It is also possible that there are more volumes with different lengths and diameters. Typically the damping is good over a wide frequency range [Ortwig], but the problem is the operation at low frequencies.

3. TEST SET-UP

For the study of dampers a test bench was built in the laboratory of Institute of Hydraulic and Automation- Tampere University of Technology. Hydraulic scheme of the test bench is depicted in figure 2.

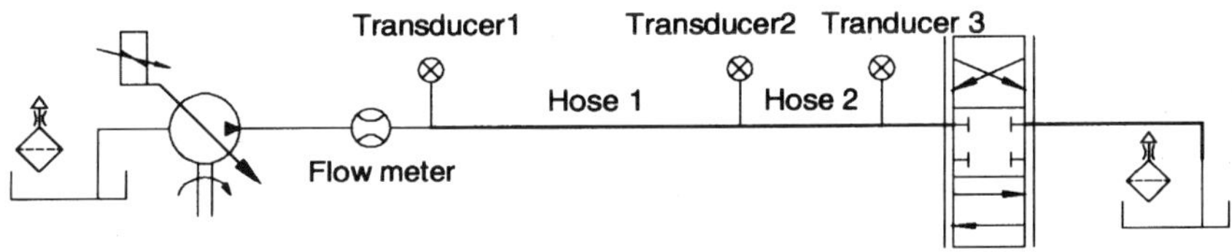

Figure 2. Hydraulic scheme of the test bench.

The hydraulic pump was variable displacement pump with electrical displacement control from Mannesman Rexroth Hydromatik. The pressure transducers were Druck PTX 1400 – 250bar. The transducer1 was immediately after the pump unit. The place of the transducer 2 was between the hose 1 and hose 2 and transducer 3 was just before the servo valve.

The length of the hose 1 was 15 metres and its nominal size was 1", length of the hose 2 was 5 metres and its nominal size was ¾", respectively. Originally the aim of the project was reduction of the pressure oscillation in the rock-drilling machine. That's why the hose arrangement was the above described. The hoses were identical to those of the actual rock-drilling machine. Position of the examined damper was between the transducer3 and the servo valve.

Control valve was 4-way directional servo valve. Hydraulic oil was Shell Tellus S46 and during measurements temperature was kept at 50°C (±1°C), when the viscosity of oil was 32cSt. The control and measurements was realised with dSpace DS1102.

4. TEST WITHOUT DAMPER

The test sequence was the same all the time. First the settings of the pump and the servo valve were adjusted so that the pressure on transducer 1 was 160 bar and oil volume flow 0.00166 m^3/s. When the steady state was achieved, the servo valve produced the flow and pressure oscillation. The servo valve was controlled with a sinusoidal signal. The frequency was swept evenly from zero to hundred hertz. The amplitude of the control signal was about $1.17*10^{-4}$ m^3/s and it was kept constant during the measurement.

The measurement result without a damper is shown in figure 3. The different resonance frequencies can be seen. The resonance seen at 15Hz, is obviously the resonance frequency of the pump displacement control system, because this resonance is observed with the pressure transducer 1. The strongest resonance is about at the frequency 42Hz. In this paper the resonance frequencies are not analysed. The main purpose of this study is to stabilise the frequency of 50 Hz in spite of some resonance frequencies in the hose system. A good damper should have good damping capability in a large frequency range.

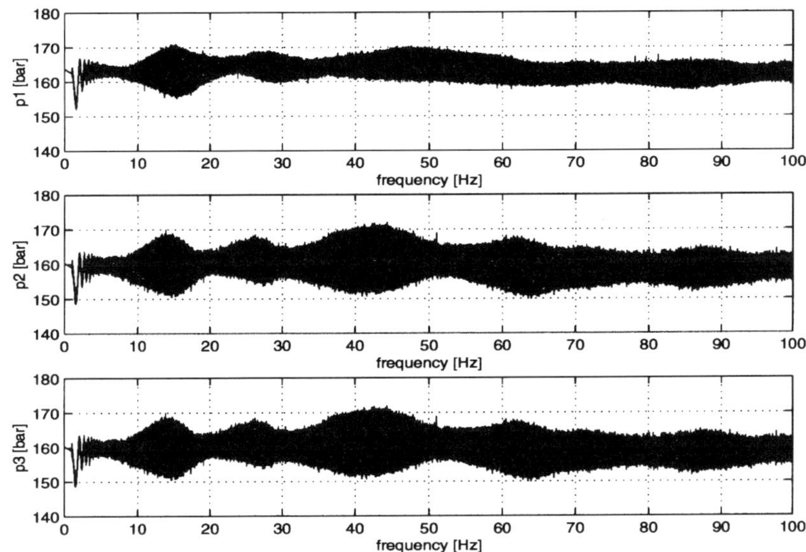

Figure 3. Pressure response of the hose arrangement.

The oscillation of the pressure at the point 3 on the frequency 50Hz was 9bar (peak-to peak value). The oscillation amplitude of the pressure was found manually, because it was difficult to analyse the measurement data automatically. Non-Linearity and hysteresis of the pressure transducer was ±0.25%, and accuracy of the manual reading was about ±1bar. In this study it was not intended to analyse the absolute size of the pressure oscillation but the aim was to study the bandwidth of the different pressure dampers.

5. ACCUMULATOR AS A DAMPER

As mention before, the hydro-pneumatic accumulator is probably the most well known damper in use. In this study it was concentrated on a traditional standard accumulator. There is a remark in the reference [1] that the standard accumulator can be used to damp pressure oscillation at frequencies lower than 100 Hz. If the frequencies are higher, special accumulators have to be used. These dampers typically have a large connection, so the damper can absorb higher frequency oscillations. There are also tube type accumulators [6], which are designed for high frequencies. They are usually called noise suppressors.

5.1 Theory

If low frequencies have to be damped (<100Hz), an accumulator can be dimensioned with the help of a basic formula. The use of this formula requires that the volume that causes the oscillations be known. The dynamics are not considered in this formula.

$$p_0 * V_0^{\kappa} = p_1 * V_1^{\kappa} = p_2 * V_2^{\kappa} \qquad (1)$$

if $\Delta V = V_1 - V_2$ the formula can be derived to the following form:

$$V_0 = \frac{\Delta V}{\left(\frac{p_0}{p_1}\right)^{\frac{1}{\kappa}} - \left(\frac{p_0}{p_2}\right)^{\frac{1}{\kappa}}} \qquad (2)$$

Another way to determine the dimensions of the accumulator and its connection to the pipe is to evaluate the natural frequency of the accumulator and connection system. Using the same notations as Garbacik [1] the following equation can be archived:

$$\omega_A = 2 * \pi * f = \frac{d_A}{\sqrt{L_A * V_G}} \sqrt{\frac{\pi * \kappa * p_A}{2 * \rho}} \qquad (3)$$

At low frequencies formula 2 is more useful [2]. In this case it is reasonable to install the accumulator near the head pipeline. In Figure 4 there are results when the effect of the different pre-charge pressures of the accumulators has been studied. The nominal size of the accumulator was $0.0012 m^3$ and it had been connected to the head pipeline with ¾" connecting pipe. The pressure ripple at the frequency 50Hz was 1bar. It can be concluded from figure 3 that the pre-charge pressure is good, when it is below about $0.8 * p_{working}$ (now 0.8*160bar).

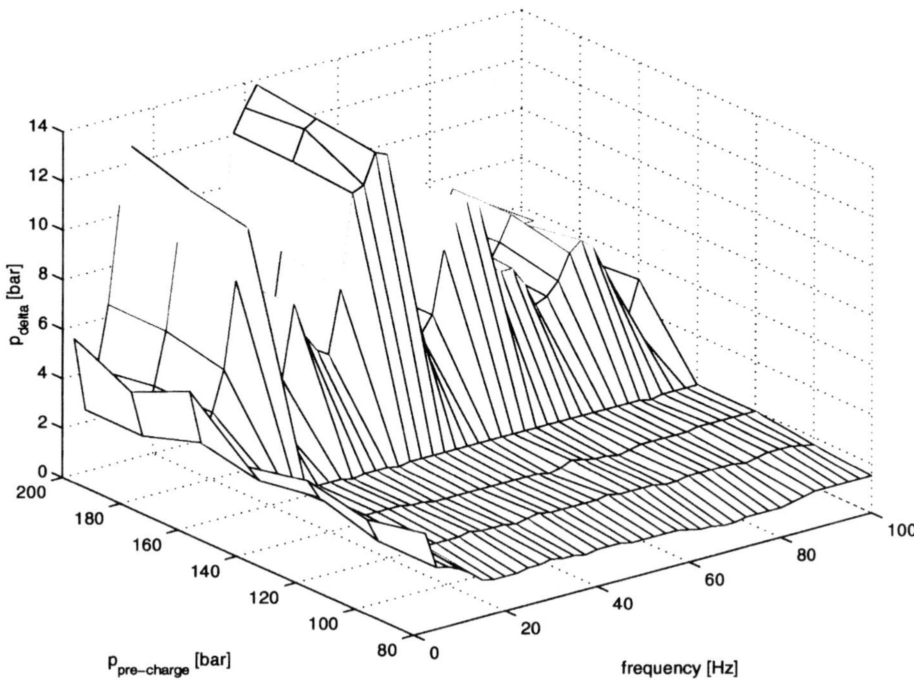

Figure 4. Pressure ripple (peak-to-peak value at transducer 3) with separate pre-charge pressure.

6. HELMHOLTZ- RESONATORS AND T-PIPES

H- and T-filter are dampers, which branch away from the main pipeline system. They are reflecting type dampers. Characteristic to them is the narrow range of operating frequencies and the requirement for careful dimensioning [3].

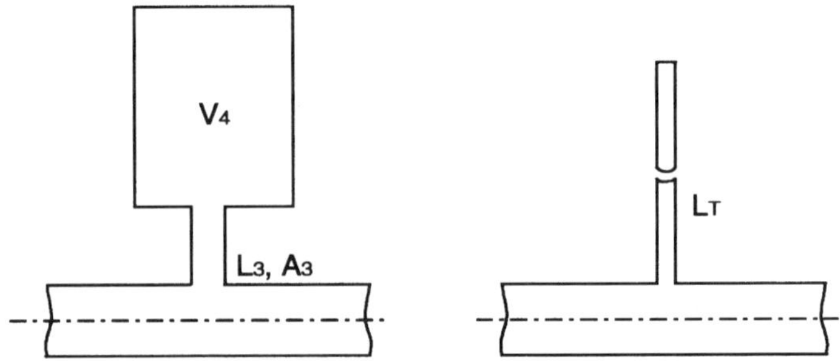

Figure 5. Schematic picture of a Helmholtz -resonator and a T-pipe.

The resonance frequency of a Helmholtz- resonator can be evaluated with the following equation [3]:

$$f = \frac{a}{2*\pi}\sqrt{\frac{A_3}{L_3 * V_4}} \qquad (4)$$

Correspondingly for T-pipe the equation (5) can be used [3]:

$$f = \frac{n*a}{4*L_T} \qquad (5)$$

In this case the important parameters were:

- a = 1200 m/s
- f = 50Hz

For the testing of the H-filter an installation, where it was possible to change volume V_4, was built. The cylinder operated as the volume V_4 and when the stroke of the cylinder was changed, the volume V_4 changed. The specifications of the cylinder were:

- L_3 = 420mm
- $A_3 = \frac{\pi * 0.01^2 m^2}{4} = 7.85*10^{-5} m^2$
- D_{cyl}=0.08m

A volume V_4 can be solve from the equation (4):

$$V_4 = \frac{A_3 * a^2}{f^2 * 4 * \pi * L_3} = \frac{7.85 * 10^{-5} * 1200^2}{50^2 * 4 * \pi^2 * 0.42} m^3 = 2.73 * 10^{-3} m^3 \qquad (6)$$

In this case the stroke of the cylinder is:

$$S = \frac{V_4 * 4}{\pi * D_{cyl}^{\ 2}} = \frac{2.73 * 10^{-3} m^3 * 4}{\pi * 0.08^2 m^2} = 0.54m$$

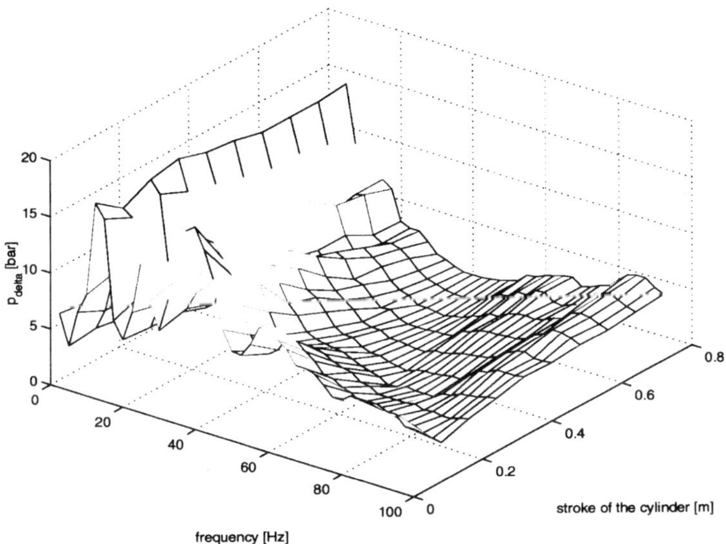

Figure 6. Function of the H-filter, when the volume V_4 is changed.

Figure 6 shows the performance of the H-filter, when the stroke of the cylinder is changed. As can be seen the experimental tests support the values calculated with equation 4. The peak to peak value p_{delta} is smallest, when the stroke of the cylinder is about 0.5m at the frequency 50Hz.

The performance of the H-filter is presented in figure 7, when the pressure level was changed. The stroke of the cylinder was 0.5m and then the volume V_4 was $2.5*10^{-3}$ m^3. As can be seen the H-filter works even when the operation point varies a little. The pressure ripple at the frequency 50Hz is now about 3 bar (working pressure 160bar).

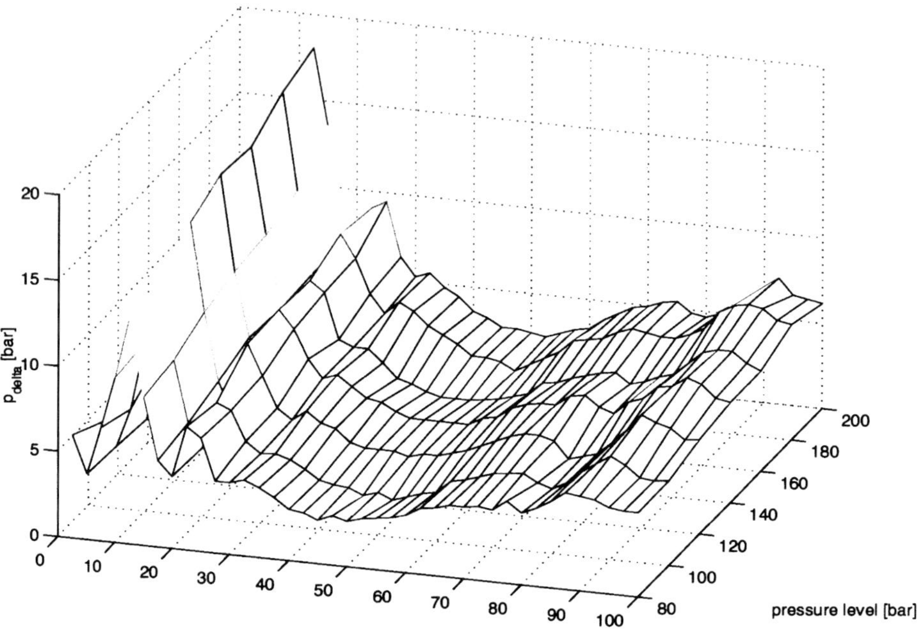

Figure 7. Function of the H-filter, when the pressure level is changed.

The T-filter is tested in the same way at different pressure levels. Length of the hydraulic T-line became:

$$L_T = \frac{a}{4 * f} = \frac{1200}{4 * 50} m = 6m$$

Figure 8 shows the results of the measurements. The measurement results validate the statements found in literature that the frequency band is narrower than with the H-filter [5]. The pressure ripple at the frequency 50Hz is 2bar (working pressure 160bar).

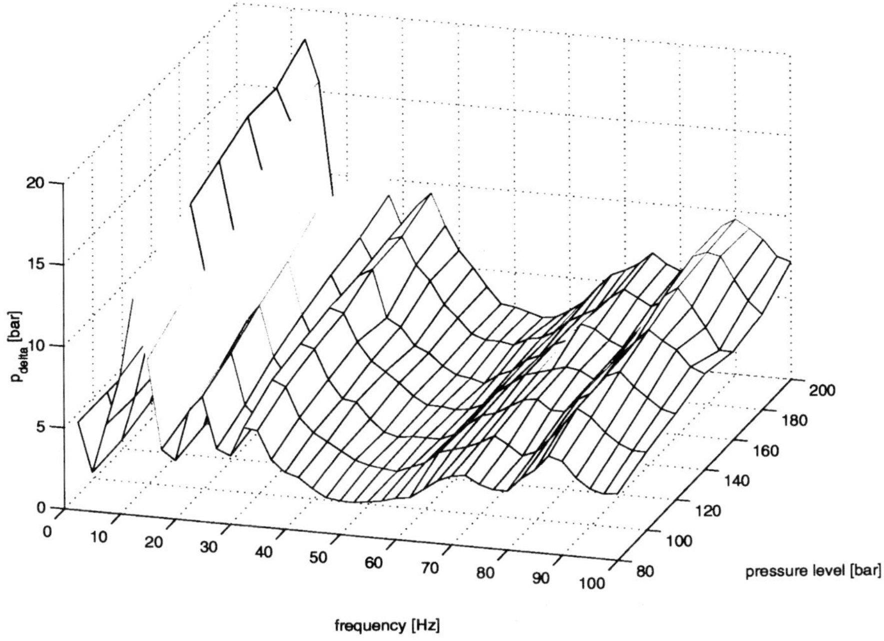

Figure 8. Function of the T-filter, when the pressure level is changed.

In principle it is easy to dimension a Helmholtz- resonator and T-pipe provided that the speed of the sound is known. It should also be ensured that there is no air in the dampers.

Earlier it was mentioned that the bandwidth of the H-filter is wider than bandwidth of the T-filter. One problem occurs when attenuating a low frequency with an H-filter. It becomes quite big. For example the best damping was reached in these tests when the length of the connection pipe was about 40cm, the diameter of the volume V_4 was 80mm and the length of the volume V_4 was 50cm. This means that the total length of the damper becomes almost 1m. It is difficult to install this kind of damper in mobile machines. T-filter, which consists of 6m hose, is easy to hide among the other hoses. It seem that the T-filter is a more practical solution in mobile machine applications.

7. HYDRAULIC MOTOR AS A DAMPER

One way to damp the oscillations of the pressure is increase the inertia of the system oil. Adding a hydraulic motor to the pipeline can do this. The hydraulic motor operates then as a flywheel. This idea was tested by installing the hydraulic motor to the pipeline and by running the test programme. The hydraulic motor was Brueninghaus Hydromatik A6VM. The displacement of 80 cm^3 was used during the test. The results are presented in the Figure 9, when there was no extra inertia load on the hydraulic motor (solid line) and when there was the effective load $J_{eff}=J_{motor}+J_{flywheel}=0.0080 kg*m^2+0.090 kg*m^2 = 0.098 kg*m^2$ (dashed line).

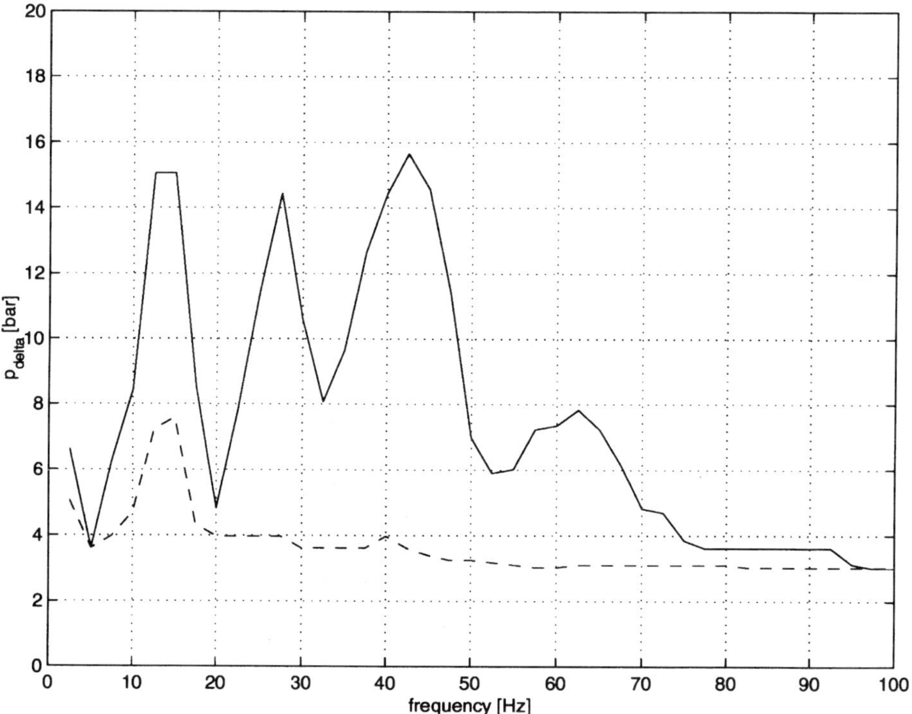

Figure 9. Hydraulic motor operates as a damper.

The extremely good damping was achieved on a wide frequency band. The difference to the previous ones is that the hydraulic motor actually does not act as a damper. It isolates the pulsating pressure line from the headline. In the test this meant that the oscillation at the servo valve increased strongly. The motor causes also the pressure losses and leakage. The motor is mainly an interface between an unstable and stable line.

8. CONCLUSIONS

The performance of different pressure dampers was experimentally tested. The excitation was a ripple of the oil flow in the downstream of the hose. The necessary simplified formulas for the dimensioning of dampers have been presented. It can be stated that all studied dampers operate sufficiently even at low frequencies. According to this study the following conclusion can be made.

- The accumulator damps well when the pre-charge pressure isn't too high and the nominal-volume is big enough.

- Helmholtz-resonator and T-pipe operates quite well but T-pipe is more practical in mobile hydraulic applications. The required total length of the Helmholtz-resonator was about 1m. The length of the T-pipe was about 6m at frequency 50Hz.

- A hydraulic motor can be operated as a damper if one pulsating pipeline is needed to be isolated from another pipeline.

9. REFERENCES

1. Garbacik A., Szewczyk K., 1995. New Aspects of Modelling of Fluid Power Control. Politechnika Krakowska. 130 p.

2. Ijas M., Virvalo T., 2000. Experimental Study of Hydraulic Pulsation Dampers for Low Frequencies. Sixth Triennal International Symposium on Fluid Control, Measurement and Visualization, Flucome 2000. August 13-17. 2000. Sherbrook. Canada

3. Larsson P., 1987. On Fluid Power Attenuators- Analysis, Measurements and Performance Optimization. Linköping Studies in Science and Technology Thesis No. 101, 1987. 94 p.

4. Ortwig H., Goebbels K., 1999. Noise Reduction in Hydraulic Circuits. The Sixth Scandinavian Conference on Fluid Power, SICFP '99. May 26-28. 1999, Tampere, Finland 1071-1082 pp.

5. Viersma, T. 1980. Analysis and design of hydraulic Servosystems and Pipelines. Amsterdam, Elsevier Scientific Publishing Company. 273 p.

6. Wilkes, R. 1998. Getting to the root of hydraulic noise problems. Hydraulics & Pneumatics June 1998. 59-60 pp.

Estimating parameters in a proportional solenoid valve

A ROSA, D MOURRE, R BURTON, and **D BITNER**
Department of Mechanical Engineering, University of Saskatchewan, Canada

ABSTRACT

Many control schemes, simple or sophisticated, utilize high performance electro-hydraulic components as an interface to mechanical hardware. Failure of an electro-hydraulic component in these applications may impact safety, maintenance schedules and/or productivity of the overall operation. A condition monitoring scheme that would measure the performance or health of critical electro-hydraulic components would be desirable in addressing the above mentioned concerns.

Present research at the University of Saskatchewan involves the feasibility of acquiring data from a proportional solenoid valve on-line for the purpose of condition monitoring. Once data have been acquired (specifically spool displacement and spool differential pressure), an algorithm can be implemented to estimate the desired valve parameters. Some of the parameters that affect the performance of the main stage valve spool which may be used in a condition monitoring scheme include area orifice gradient, spring rates, spring pretensions and friction.

This paper considers the use of individual 'neuron-like' structures to estimate each valve parameter. Each neuron trains to a specific valve parameter with its output being the estimated parameter. The individual neurons act and train independently; however, their outputs are integrated to compute the main spool differential pressure. This differential pressure calculation is compared to the measured value and the difference is used to train each neural structure. Experimental results are presented which demonstrate the accuracy and feasibility of the procedure.

1 NOMENCLATURE

P_i	Pilot spool pressure on side i (Pa)
P_L	Pressure across load ports A and B (Pa)
A	Spool area (m^2)
K_i	Valve spring constant on side i (N/m)
K	Effective spring constant (N/m)
x_v	Spool position (m)
$F_{pretension}$, F_{pti}	Spool pretension in spring i (N)
F_c	Spool coulomb friction (N)
F_f	Steady state flow force (N)
B_s	Equivalent spool and LVDT viscous damping coefficient (N s/m)
M	Mass of main spool (kg)
P_s	Supply pressure (constant) (Pa)
w	Spool orifice area gradient (m)
ΔP	Pressure differential across the spool (Pa)
C_d	Orifice discharge coefficient
C_v	Orifice velocity coefficient
θ	Orifice jet discharge angle (degrees)

2 INTRODUCTION

High performance electro-hydraulic components are becoming commonplace in industrial applications partly because of improved computer technology. Increased CPU speed and memory capacities have allowed the implementation of rigorous and sophisticated control algorithms that demand higher performance from electro-hydraulic hardware. As these control systems evolve and their reliance on electro-hydraulic hardware increases concerns for safety, maintenance and productivity must also be addressed. Failure of an electro-hydraulic component may affect some or all of these concerns to varying degrees. Hence, the need to detect or predict failure is becoming increasingly important and the ability to monitor the health of electro-hydraulic components using a simple and reliable method has become an issue of interest. Once the component's health has been established and the change of this health over time is monitored (condition monitoring), a time to failure may be predicted. The goal of such a condition monitoring program is to detect when a component should be replaced before a catastrophic failure can occur. Condition monitoring may also include modifying the component's controller settings to compensate for changes in the component's operating characteristics. The goal is of course to optimize performance of the component.

In a recent paper by Burton et al (1) the fundamentals of a unique condition monitoring technique were established. The presented results indicated that the proposed condition monitoring scheme is in fact feasible. The basis behind the proposed method is to acquire information from a system on-line during start up or shut down procedures so as to minimize interference with the system's duty cycle. This information can then be analyzed off-line to determine the component's health while the system returns to normal operation. Three very important conditions need to be met for this type of approach to work:

1. the progress of a fault must be relatively slow in its initial stages. That is, the fault cannot reach catastrophic conditions in a normal operating cycle (the time between startup and shut down),

2. the method requires that a start up or shut down cycle exists so that information from the system or component being monitored can be extracted, and
3. the information attained during start up (or shut down) must be acquired safely and the process must be transparent to any human operator/s of the equipment.

The initial research [1] was focused towards condition monitoring of solenoid proportional valves. Valve condition monitoring requires that health (performance of some type) be monitored over time. In this case, specific main spool valve parameters were chosen to indicate valve performance. Some of these included: spring constants, viscous damping coefficient, orifice area gradient, and spool mass. Valve parameters were calculated using graphical methods via the data acquired at a specific operating "norm" (a prescribed reproducible system operating condition). Even though the graphical method is feasible, practical implementation would require automation. Graphical based methods for parameter identification are not feasible for an automated system requiring frequent computer based data manipulation. Therefore, there is a need to upgrade the graphical method and automate the estimation process of the proportional solenoid valve parameters. The automated condition monitoring scheme should be flexible, robust and easily adapted to different manufacturer's valves. The methodology in performing the data acquisition of appropriate states must be simple and reliable. As well, the algorithms used to obtain the parameter estimations must yield accurate and repeatable results. In this paper, the graphical method is replaced by numerical algorithms that in effect automate the procedures presented in (1). The objectives of this paper are then:

1. to present a practical procedure for extracting specific data from an electro-hydraulic proportional valve on-line,
2. to present algorithms that may be used in off-line data manipulation for parameter estimation, and
3. to demonstrate the accuracy and hence feasibility of these algorithms in estimating valve parameters.

This paper attempts to further current condition monitoring technology and more specifically aid in the development of a practical condition monitoring system for electro-hydraulic proportional valves.

3 EXPERIMENTAL IMPLEMENTATION

A schematic of the test setup used to acquire measurable states (temperature, pressure and spool position) from the proportional solenoid valve is shown in Figure 1. The setup includes three electro-hydraulic on/off solenoid valves (Valves A, B and C in Figure 1). Valves A and B provide a means for isolating the pilot operated proportional solenoid valve from the load. Valve C allows flow to bypass freely across the proportional valve's load ports. A differential pressure transducer is installed to measure the end cap pressure difference across the main stage spool, an LVDT (integral to the valve) is used to measure the main stage spool position and a temperature sensor is placed downstream of the valve to measure the fluid temperature. The two signals (spool position and differential pressure) are required by the condition monitoring scheme to estimate the valve parameters. A PC with a 12 bit multifunction data acquisition board is used to collect data from the sensors and provide a desired spool position control signal to the valve electronics.

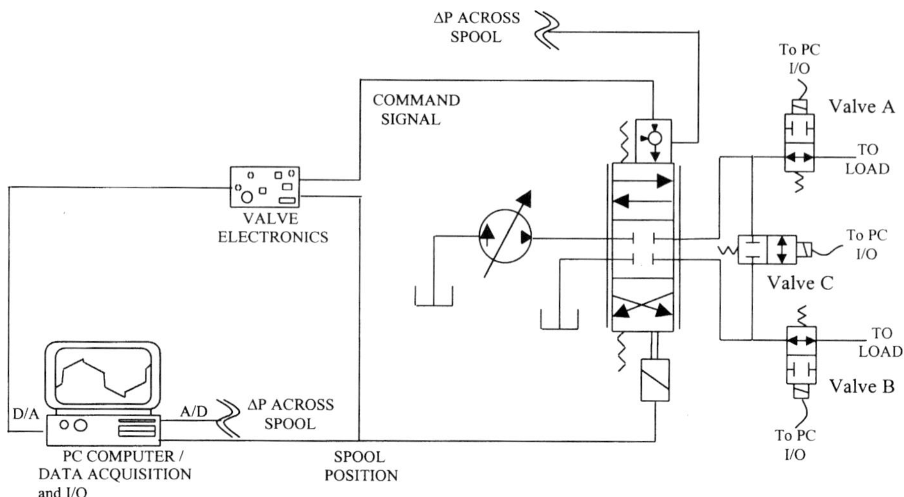

Figure 1: Schematic of Valve and Associated Instrumentation

A specific procedure was implemented in [1] so that particular parameters of the valve could be estimated. The procedure requires data obtained from two different operating conditions. The first operating condition is a shorted port scenario where the load ports are blocked (Valves A and B closed) and fluid is allowed to flow unrestricted across the load ports (Valve C open). In this case flow is allowed to pass through the proportional valve while the main spool of the valve is forced (using position feedback) to follow a desired input displacement.

In the second scenario, the load ports are again blocked (Valves A and B closed) and no flow is permitted between load ports (Valve C now closed). The spool is forced to follow the same desired input displacement as in the first scenario. A summary of the procedure used to obtain data is as follows.

1. The load is first isolated from the proportional solenoid valve by blocking the outlet ports via Valves A and B.
2. The pump supplying fluid to the valve is started.
3. Fluid is allowed to flow through the proportional valve via the shorted load ports (Valve C is open) causing a local increase in the system temperature.
4. When the desired operating "norm" has been achieved, data acquisition in the shorted port scenario is initiated.
5. The blocked port (close Valve C) or "no flow" scenario is then executed and the appropriate data collected.
6. Once all data has been obtained, the valves are returned to their normal positions (Valves A and B open, Valve C closed) and the proportional solenoid valve is allowed to resume its normal duty cycle.

4 VALVE SPOOL CONFIGURATION AND BASIC EQUATIONS

A schematic of the main stage of the proportional solenoid valve is shown in Figure 2. A differential pressure (P_2-P_1) across the main spool causes a displacement (x_v) which is monitored by an LVDT. The spool is overlapped and two springs (1 and 2) with spring constants K_1 and K_2 respectively, act on either ends of the spool.

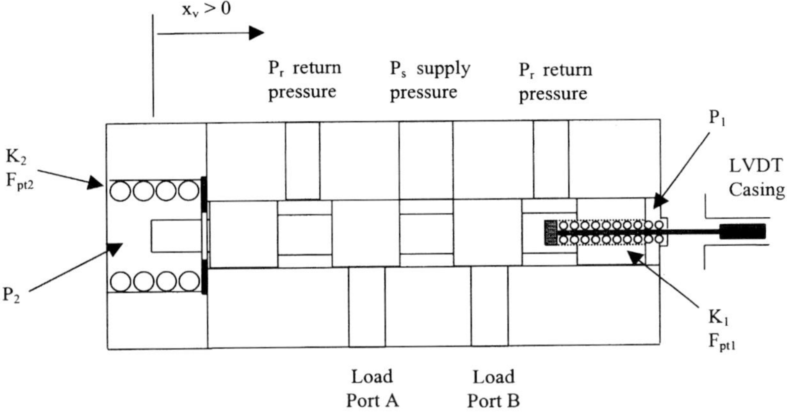

Figure 2: Proportional Valve's Main Stage

Spring 1 is in compression for both positive and negative x_v. Spring 2 is also in compression, but only comes in contact with the spool when $x_v < 0$. Therefore, for $x_v < 0$, the effective spring constant is $K_1 + K_2$. Other forces that act on the spool include spring pretensions (F_{pt1} and F_{pt2}), coulomb friction (F_c), viscous friction ($B_s \times \dot{x}_v$) and the force resulting from acceleration of the spool mass ($M \times \ddot{x}_v$). If the load ports (Ports A and B) are blocked there can be no flow through the valve and hence no flow forces acting on the spool. Under these conditions the force balance equation of the spool can be expressed as:

$$(P_2 - P_1)A - B_s \frac{dx_v}{dt} - Kx_v - F_{pretension} - F_c \text{sign}\left(\frac{dx_v}{dt}\right) = M\frac{d^2x_v}{dt^2} \quad (1)$$

for, $x_v \geq 0$, $K = K_1$; $F_{pretension} = F_{pt1}$

$x_v < 0$, $K = K_1 + K_2$; $F_{pretension} = F_{pt1} + F_{pt2}$

The parameters of interest in Equation (1) which could be used as an indication of the health of the valve are B_s, F_c, F_{pt1}, F_{pt2}, K_1, K_2 and M. When considering the shorted port scenario (Load Ports A and B connected), the steady state flow reaction force, F_f, becomes significant and can be expressed as:

$$F_f = 2C_d C_v \cos\theta (P_s - P_L) w x_v \quad (2)$$

Including F_f in the spool force balance and neglecting transient flow forces, Equation (1) becomes:

$$(P_2 - P_1)A - B_s \frac{dx_v}{dt} - Kx_v - F_{pretension} - F_c \text{sign}\left(\frac{dx_v}{dt}\right) - 2C_d C_v \cos\theta (P_s - P_L) w x_v$$

$$= M\frac{d^2x_v}{dt^2} \quad (3)$$

If the supply pressure (P_s in Equation (3)) is assumed constant and the valve's orifices are matched and symmetrical, then the load pressure (P_L in Equation (3)) is also constant and equal to half the supply pressure ($P_L=P_s/2$). C_d, C_v and $\cos\theta$ are constants [2]. The parameter of interest in Equation (3) is w, the orifice area gradient. This parameter could possibly be used as an indication of wear on the spool or lands if it could tracked over time. The parameter, C_d, is most likely to change with metering edge wear. However, instead of looking at a $C_d \times w$ term, the process is simplified by maintaining a constant discharge coefficient. Since this overall analysis is dependant on parameter changes, errors due to a changing C_d in the orifice area gradient term are not important as long as information regarding the change of wear in the valve is obtainable. The effect of metering notches on parameter estimations is not known as the valve in study did not have notches.

The technique used in [1] to obtain the unknown parameters given in Equations (1) and (3) dictated that a specific spool displacement be used. For both scenarios, a triangular input signal (see Figure 3a) is fed into the valve electronics. Position feedback on the valve spool via the LVDT ensures that spool motion follows the input signal. Utilizing differentiation and filtering (4th order Bessel filter, f_c = 14.3 Hz), the corresponding velocity and acceleration profiles of the spool motion can also be determined.

The corresponding differential pressure profiles ($P_2 - P_1$) for the blocked and shorted port scenarios are shown in Figure 3b. These profiles represent typical outputs in the blocked and shorted port scenarios based on Equations (1) and (3) respectively for the triangular spool displacement shown in Figure 3a. The profiles contain pertinent information that is required to estimate the valve parameters.

The differential pressure waveforms are divided into four quadrants. Each quadrant represents a different direction in either spool velocity or spool position resulting in a different form of Equation (1) and Equation (3) in each quadrant.

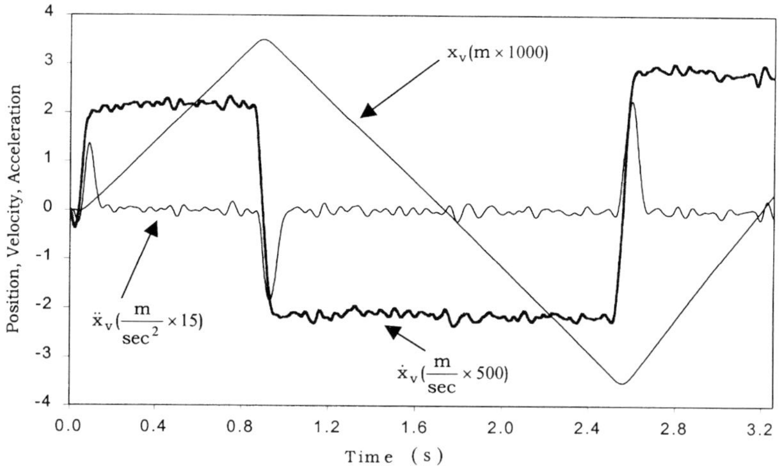

a) Triangular Spool Position verses Time

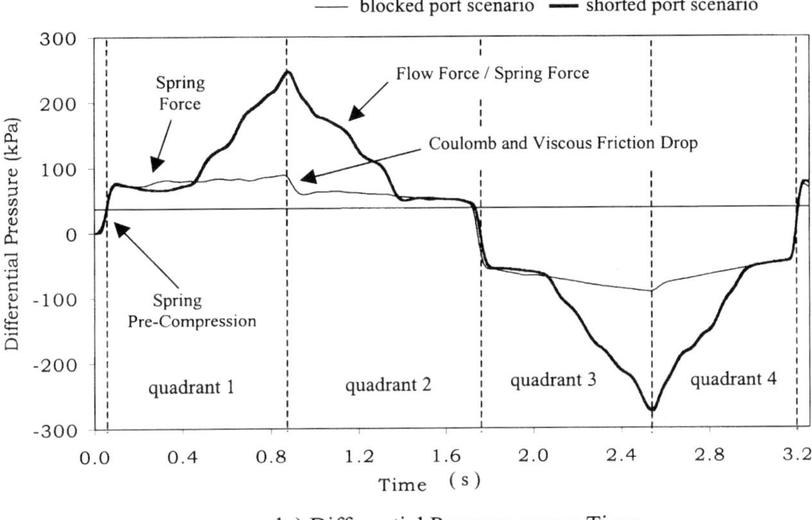

b.) Differential Pressure verses Time

Figure 3: Typical Closed Port Waveforms of Spool Displacement and Differential Pressure versus Time

At the start of quadrant 1, the pressure in both blocked and shorted port scenarios rapidly increases. This increase is needed to overcome the initial pre compression of spring 1. Following this increase, a gradual and approximately linear rise in differential pressure occurs as a result of spring 1 being compressed. In the blocked port scenario, an instantaneous drop in differential pressure occurs at the transition from quadrant 1 to quadrant 2. This drop is due to the sign change in velocity that causes the friction forces to change their direction of application. The magnitude of this drop, in terms of a net force on the spool, is equal to twice the sum of coulomb and viscous friction forces. In order to calculate the coulomb and viscous friction parameters (F_c and B_s), pressure profile curves at two different velocities (triangular input signal at two different frequencies) are required. Coulomb friction is assumed constant and therefore the difference in magnitude of the above mentioned drop from one velocity to the next can be solely attributed to viscous friction, thereby simplifying the evaluation of B_s.

In the shorted port scenario, the effect of deadzone and the effect of flow reaction forces on the resulting pressure curve (Figure 3b) are quite evident. As the spool opens, the pressure profile initially follows the closed port profile closely. In this region, the deadzone in the valve does not allow flow through the valve, effectively mimicking the closed port scenario. Upon leaving the deadzone region, flow begins to pass through the valve and a steep linear rise in differential pressure occurs. The increased slope in this region is related to the flow reaction force which is linearly related to spool displacement as stated in Equation (2). The pressure profiles in Quadrants 3 and 4 and how they relate to negative deadzone and flow forces are similar to quadrants 1 and 2.

5 PARAMETER ESTIMATION

Figure 4, is a schematic of the neural-based training configuration used to estimate the proportional valve's parameters. SIMULATION MODEL 1 (SM1 for short) is a block diagram representation of Equation (1) that contains predetermined valve parameters that reflect the parameters of the actual valve which were obtained using the methods given in (1). The NEURAL MODEL 1 (NM1 for short) is structurally the same as SM1 except that the parameters have been replaced by individual neuron-type functions.

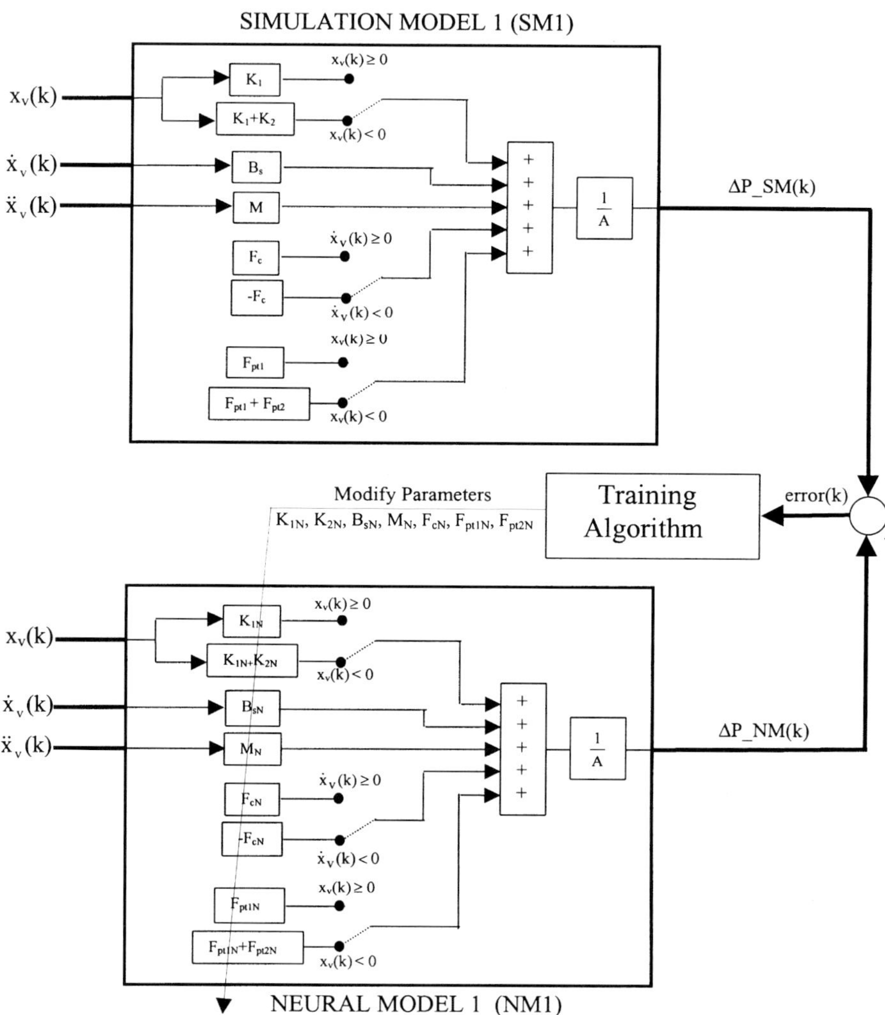

Figure 4: Schematic of the Neural Training Configuration

The object of this configuration is to modify an adjustable weight and gain in each individual neuron until ΔP_NM(k) approaches ΔP_SM(k). Under this condition, the neuron-type parameters should mimic the predetermined plant parameters B_s, F_c, F_{pt1}, F_{pt2}, K_1, K_2 and M. NM1 will then essentially contain the estimated parameters of SM1.

Although the technique presented is intended for determining parameters of an actual valve, SM1 was initially chosen. The advantage being that the data source is free of sensor noise and unknown valve characteristics. Also, the valve parameters are known and have exact values, which is important in determining the accuracy of the trained neural model and hence the effectiveness of the training algorithm.

In the configuration shown in Figure 4, a sampled data set containing spool position $x_v(k)$, spool velocity $\dot{x}_v(k)$, and spool acceleration $\ddot{x}_v(k)$ are used as inputs into the two models. The input $x_v(k)$, is a multi-cycle triangular wave with constant amplitude and varying frequency. The corresponding difference between the outputs, ΔP_SM(k) and ΔP_NM(k), is passed to the training algorithm that adjusts the weight and gain in each neural structure. The generic structure of these neural parameters is depicted in Figure 5.

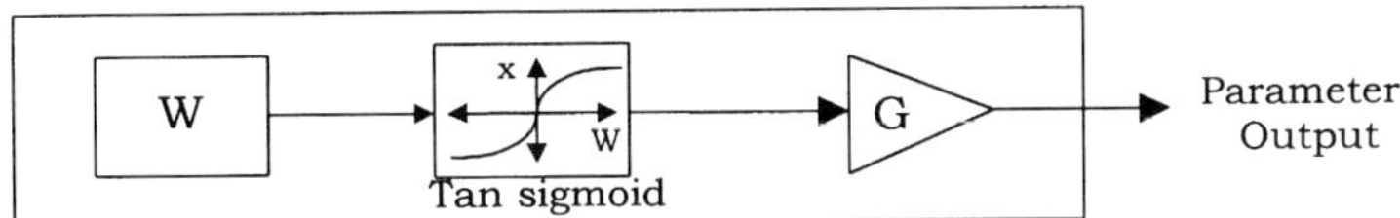

Figure 5: Generic Neural Structure used in Neural Model 1

The training algorithm in this configuration implements batch training [3] with data sets that contain 4162 points sampled at 500 Hz. An error, ΔP_SM(k) - ΔP_NM(k), is first calculated at each point, k, in the data set. Next, a performance index at each point based on the square of the error at each point is calculated [4]. Following this, a gradient of the performance index with respect to each weight (W in Figure 5) and a performance index with respect to each gain (G in Figure 5), of each individual neuron, is evaluated for each point in the data set. The average gradient with respect to each weight and the average gradient with respect to each gain are then evaluated over the entire data set. New values of weight and gain for each neuron are then calculated by multiplying the average gradient value by a pre-set learning rate and subtracting the resultant from the previous weight or gain value. This process is repeated until the average performance index over the entire data set is reduced to an acceptable level or reaches a minimum value. The neural model uses an individual neuron, not a layer of neurons, to estimate each valve parameter.

The estimated parameters from the blocked port scenario, are then used in the shorted port scenario to facilitate the determination of the remaining unknown parameter which is the orifice area gradient, w. However, before w can be estimated the deadzone locations of the spool must be determined.

A simple algorithm has been implemented to determine spool deadzone. Using the shorted port scenario, a single cycle of triangular displacement waveform, which is divided into 1600 data points, is input into a new simulation model SM2 which is similar to SM1 but based on Equation (3). The algorithm uses the resultant differential pressure output, as shown in Figure 6, to find the two points of maximum and minimum differential pressure. Starting at the maximum ΔP and working towards the right, a slope of the differential pressure profile is calculated by taking a moving average of 12 consecutive data points.

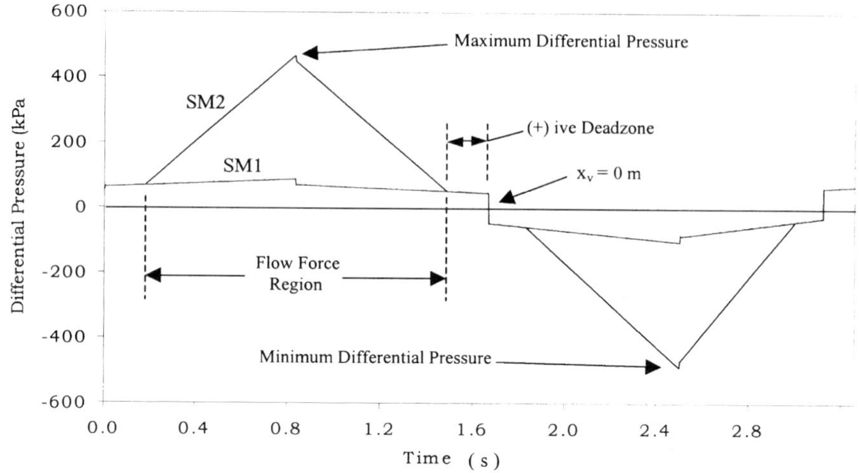

Figure 6: Simulation Results of the Blocked and Shorted Scenarios

The slope calculation is moved along the line data point by data point until it reaches a value corresponding to spring constant K_1. The spool location at this point indicates the onset of deadzone. Since the ports are now effectively blocked, the flow reaction forces disappear. The distance from this point to x_v=0 is the magnitude of the deadzone in the positive spool direction. Similarly, the deadzone in the negative spool direction is calculated.

With the deadzone locations now known, the region of spool displacement where the flow forces act is now also known. For a new NEURAL MODEL (NM2) which trains to SM2, all the estimated neural parameters previously determined in NM1 are used as fixed parameters. Note that the configuration of SM2 and NM2 (not shown) is similar in structure to SM1 and NM1, but now includes the additional parameter w, which is to be estimated. The orifice area gradient w is represented by the neural structure shown in Figure 5 and estimated in a similar fashion to the other neural valve parameters previously determined in NM1.

6 RESULTS AND DISCUSSION

This paper has considered the implementation of several algorithms designed to estimate proportional valve parameters. These algorithms have been designed to operate on an actual system with information acquired on-line from a proportional solenoid valve. However, the algorithms were initially tested using simulated data generated from SM1 and SM2. In the creation of simulated data, parameters were set to resemble true valve parameters and true experimental conditions. The algorithms were then tested using experimental data collected on-line from a proportional valve. For both the simulated and experimental cases, sampling frequencies of 500 Hz were used. Also, training arrays of identical size and identical spool displacement inputs were used.

6.1 Simulation Results

In Table 1 a typical set of estimated parameter values is presented alongside preset values.

Table 1: Simulation Results Compared to Preset Parameter Values

Valve Parameter	Preset Parameter Value	Parameter Estimation Triangular Input	% Difference
Mass (kg)	0.156	0.156	0.00
Bs (Ns/m)	100.0	101.2	1.25
Fc (N)	2.00	1.99	-0.45
K1 (N/m)	1800	1801	0.08
K2 (N/m)	2300	2299	-0.06
Fpt1 (N)	15.50	15.50	0.00
Fpt2 (N)	26.50	26.50	0.00
w (m)	0.00610	0.00610	0.00
Pos. Deadzone (m)	0.000850	0.000855	0.60
Neg. Deadzone (m)	0.000780	0.000772	-1.03

The results show that the estimated parameters obtained are in good agreement with the corresponding preset values used in SM1 and SM2. The largest difference (Table 1) between estimated and preset values is the viscous damping coefficient, B_s - at 1.25 %. The smallest difference (Table 1) between estimated and preset values is demonstrated by the mass, spring pretensions and orifice area gradient which are exact estimations.

6.2 Experimental Results

Using the experimental setup illustrated in Figure 1, four sets of actual valve data were collected. Each set includes measurements of x_v and ΔP for a triangular input signal of varying frequency for both the blocked and shorted port scenarios. To demonstrate repeatability, all four data sets were obtained using identical inputs and the same operating conditions (temperature and source pressure). The data sets were then processed through the algorithms, estimating the valve parameters for each of the four data sets.

Table 2 summarizes the actual parameter values of the valve and the estimated parameters obtained for Data Set #1. It is apparent from Table 2 that all the estimated parameters deviate from their actual counterparts by significant amounts. In quadrants 1 and 2, where the spring constant K_1 acts (see Figure 3b which illustrates Data Set #1), the waveform is not ideal (not straight) and therefore is a source of error influencing the estimated value of K_1. Unfortunately, an error in the estimated value of K_1 carries over into the estimation accuracy of the other parameters and vice versa, as they are all mathematically intertwined in the neural models NM1 and NM2.

It is apparent that the mass estimation is in error by close to 5000% when compared to the actual value (Table 1). Upon further investigation it became evident that the magnitude of force resulting from the acceleration of the spool mass was an order of magnitude 1000 times less than the forces resulting from spring constants, friction, etc. Therefore, an error in the prediction of mass affects the force balance on the spool very little and hence makes its value difficult to estimate for this kind of input signal.

Table 2: Actual and Estimated Parameter Values for Data Set #1

Valve Parameter	Actual Parameter Value	Algorithm Parameter Estimation (Data Set #1)	Percent Difference
Mass (kg)	0.156	7.65	4804
Bs (Ns/m)	NA	99.5	NA
Fc (N)	NA	1.92	NA
K1 (N/m)	1294	1859	43.7
K2 (N/m)	1723	2521	46.3
Fpt1 (N)	16.76	15.54	-7.28
Fpt2 (N)	31.70	26.40	-16.7
w (m)	0.00520	0.00661	27.1
+ deadzone (m)	NA	0.000890	NA
- deadzone (m)	NA	0.000687	NA

In Table 2, several parameters are not known and a comparison can be made only to those readily measurable off line. Table 3 summarizes the repeatability of the parameter estimations for Data Sets #1 to #4. Repeatability in this case is defined as the percent difference between maximum and minimum values. The mass estimations in Data Sets #2, #3 and #4 are also in error with respect to the actual, as was the estimation from Data Set #1. Ideally, the algorithms should be able to estimate valve parameters numerically close to the actual parameters of the valve. Obviously, this is not the case. This would imply that the model used for estimation does not adequately define the dynamics of the spool; indeed, the technique tends to provide an estimation of parameters which have similar effects on the system behavior.

Table 3: Comparison of Experimental Results

Valve Parameter	Parameter Estimation				Parameter Range	% Diff. In Max and Min
	Data Set #1	Data Set #2	Data Set #3	Data Set #4		
Mass (kg)	7.65	7.93	8.53	7.90	0.88	11.5
Bs (Ns/m)	99.5	68.6	102.9	76.8	34.3	49.9
Fc (N)	1.92	1.95	1.73	1.84	0.22	12.9
K1 (N/m)	1859	1946	2031	1983	173	9.3
K2 (N/m)	2521	2392	2255	2309	266	11.8
Fpt1 (N)	15.54	15.30	15.08	15.18	0.46	3.0
Fpt2 (N)	26.40	26.32	26.24	26.31	0.17	0.6
w (m)	0.00661	0.00649	0.00643	0.00643	0.00018	2.8
+deadzone (m)	0.000890	0.000867	0.000860	0.000860	0.000030	3.5
-deadzone (m)	0.000687	0.000622	0.000623	0.000628	0.000065	10.4

In terms of condition monitoring, a more important issue is the ability of the training algorithms to repeatedly estimate consistent valve parameter values. In this respect, a relative change in the estimated parameters from one duty cycle to the next could be used as an indication of a slowly progressing fault. Table 3 shows that the repeatability of the parameter estimation for Data Sets #1 to #4 is less than 12%. Estimation of the viscous friction coefficient, however, demonstrates the worst repeatability out of the 8 parameters at approximately 50%. This repeatability is a cause of some considerable concern and indicates that the experimental procedure which enhances the region in which viscous damping dominates must be reassessed. On the other hand, the parameter which has been shown to be a good measure of wear in a valve is the area gradient, w, and is repeatable to within 3%.

7 SUMMARY and CONCLUSIONS

1. This paper outlined a practical procedure for acquiring data from an electro-hydraulic proportional valve for the purpose of condition monitoring. More specifically, the proposed procedure involves isolation of the load from the proportional valve during a start-up or shut-down procedure using external on/off valves. Utilizing appropriate instrumentation and computer hardware, the state variables (P_2-P_1, x_v, $\dot{x}_v$, and $\ddot{x}_v$) can be acquired and stored in memory for post-processing. The data acquisition procedure is concluded to be feasible for implementation in an industrial application because of its simplicity and transparency to operator/s actions.
2. A unique neural based algorithm was presented in this paper which was developed for the intent of estimating parameters of the main spool of a proportional valve. The algorithm employs a back-propagation based training algorithm that modifies a weight and a gain in neuron-like functions that are embedded within a mathematical model of the valve. The algorithm first requires a blocked port operating condition of which the parameters K_1, K_2, B_s, M, F_c, F_{pt1} and F_{pt2} can be estimated. Secondly, an open port scenario is required to estimate deadzone values and the orifice area gradient, w. The algorithm is autonomous from start to finish. Therefore, it is feasible to implement this algorithm in an industrial environment.
3. The ability of the neuron-based algorithms to estimate valve parameters was first investigated using simulated valve data. In this study, the form of the neural based models NM1 and NM2 were structurally identical to the simulation models SM1 and SM2. The training algorithms estimated all preset valve parameters with a high degree of accuracy – less than 2% in all cases. Next, the algorithms were executed using data acquired from a proportional valve operating on-line. The ability of the algorithms to estimate accurate values of some parameters (mass for example) was in question. However, for some critical indicators of valve wear, (the area gradient, for example), the repeatability of the estimation was less than 3% which is quite acceptable. A future consideration will be to assume that the mass, spring constants and spring pretensions do not change significantly over time (or at least with respect to changes in other parameters); as such, nominal values can be used in the model and need not be estimated. This is now being investigated.
4. The research is in its preliminary stages. The application of the estimation technique to a simulated model showed great potential. The algorithms are computationally simple which is very important from an implementation point of view. However, the application of the technique to experimental data indicated problems. An important conclusion to be drawn from this work is the need to further explore waveforms (input signals and testing procedures) that can enhance the effects certain parameters have in a particular region. In addition, the ability of the technique to track a fault must also be ascertained using experimental data. This is now being done using a special instrumented test facility. The initial research was completed with a used valve which had visible evidence of wear on the main spool. This is believed to be a large contributing factor to the problems associated with parameter estimations. Recently, a new valve has been implemented revealing a far more linear reaction force profile. Testing is in progress utilizing the new valve with results showing improved repeatability and accuracy to known valve parameters.

8 ACKNOWLEDGEMENTS

The project was funded from the IRIS Phase III Network of Centers of Excellence, Intelligent Decision Support Theme

9 REFERENCES

(1) R.T. Burton, D.P. Mourre, D. Bitner, P. Ukrainetz. "A Technique to Estimate Some Valve Parameters in a Proportional Valve". Bath Workshop on Power Transmission and Motion Control, 8-10 Sept., 1999.

(2) W. E. Merritt, Hydraulic Control Systems, John Wiley &Sons, 1967.

(3) K. S. Narenda and K. Parthasarathy. "Identification and Control of Dynamical Systems Using Neural Networks". IEEE Trans. Neural Networks. vol. 1, 1990, pp 4-27.

(4) S. Chow, C. F. N. Cowan, S.A. Billings and P.M. Grant. "Parallel Recursive Error Algorithm for Training Layered Neural Recursive". International Journal of Control. vol. 55, no. 6, 1990, pp 1215-1228

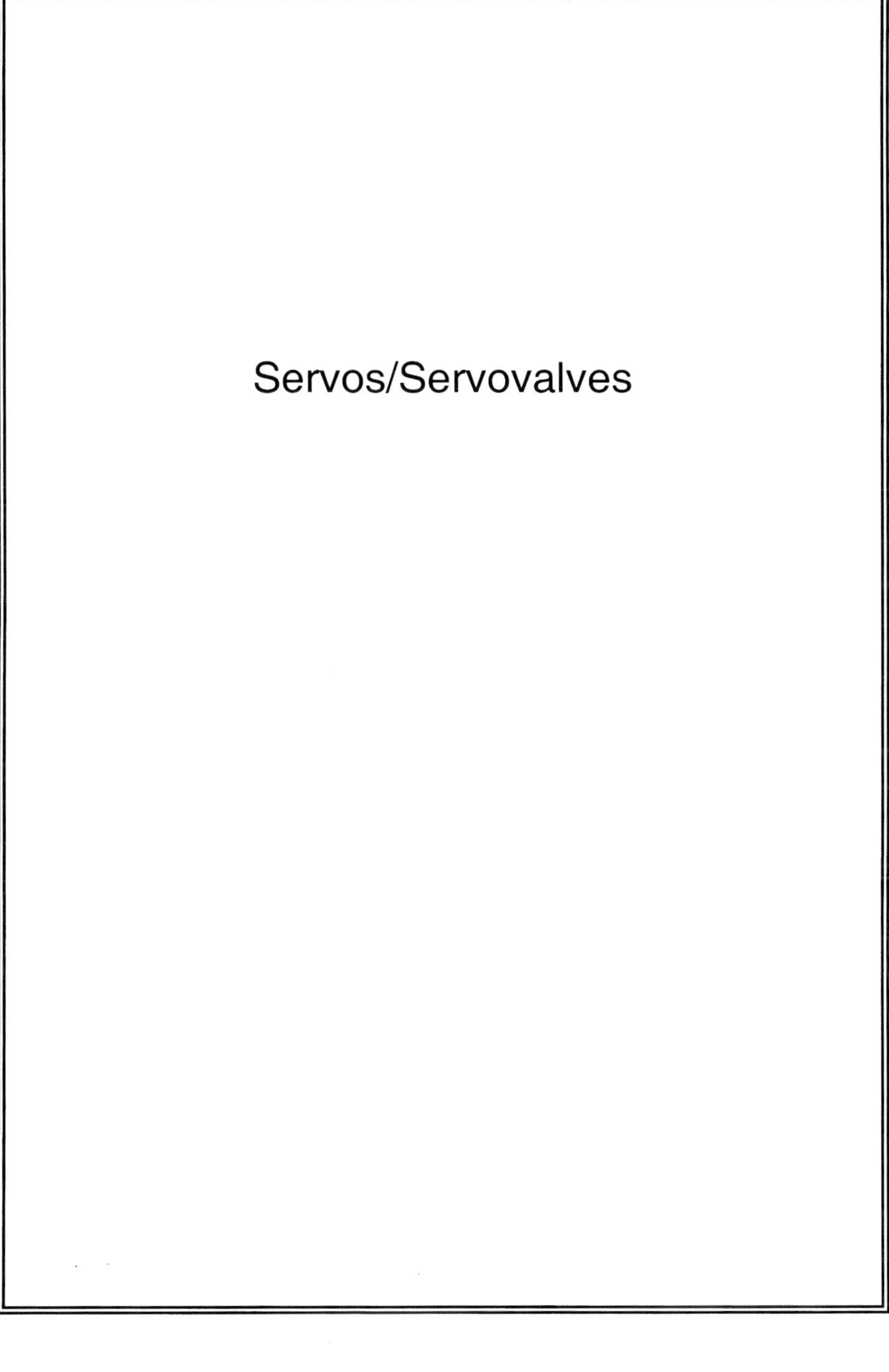

Servos/Servovalves

Accurate positioning of a pneumatic servosystem with air bearings

T KAGAWA, L R TOKASHIKI, and **T FUJITA**
Precision and Intelligence Laboratory, Tokyo Institute of Technology, Japan
K SAKAKI and **F MAKINO**
Sumitomo Heavy Industries Limited, Japan

ABSTRACT

The semiconductor processing industry has been challenged to define and position images to more exacting tolerances. Responding to this need, steppers (reduction-projection exposure systems) have reduced the wavelength of its light source. One of the main concerns in such systems is the need for an accurate and fast positioning system that can operate in a vacuum environment without heat generation.
This paper presents a pneumatic servo systems for positioning control. To reduce the effects of friction force, a pneumatic actuator prototype has been constructed. Seals are removed and the slider of the actuator is mounted with externally pressurised air bearings. For the positioning control, a state feedback controller with disturbance observer is used. Experimental results show that accuracy of ± 0.2 [μm] can be obtained.

1. INTRODUCTION

Microlithography have grown over the last twenty years into one of the most sophisticated aspects of the entire integrated circuit technology (1). To define and position images to more exacting tolerances, steppers have reduced the wavelength of its light source; going from the g-line to the i-line, that uses high-pressure mercury lamps as the light source. Now, microlithography has made a transition from using mercury lamps to excimer laser as the illumination source for large-scale production. In an attempt to print even smaller feature sizes on the wafer with high productivity, steppers using electron beam exposure systems are currently under research.
An exposure system for microlithography consists of three parts: a lithographic lens, an illumination system and a wafer positioning system. In the wafer positioning system, the wafer is held by a vacuum chuck on an ultraprecision x-y stage. The stage position is determined by laser interferometer. Traditionally, linear motors have been used in steppers. However, electrical motors become a heat source that can have serious consequences (2),

especially in a vacuum environment needed for electron beam lithography.
Lithographic exposure equipment is quite sensitive to temperature changes. Some of the problems brought by temperature changes are:

a) Change of the index of refraction of the lens that alters the optical behaviour of lithographic projection lens.
b) Change of the index of refraction of the air that affects the performance of optical measurement instruments like stage interferometers.
c) Thermal expansion of the structural materials that brings errors in measurements.
d) Thermal expansion of the wafer that can be significant.

Therefore, heat generation may become a serious problem in large-scale production. Actuators with fast and accurate positioning capabilities, non-magnetic and without heat generation to operate in a vacuum environment are highly desired. This study presents a pneumatic servo system to accomplish fast and accurate positioning.
Despite its many benefits, pneumatic servo systems experience a lack of attention due to their inherent non-linear nature that demands sophisticated control algorithms (3). One of the most critical problems in servo pneumatic positioning systems is that friction force affects convergence performance near the reference position, decreasing the positioning accuracy (4)(5).
To reduce the effects of friction force, a pneumatic actuator prototype has been constructed. Seals used in normal pneumatic cylinders are removed and the slider of the pneumatic actuator is mounted with externally pressurised air bearings. An air film separates the slider and its guide, attaining a non-contact drive.
For the design and tuning of the controller, a mathematical model of the system is linearised and used as a basis for a state feedback controller design. To compensate for parameter variations and non-linear characteristics, a disturbance observer is added.
Experimental results are shown for continuous step responses of 1[μm] and velocity control of 1[mm/s]. Accuracy of ±0.2[μm] is obtained with good repeatability.

2. NOMENCLATURE

A	Pressurised area	[m^2]
b	Critical pressure ratio	[·]
G	Mass flow rate	[kg/s]
K_a	Acceleration feedback gain	[Vs^2/m]
K_f	Constant (Eq.(3))	[s/m]
K_n	Parameter (Eq.(17))	[m/(sV)]
K_p	Proportional gain	[V/m]
K_{sn}	Flow gain	[m^2/V]
K_v	Velocity feedback gain	[Vs/m]
M	Slider mass	[kg]
P	Pressure	[Pa]
R	Gas Constant	[J/(kgK)]
s	Laplace operator	[s^{-1}]
S_e	Effective area	[m^2]
t	Time	[s]

u	Controller output	[V]
x	Slider displacement	[m]
V	Volume	[m^3]
φ	Parameter (Eq.(9))	[·]
κ	Specific heat ratio	[·]
θ	Temperature	[K]
ω_n	Natural frequency	[rad/s]

Subscripts

1, 2	Actuator chambers
a	Atmosphere
o	Equilibrium point
ref	Reference
s	Supply

3. COMPONENTS OF THE PNEUMATIC SERVO SYSTEM

Fig.1 shows the main components of the servo system: pneumatic actuator and two servo valves. The slider of the pneumatic actuator is mounted with externally pressurised air bearings. Details of the pneumatic actuator are given in Section 4. The pneumatic actuator is driven by two high-speed 3-port, 3-position servo valves.

The displacement of the pneumatic actuator is measured by a linear transducer of resolution 0.05[μm]. Values of displacement are passed to a personal computer through a counter. The personal computer acts as a controller and sends the control signals to the servo valves.

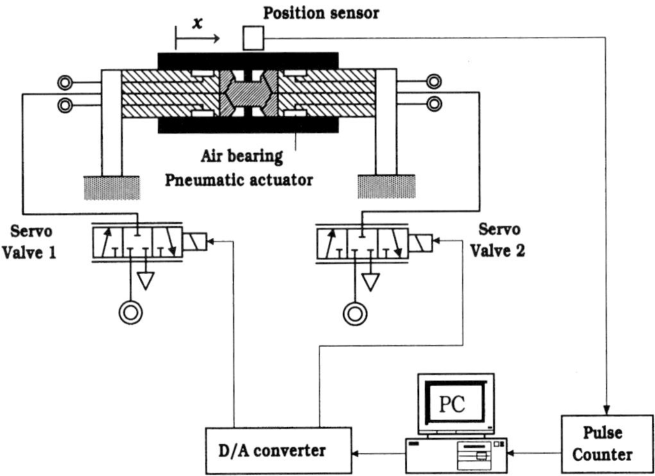

Fig.1 Pneumatic servo system

4. PNEUMATIC ACTUATOR WITH AIR BEARINGS

Fig.2 shows a schematic view of the pneumatic actuator prototype. The slider is the moving part of the actuator and the guide is fixed at its extremes. The pressured wall is attached to the slider; consequently, the slider moves because of the pressure difference at both sides of the pressured wall. The pressured wall forms four chambers between guide and slider. Each of the two chambers at each side of the pressured wall are charged or discharged through the same pipe.

The slider is mounted in the guide through externally pressurised air bearings. The air bearings act through holes in the surface of the guide, attaining a non-contact drive between slider and guide during movement. Air for the bearings is supplied and discharged from both extremes of the guide. The distance between slider and guide is 5[μm]. The material of the pneumatic actuator is ceramics.

Table 1 shows the main features of the pneumatic actuator.

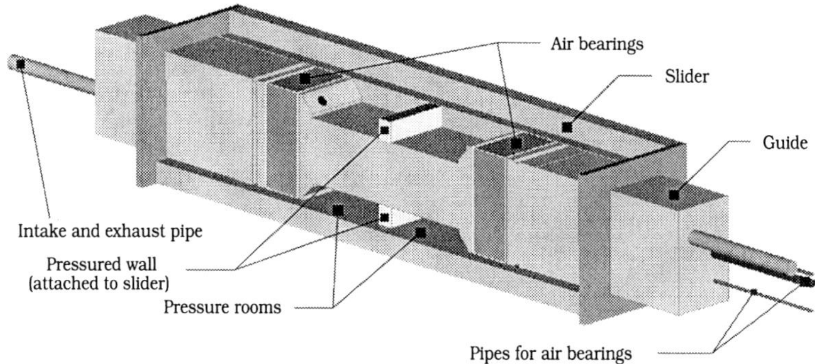

Fig.2 Pneumatic actuator with air bearings

Slider stroke	0.24 [m]
Pressured wall area	0.00486 [m^2]
Slider mass	72.3 [kg]

Table 1: Main features of the pneumatic actuator

Table 2 shows straightness, angular motion and stiffness characteristics of the pneumatic actuator. From Table 2, it can be seen that the pneumatic actuator shows good characteristics.

Performance	Direction	Experimental result
Straightness [μm] / 200 [mm]	Horizontal	0.5
	Vertical	0.5
Angular motion [Second] / 200 [mm] ([μrad] / 200 [mm])	Yawing	23 (11.2)
	Pitching	21 (10.2)
	Rolling	0.9 (4.4)
Stiffness [N / μm]	Horizontal	120
	Vertical	200

Table2: Guide performance

5. MATHEMATICAL MODEL

5.1 Basic equations

The pneumatic servo system consists of two servo valves, a pneumatic actuator and the connecting pipes. In this study, the influence of the pipes is neglected. Assuming isothermal conditions, three main variables have to be known to model the pneumatic servo system: air mass flow through the servo valves, pressure at each chamber of the actuator and piston velocity. The three equations needed to know these variables are the flow equation, state equation of ideal gases and motion equation. These equations are written below under the following assumptions:

a) Influence of pipes is neglected.
b) State change is isothermal.
c) Chamber 1 is charged while chamber 2 is discharged.
d) Air mass flow is choked in chamber 1 and is not choked in chamber 2.
e) Friction force is negligible because of the air bearings.
f) Air leakage is negligible.

[Flow equation]

$$G_1 = K_f S_{e1} P_s \sqrt{1 - \left(\frac{\frac{P_1}{P_s} - b}{1 - b} \right)^2} \quad , \quad \frac{P_1}{P_s} > b \tag{1}$$

$$G_2 = K_f S_{e2} P_2 \quad , \quad \frac{P_a}{P_2} \leq b \tag{2}$$

Where: $$K_f = \sqrt{\frac{\kappa}{R\theta_a} \left(\frac{2}{\kappa + 1} \right)^{\frac{\kappa+1}{\kappa-1}}} \tag{3}$$

[State equation of ideal gases]

$$V_1 \frac{dP_1}{dt} = R\theta_a G_1 - AP_1 \frac{dx}{dt} \tag{4}$$

$$V_2 \frac{dP_2}{dt} = R\theta_a G_2 + AP_2 \frac{dx}{dt} \tag{5}$$

[Motion equation]

$$M \frac{d^2x}{dt^2} = (P_1 - P_2)A \tag{6}$$

5.2 Linearised equations

Linearising equations for the mass flow around the equilibrium point, the following equations are obtained (6):

$$G_1 = \varphi K_f P_o S_{e1} \tag{7}$$

$$G_2 = K_f P_o S_{e2} \tag{8}$$

Where

$$\varphi = \frac{P_s}{P_o}\sqrt{1-\left(\frac{\frac{P_o}{P_s}-b}{1-b}\right)^2} \tag{9}$$

Assuming a linear relationship between effective area and servo valve input:

$$S_{e1} = K_{sn}u \tag{10}$$

$$S_{e2} = -K_{sn}u \tag{11}$$

From Eq.(7), Eq.(8),Eq.(10) and Eq.(11):

$$G_1 = \varphi K_f K_{sn} P_o u \tag{12}$$

$$G_2 = -K_f K_{sn} P_o u \tag{13}$$

Linearising the state equation of gases, Eq.(4) and Eq.(5), around the equilibrium point and taking the central position as a linearisation point:

$$V_o \frac{dP_1}{dt} = R\theta_a G_1 - AP_o \frac{dx}{dt} \tag{14}$$

$$V_o \frac{dP_2}{dt} = R\theta_a G_2 + AP_o \frac{dx}{dt} \tag{15}$$

5.3 Open loop transfer function

From the linearised equations of flow, Eq.(12) and Eq(13), linearised state equations of ideal gases, Eq.(14) and Eq.(15), and equation of motion, Eq.(6); the open loop transfer function from the servo valve input u to displacement of the actuator x is given by Eq.(16):

$$P_n(s) = \frac{x(s)}{u(s)} = \frac{K_n \omega_n^2}{s(s^2 + \omega_n^2)} \tag{16}$$

Where coefficient K_n and natural frequency ω_n are given by:

$$K_n = \frac{K_f K_{sn} R\theta_a (1+\varphi)}{2A} \tag{17}$$

$$\omega_n = \sqrt{\frac{2P_o A^2}{MV_o}} \tag{18}$$

The block diagram from effective area of servo valve S_e to actuator velocity $\dot{x}$ is shown in Fig.3.

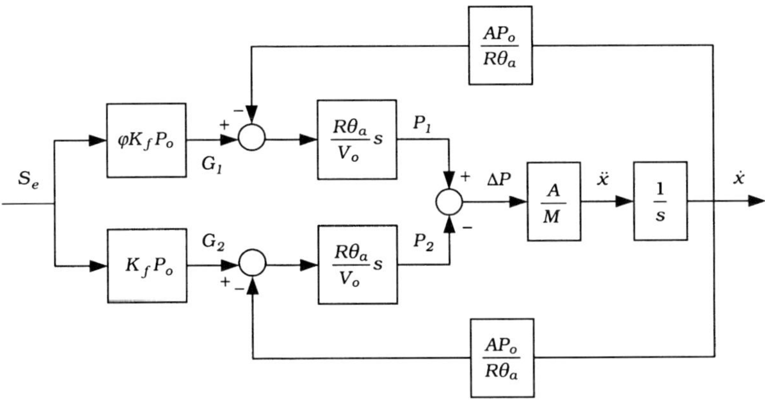

Fig.3: Block diagram from effective area of servo valve S_e to actuator velocity $\dot{x}$

6. DISPLACEMENT CONTROL

6.1 Controller gains

The linear model derived in Section 5 is used as a basis for a controller design. A simple controller that has proven to stabilise pneumatic servo systems is the state feedback controller (7)(8). Position, velocity and acceleration of the system to be controlled are used to compute the controller output:

$$u = K_p(x - x_{ref}) - K_v \dot{x} - K_a \ddot{x} \tag{19}$$

The closed loop transfer function from x_{ref} to x becomes:

$$G_c(s) = \frac{x(s)}{x_{ref}(s)} = \frac{K_p K_n \omega_n^2}{s^3 + K_a K_n \omega_n^2 s^2 + (K_v K_n + 1)\omega_n^2 s + K_p K_n \omega_n^2} \quad (20)$$

Information on velocity and acceleration can be obtained by differentiating the position signal but the differentiation increases the noise present in the position signal. Therefore, velocity and acceleration are obtained using a Kalman filter.
For the controller tuning a gain-phase margin diagram of 3rd order is used to estimate the dynamic control performance. By introducing dimensionless parameter α and β in Eq.(20):

$$G_c(s) = \frac{x(s)}{x_{ref}(s)} = \frac{1}{s^3 + \alpha s^2 + \beta s + 1} \quad (21)$$

From Eq.(20) and Eq.(21):

$$K_v = \frac{1}{K_n}\left\{\beta\left(\frac{K_n K_p}{\omega_n}\right)^{\frac{2}{3}} - 1\right\} \quad (22)$$

$$K_a = \frac{\alpha K_p^{\frac{1}{3}}}{(K_n \omega_n^2)^{\frac{2}{3}}} \quad (23)$$

Gains K_v and K_a have been expressed as functions of the proportional gain K_p and the dimensionless parameters α and β. Consequently, gains K_v and K_a are obtained by selecting gain K_p and choosing suitable values of α and β according to desirable gain and phase margin.

6.2 Disturbance compensation

The robustness of a state feedback controller is weak since feedback gains are based on a simplified linear model. To compensate for nonlinear characteristics of the servo valves and parameter variations, a disturbance observer is introduced (9). By using a disturbance observer, robustness is added to the controller against disturbances. The observer filter used in this study is given by Eq.(24):

$$F(s) = \frac{1}{Ts + 1} \quad (24)$$

Fig.4 shows the block diagram of the position control system with state feedback, Kalman filter and disturbance observer. The closed loop transfer function has a steady state velocity error. To compensate for this steady state error, an inverse model has been added.

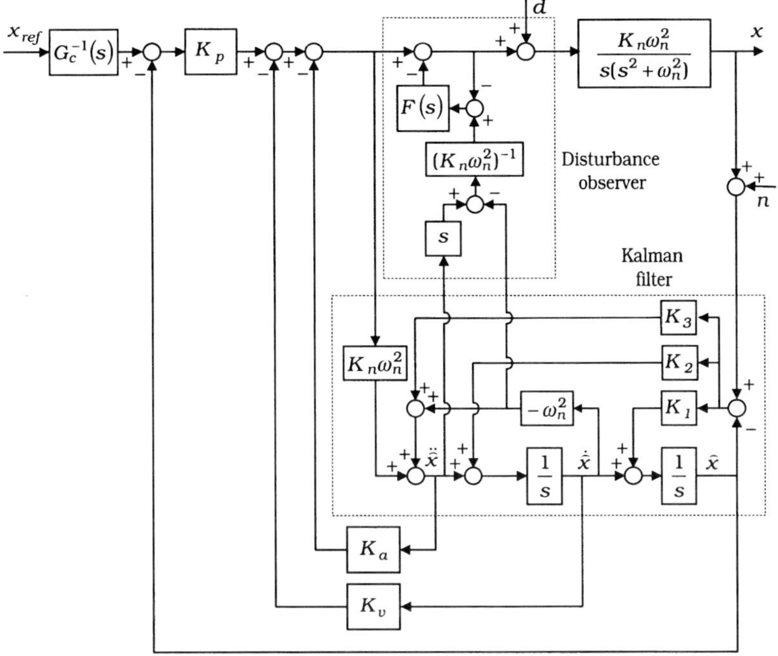

Fig.4: Block diagram of position control system with state feedback, disturbance observer and inverse model.

7. POSITIONING RESULTS

To study the performance of the positioning control system, tests of continuous step response and constant velocity control were carried out.

Fig.5 shows the result for continuous step responses of ±1[µm]. The upper figure shows slider displacement from –5[mm] to 5[mm]. The lower figure shows an enlargement of the upper figure from 8[s] to 16[s]. The error average is below 0.01[µm] and the standard deviation is 0.08[µm]. From the various step responses, it can be seen that repeatability is good. Although good accuracy can be obtained, fluctuations can be seen around the reference position that are in the range of ± 0.2[µm].

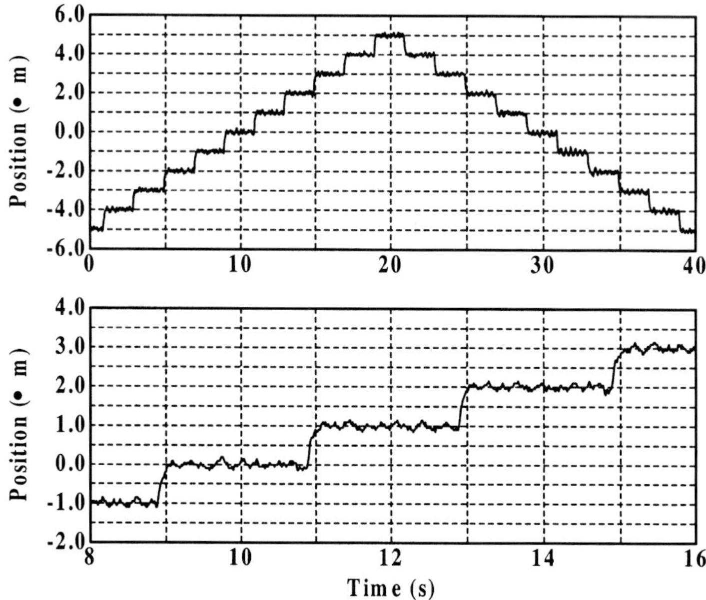

Fig.5: Continuous step response test

Fig.6 shows the slider displacement, position error and velocity response when ±1[mm/s] constant velocity control is performed. The actuator is driven from –4[mm] to 4[mm] and vice versa. In the positions ±4[mm], the actuator is stopped for 2[s]. The positioning error in the velocity control is in the same range as in the continuous step response test of ±0.2[μm]. It can be seen that although a stable velocity is obtained, there are overshoots in the velocity response at the beginning and end of the motion. These overshoots are attributed to the output of the values of velocity and acceleration of the Kalman filter.

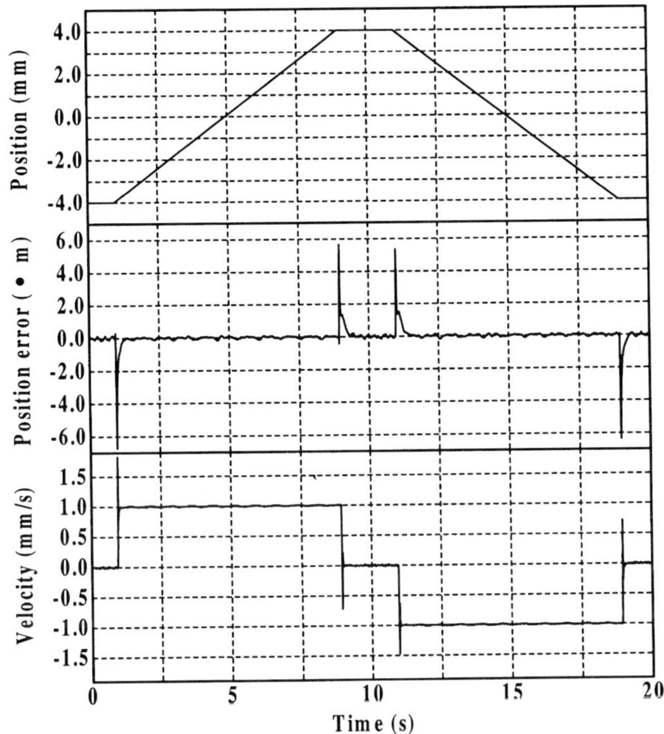

Fig.6: Constant velocity control test

8. CONCLUSIONS

a) With the purpose to design a fast and precise positioning system, this study presents results using a pneumatic actuator prototype in which common seals are removed and externally pressurised air bearings are introduced to reduce friction. Experimental continuous step response and constant velocity tests show that an accuracy of ±0.2[μm] can be obtained.

b) A simple controller consisting of a state feedback, disturbance observer and inverse model is designed for the control of the pneumatic actuator.

c) The next steps of this research will try to reduce the ±0.2[μm] fluctuations around the reference position. Compensation of the non-linear characteristics of servo valves and hysterisys will be introduced in an attempt to improve positioning accuracy.

REFERENCES

(1) Sheats J., Smith B.: Microlithography, science and technology. Marcel Dekker, Inc. (1998)

(2) Makinouchi S., Hayashi Y., Kamiya S.: New stage system for step and repeat scanning stepper. Journal of Precision Engineering, 61-12, 1676/1680 (1995) (In Japanese)

(3) Uebing M., Vaughan N.D., Surgenor B.W.: On linear dynamic modelling of a pneumatic servo system. The 5th Scandinavian International Conference on Fluid Power 2, 363/378 (1997)

(4) Jang J., Kagawa T., Fujita T., Ohsumi Y.: Development of a pneumatic servo table mounted with externally pressurised air bearings for accuracy and speedy positioning". Journal of the Japan Hydraulics and Pneumatics Society, 28-4, 451/457 (1997) (In Japanese)

(5) Brun X., Sesmat S., Scavarda S., Thomasset D.: Simulation and experimental study of the partial equilibrium of an electropneumatic positioning system, Cause of the sticking and restarting phenomenon. Proceedings of the 4th JHPS International Symposium on Fluid Power, 125/130 (1999)

(6) Sakaki K., Shibukawa T., Kagawa T.: Theoretical approach to optimum effective area for pneumatic servo system. Proceedings of 7th Symposium on Fluid Control and Measurement, The Society of Instrument and Control Engineering, 99/102 (1999) (In Japanese)

(7) Backé W., Eschman R.: SSP-A simulation program for pneumatics. 6th Bath International Fluid Power Workshop, Modelling and Simulation, 64/79 (1993)

(8) Pu J., Moore P.R., Harrison R., Weston R.H.: A study of gain-scheduling method for controlling the motion of pneumatic servos. 6th Bath International Fluid Power Workshop, Modelling and Simulation, 193/209 (1993)

(9) Yokota S., Morimoto S., Toratani T.: Control of electro-hydraulic manipulators for electric distributed power line task using disturbance observer. Journal of The Japan Hydraulics and Pneumatics Society, 25-1, 113/117 (1994) (In Japanese)

Study of magnetic circuits for servovalve torque motors

E URATA
Kanagawa University, Japan

The moment generated by magneto-motive force acting on an armature in a servovalve torque motor is studied. Basic equations describing flux density in an air gap are presented and verified by experiments using a permanent magnet circuit. The derived basic equations are applied to another permanent magnet circuit in a torque motor. Average differences between theoretical and experimental results for the flux density and flapper deflection of the torque motor were less than about 10 percent.

1 INTRODUCTION

Purpose of this paper is to evaluate basic parameters of a servovalve torque motor containing permanent magnets. Figure 1 shows the torque motor. Forces and moments of magnetic, elastic and hydraulic origins are acting on the assembly composed of a flapper, a flexure tube and an armature. The equilibrium of these moments and forces determine the flapper position. Our discussion in this paper focuses on the magnetic torque acting on the armature.

Until now, only a few authors studied inherently nonlinear characteristic of the armature torque[(1), (2), (3)]. On the other hand, a linear approximation of the torque motor,

$$T_A \approx K_B i + K_G \psi ,$$

is widely accepted [(4), (5)] and valid for commercial servovalves. However, few studies have been made for derivation of the constants K_G and K_B. For designing of servovalves and precise modeling of electrohydraulic servosystems, we need a mathematical model by which we can deduce these parameters from dimensions and characteristics of magnetic elements. Therefore, we have developed a new mathematical model, derived the above constants and examined them through experiments.

In section 2 of this paper we establish a basic equation for permanent magnet circuits and derive parameters to be used in other circuits. In section 3, we apply the result of section 2 to the torque motor. In section 4, we show experiments on a basic permanent magnet circuit and a prototype torque motor and then verify validity of the theory.

Notation

A_g: Cross sectional area of air gap, A_p: Cross sectional area of permanent magnet, B: Flux density, B_r: Residual flux density, b_p: Width of pole piece, F_A: Magnetic attractive force acting on armature end, H: Magnetic field intensity, H_g: Magnetic field in air gap, i: Electric current to coil, K_E: Bending rigidity of flexure tube, L_A : Length of armature, l_0: Normal air gap of torque motor, l_F: Flapper length, l_g: Air gap, l_p: Length of permanent magnet, l_t: Length of flexure tube, n: Coil turns, R_g: Reluctance of air gap, R_p: Reluctance of permanent magnet, T_A: Armature moment exerted by magnetic attraction, V_P: Magneto-motive force of permanent magnet, X: Displacement of armature end, Y: Displacement of flapper end, μ: Recoil permeability of permanent magnet, μ_0: Permeability of vacuum, μ_r: Relative permeability $=\mu/\mu_0$, Φ: Flux, Φ_g: Flux through air gap, Φ_l: Leakage flux, Φ_P: Flux through permanent magnet, ψ: Armature deflection angle.

2 BASIC PERMANENT MAGNETIC CIRCUIT

When we calculate flux density in an air gap of a permanent magnet circuit using traditional methods, the result often shows a large discrepancy from measured values. In this section, we will try to improve mathematical expression for a basic permanent magnet circuit shown in Fig. 2. The circuit is composed of a permanent magnet, a pair of yokes, a coil and an air gap. Figure 3 illustrates an image of flux lines around the basic magnetic circuit shown Fig.2. Applying Ampere's circuit theorem [6] to the closed curve C_1 in Fig.3, we have

$$\int_{\mathrm{PM}} H dz + H_g l_g = ni. \qquad (1)$$

The notation "PM" under the integral in (1) indicates the permanent magnet. In the above equation, we ignored reluctance of the yokes because it is usually far smaller than the air gap reluctance. Referring to the demagnetization curve illustrated in Fig.4, the relation between the magnetic field and the flux density at a point in a permanent magnet is expressed by

$$B(x, y, z) = B_r + \mu H(x, y, z). \qquad (2)$$

For convenience, the coordinate is taken as the planes of z=0 and $z=l_p$ coincide with the planes of S-pole and N-pole of the permanent magnet, respectively. Although magnetic field and flux density are vector quantities, we assume that x- and y-components are small and negligible within the permanent magnet. Then, a scalar expression (2) represents the relation between the z-components of the magnetic field and the flux density.

Before estimating the integral in (1), we examine the boundary condition for the permanent magnet. At just inside of N-pole surface, namely at $z=l_p-0$, we have

$$B_s(x,y)=B(x,y,l_p-0),\quad H_s(x,y)=H(x,y,l_p-0). \tag{3}$$

Then, (2) and (3) yield

$$B_s(x,y)=B_r+\mu H_s(x,y). \tag{4}$$

Since the flux density component normal to a permanent magnet surface is continuous,

$$B_s(x,y)=B(x,y,l_p-0)=B(x,y,l_p)=B(x,y,l_p+0). \tag{5}$$

Now, let us consider an integral curve that becomes parallel to z-axis in the permanent magnet. Because the end poles reduce the magnetic field in the permanent magnet [(6), (7)], the magnetic field and the flux density along the integral path in the permanent magnet become as illustrated in Fig.4. Integrating H from 0 to l_p with respect to z, we have

$$\int_{\mathrm{PM}} H dz=\int_0^{l_p} H(x,y,z)dz=l_p H_m(x,y). \tag{6}$$

In this expression, $H_m(x, y)$ indicates the average of the magnetic field along the integral path. As observed from Fig.4, this average (H_m) is different from the magnetic field (H_s) at the point just inside of the intersection of the integral path and the magnet surface. Therefore, we introduce a new function $\Lambda(x, y)$ that satisfies

$$H_s(x,y)=\Lambda(x,y)H_m(x,y). \tag{7}$$

Substituting (4)-(7) into (1), we have

$$\frac{l_p}{\mu}B_s(x,y)=\left(ni-H_g l_g\right)\Lambda(x,y)+\frac{l_p}{\mu}B_r. \tag{8}$$

Integrating both sides of (8) on the magnet surface $z=l_p$, we have

$$\frac{l_p}{\mu}\int_{\mathrm{A}} B_s(x,y)dxdy=\left(ni-H_g l_g\right)\int_{\mathrm{A}} \Lambda(x,y)dxdy+\frac{A_p l_p}{\mu}B_r. \tag{9}$$

The integral of flux density over the magnet surface gives the flux through the surface, namely,

$$\int_{\mathrm{A}} B_s(x,y)dxdy=\Phi_p. \tag{10}$$

Here we introduce a constant λ defined by

$$\int_{\mathrm{A}} \Lambda(x,y)dxdy=\lambda A_p. \tag{11}$$

Thus, λ refers to the average of $\Lambda(x, y)$ over the surface. Equation (9), (10) and (11) yield

$$\frac{l_p}{\mu}\Phi_p=\left(ni-H_g l_g\right)\lambda A_p+\frac{A_p l_p}{\mu}B_r. \tag{12}$$

Reluctance R_p and magneto-motive force V_p of the permanent magnet are defined as

$$R_p=\frac{l_p}{A_p\mu},\quad V_p=\frac{l_p}{\mu}B_r. \tag{13}$$

On the other hand, flux density and magnetic field in the air gap is almost uniform:

$$H_g l_g = B_g l_g / \mu_0 = R_g \Phi_g, \tag{14}$$

where

$$R_g = l_g / (\mu_0 A_g), \quad \Phi_g = A_g B_g. \tag{15}$$

Substitution of (13) and (15) into (12) yields

$$\frac{R_p}{\lambda}\Phi_p + R_g \Phi_g = ni + \frac{V_p}{\lambda}. \tag{16}$$

Let us compare (16) with the following conventional basic equation for the basic magnetic circuit appearing in texts [(8), (9)]:

$$R_p \Phi_p + R_g \Phi_g = ni + V_p. \tag{16A}$$

The above equation can be deduced if we assumed $H_m(x,y)=H_s(x,y)$. However, this assumption does not reflect the physical fact. As seen in Fig.4, the mean flux density (B_m) corresponding to Ampere's rule is not the real flux density (B_s) that is integrated according to Gauss's rule. Apparent difference between (16) and (16A) is that the magneto-motive force and the reluctance of the permanent magnet are divided by λ in (16). The experiment in Section 5 coincides with (16) but not with (16A).

Next, let us consider leakage flux. The flux lines through the air gap do not occupy all part of the permanent magnet surface. Flux lines that do not pass the air gap occupy the other part of the permanent magnet surface. These conditions are modeled as shown in Fig.5 and expressed by the following simple equation:

$$\Phi_g = r\Phi_p. \tag{17}$$

Contrary to the above assumption, traditional magnetic circuit models assume additional leakage reluctance and estimate its value. The above expression is equivalent to assume leakage reluctance, mathematically. However, the procedure to evaluate flux passing the air gap becomes simpler when we use (17).

Substitution of (15) and (17) into (16) yields

$$B_g = \frac{V_p \mu_0}{l_g} \cdot \frac{1}{\lambda + \alpha / r} + \frac{ni\mu_0}{l_g} \cdot \frac{\lambda}{\lambda + \alpha / r}, \tag{18}$$

where

$$\alpha = R_p / R_g. \tag{19}$$

An electric circuit analogy corresponding to (16) and (17) is illustrated in Fig.6, together with the traditional analogy [(8), (9)]. The magnet reluctance cannot be ignored because α is usually larger than one. The essential difference between the new and the traditional analogies is that the magneto-motive force and reluctance of the permanent magnet is divided by λ in the new analogy. The unknown parameters, λ and r, are determined using relating physical quantities.

We will find these values by the experiment in Section 4.

3 MAGNETIC CIRCUIT OF TORQUE MOTORS

Structure of the magnetic circuit of the torque motor and the equivalent electric circuit diagram are illustrated in Fig.7. The permanent magnet and the coil generate magnetic flux. The electric current to the coil generates unbalance of the fluxes through the four air gaps. Consequently, the four magnetic poles attract the armature with different forces. These four forces result in a moment to deflect the armature and flapper assembly. We calculate, in present section, the moment generated by the magnetic attractive forces.
Structural condition gives the following distribution of air gaps.

$$\left.\begin{array}{ll} l_1 = l_0 + X, & l_2 = l_0 - X, \\ l_3 = l_0 + X, & l_4 = l_0 - X \end{array}\right\}. \tag{20}$$

Practically, the four gaps are not equal at *X*=0 because the manufacturing and constructing inaccuracy are not avoidable. However, we develop here a theory for the ideal symmetric structure. The reluctance between the poles and the armature ends, namely, the air gap reluctance is

$$R_i = l_i / (\mu_0 A_g), \quad (i = 1, 2, 3, 4). \tag{21}$$

The reluctance of the permanent magnet is

$$R_p = l_p / (\mu_0 \mu_r A_p). \tag{22}$$

Conservation of flux results in

$$r\Phi_p = \Phi_1 + \Phi_4, \quad \Phi_2 = \Phi_a + \Phi_1, \quad \Phi_4 = \Phi_a + \Phi_3. \tag{23}$$

The relations between the flux and the magneto-motive forces are

$$\left.\begin{array}{l} V_p / \lambda = \Phi_1 R_1 + \Phi_2 R_2 + \Phi_p R_p / \lambda \\ ni = \Phi_2 R_2 - \Phi_3 R_3 \\ ni = \Phi_4 R_4 - \Phi_1 R_1 \end{array}\right\}. \tag{24}$$

Since the flux density in the air gap is practically uniform, there follows

$$B_i = \Phi_i / A_g, \quad (i = 1,2,3,4). \tag{25}$$

Let us introduce the following nondimensional variables

$$\xi = X / l_0, \quad \theta = ni / V_p. \tag{26}$$

Substitution of (25) and (26) into (23) and (24) yields

$$\left.\begin{aligned}B_1 = B_3 &= \frac{V_p\mu_0}{2l_g}\cdot\frac{\left[r(1-\xi)(1-\lambda\theta)-\alpha\theta\right]}{\left(1-\xi^2\right)r\lambda+\alpha}\\ B_2 = B_4 &= \frac{V_p\mu_0}{2l_g}\cdot\frac{\left[r(1+\xi)(1+\lambda\theta)+\alpha\theta\right]}{\left(1-\xi^2\right)r\lambda+\alpha}\end{aligned}\right\}. \tag{27}$$

Since Maxwell stress generated by the flux density in the each air gap is

$$B_i^{\,2}/(2\mu_0), \quad (i = 1, 2, 3, 4)\,, \tag{28}$$

the force acting on the right end of the armature is

$$F_A = (B_2^{\,2} - B_1^{\,2})A_g/(2\mu_0)\,. \tag{29}$$

Another force of the same magnitude but inverted direction acts on the left end of the armature. Consequently, the armature receives the following moment:

$$T_A = (B_2^{\,2} - B_1^{\,2})(L_A - b_p)A_g/(2\mu_0)\,. \tag{30}$$

Substitution of (27) into (30) yields

$$T_A = U_0\eta\cdot\frac{r(1+\lambda\theta\xi)\left[r\xi+(\alpha+r\lambda)\theta\right]}{\left[\left(1-\xi^2\right)r\lambda+\alpha\right]^2}, \tag{31}$$

where

$$U_0 = \frac{\mu_0 A_g V_p^{\,2}}{l_g}, \quad \eta = \frac{L_A - b_p}{2l_g}, \quad \xi = \eta\psi. \tag{32}$$

Partial differentiation of (31) with respect to i and ψ yields

$$\left.\frac{\partial T_A}{\partial i}\right|_{i=0,\psi=0} = K_B = \frac{nU_0\eta r}{V_p(\alpha+r\lambda)} \tag{33}$$

and

$$\left.\frac{\partial T_A}{\partial \psi}\right|_{i=0,\psi=0} = K_G = \frac{U_0\eta^2 r^2}{(\alpha+r\lambda)^2}\,. \tag{34}$$

Thus, we obtain a linear approximation of the moment:

$$T_A \approx K_B i + K_G\psi\,. \tag{35}$$

Equation (31) reveals inherent nonlinearity in the torque motor dynamics. However, its linear approximation (35) becomes accurate for small values of ψ and i, when design parameters are selected properly.

4 EXPERIMENTS

4.1 Experiment on the basic model

A physical model for a basic magnetic circuit is made and its characteristic is investigated. This physical model corresponds to a quarter branch of the torque motor magnetic circuit. Using this basic model, we study the length parameter λ and area parameter r introduced in Section 2. The model is, as already shown in Fig.2, composed of a pair of yokes, a permanent magnet block, a coil and an air gap. One of the yokes passes through the coil. The size of the air gap is fixed and various magnets and magnet blocks are inserted interchangeably. The yokes and the iron spacers in the magnetic blocks are made of soft magnetic iron, SUYP1 (JIS C 2504). Figure 8 illustrates shapes of the permanent magnets and the magnet blocks. Figure 9 shows the demagnetization characteristics of the permanent magnets used for the experiment. Dimensions and characteristics of the magnets are summarized in Table 1.

The flux density in the air gap is measured inserting Hall effect sensor made of Gallium Arsenide. The dimension of the sensing tip is 75-μm×75-μm with 70-μm height. To begin, flux density distribution in the gap is measured moving the sensor in a plane that includes the gap. Thus, we find that the flux distribution in the gap is almost uniform. Next, by exciting the coil with direct current ranging from –0.12 A to 0.12 A, the flux density in the gap is measured. Figure 10 shows an example of the measured *i-B* characteristic. From this series of data, we deduce regression lines for each magnetic block:

$$B_g = B_{g0} + Ci\,. \tag{36}$$

The coefficients appearing in (36) are also noted in Table 1. Comparing (36) and (18), we have

$$B_{g0} = \frac{V_p \mu_0}{l_g} \cdot \frac{1}{\lambda + \alpha / r}, \quad C = \frac{n\mu_0}{l_g} \cdot \frac{\lambda}{\lambda + \alpha / r}. \tag{37}$$

Rearranging (37), we have the length and area factors:

$$\lambda = \frac{CV_p}{B_{g0}}, \quad r = \frac{\alpha}{V_p \mu_0 / (l_g B_{g0}) - \lambda}. \tag{38}$$

Figures 11 and 12 are scatter diagrams of these factors vs. α, respectively. Correlation coefficient between λ and α is 0.95. This fact suggests that λ is expressed as a function of α. No reasonable correlation is found between r and nondimensional parameters. Therefore, we assume a simple average as estimation of r. Thus, our estimation for λ and r based on the basic magnetic circuit shown in Fig.2 is

$$\lambda = 2.3 + 1.8\alpha, \quad r = 0.27\,. \tag{39}$$

Naturally, we are not sure whether this estimation is valid for other permanent magnet circuits. To ensure the above result, and to examine the validity of the parameter estimation by (33) and (34), we make other experiments using a prototype servovalve torque motor.

Table 1 Magnetic materials used for experiment

Code	Material*	Type	l_p [m]	A_p[m^2]	B_r[T]	μ_r	α	B_{g0}[T]	C [T/A]	λ	r
a	M1(Sr)	1	0.028	2.52E-04	0.37	1.1	3.03	0.905	0.965	7.01	0.293
b	M2(Sr)	1	0.028	2.52E-04	0.42	1.1	3.03	1.032	1.033	7.47	0.309
c	M3(Ba)	1	0.028	2.52E-04	0.24	1.05	3.17	0.585	1.021	7.79	0.304
d	M4(Ba)	1	0.028	2.52E-04	0.36	1.1	3.03	0.904	1.001	7.08	0.309
e	M5(Sr)	2	0.0191	1.44E-04	0.42	1.0	3.98	0.512	0.965	10.6	0.256
f	M5(Sr)	2	0.0101	1.44E-04	0.42	1.0	2.10	0.493	1.122	6.74	0.277
g	M5(Sr)	2	0.0051	1.44E-04	0.42	1.0	1.06	0.412	1.238	4.49	0.255
h	M6(Ba)	2	0.0152	1.44E-04	0.23	1.2	2.64	0.238	0.899	7.68	0.207
i	M6(Ba)	2	0.0031	1.44E-04	0.23	1.2	0.54	0.185	1.378	3.09	0.238

*Sr: Strontium ferrite, Ba: Barium ferrite l_g=0.6 mm, A_g=18 mm^2, n=1140

4.2 Experiment on a torque motor

Static characteristics of the torque motor shown in Fig.1 were measured and compared with the theoretical results in the previous sections. Dimensions and physical parameters of the servovalve are listed in Table 2. Experimental setup is shown in Fig.13. The magnets used in the experiment of the previous section are mounted in the torque motor. The flux densities in the four gaps and the flapper displacement are measured, by exciting the coil with electric current. Measuring method for the flux density is the same as the previous section. A measured example of the four flux densities for a magnetic material is shown in Fig.14. The displacement of the flapper is measured using a non-contact-type position sensor.

Flux densities in the four gaps in the torque motor are not equal even at its null-position. This comes from manufacturing and construction inaccuracies that are practically unavoidable.

Therefore, as the flux density of air gap at the null-position, we use the average of these four values. As estimation of rate of change of B, i.e. dB/di, the average of the absolute values of four measured values is used. These estimated values are compared with theoretical values.

Theoretical estimation of B_0 and C is as follows. First, substituting ξ=0 and θ=0 into (27), we have the flux density at i=0:

$$B_{g0} = \frac{V_p \mu_0}{2l_g} \cdot \frac{1}{\lambda + \alpha / r}. \tag{40}$$

The numerical value of α becomes half of the value in Section 4.1 because two similar gaps exist in a closed curve on that Ampere's rule is applied. The empirical values λ and r obtained in previous section are substituted into (40). Figure 15 compares the experimental and theoretical values of the flux density at the null-position. The experimental values are about 9 % less than the theoretical values; standard deviation of the ratio of experimental and theoretical values is about 9 %. This agreement is very good compared with traditional theory in which the difference becomes often several hundred percent.

Next, the rate of change of flux density, C, for the experimental rig is as follows. Since the flexure tube supports the armature, the armature moment (31) balances with the elastic moment generated in the flexure tube:

Table 2 Torque motor specifications and experimental results

Code	Material	Type	l_p/l_g	A_g/A_p	μ_r	B_r	l_g	α	η	B_{go} (exp.)	dB/di (exp.)	dY/di (exp.)
Ta	M1	1	44.1	3.57E-02	1.1	0.37	5.6E-04	1.43	32.3	0.649	1.64	2.68E-04
Tb	M2	1	49.1	3.57E-02	1.1	0.42	5.2E-04	1.59	36.0	0.836	2.17	5.15E-04
Tc	M3	1	43.1	3.57E-02	1.05	0.24	6.0E-04	1.47	31.5	0.473	1.39	1.75E-04
Td	M4	1	44.8	3.57E-02	1.1	0.36	5.5E-04	1.45	32.8	0.742	1.80	3.29E-04
Te	M5	2	26.9	6.25E-02	1.0	0.42	6.5E-04	1.68	28.9	0.387	1.25	1.34E-04
Tf	M5	2	14.2	6.25E-02	1.0	0.42	6.4E-04	0.89	28.9	0.359	1.30	1.60E-04
Tg	M5	2	6.99	6.25E-02	1.0	0.42	6.8E-04	0.44	28.1	0.295	1.28	9.42E-05
Th	M6	2	21.0	6.25E-02	1.2	0.23	6.8E-04	1.09	28.3	0.202	1.21	6.27E-05
Ti	M6	2	4.22	6.25E-02	1.2	0.23	7.1E-04	0.22	28.0	0.122	1.16	3.68E-05
Unit	-	-	-	-	-	T	m	-	-	T	T/A	m/A

K_E=85 Nm, $A_g=18\times10^{-6}\text{m}^2$, n=1140

$$K_E\psi = T_A. \quad (41)$$

Derivative of B_g (B_g is equivalent to B_2, B_4 or $-B_1$, $-B_3$) with respect to i,

$$C = \frac{dB_g}{di} = \frac{\partial B_g}{\partial i} + \frac{\partial B_g}{\partial \psi}\frac{d\psi}{di} \quad (42)$$

and differentiation of (41) with respect to i,

$$K_E\frac{d\psi}{di} = \frac{\partial T_A}{\partial i} + \frac{\partial T_A}{\partial \psi}\frac{d\psi}{di} \quad (43)$$

yield

$$\frac{dB_g}{di} = \frac{\partial B_g}{\partial i} + \frac{\partial B_g}{\partial \psi}\frac{\partial T_A}{\partial i}\Bigg/\left(K_E - \frac{\partial T_A}{\partial \psi}\right). \quad (44)$$

Substituting (33) and (34) into (44), the gradient of B against i yields

$$C = \frac{dB_g}{di} = \frac{n\mu_0}{2l_g}\cdot\frac{(\alpha/r+\lambda)^2}{(\alpha/r+\lambda)^2-(U_0/K_E)\eta^2}. \quad (45)$$

The above theoretical values and the experimental values are compared in Fig.16. The theoretical values are about 5 % larger than the corresponding theoretical values; standard deviation of the ratio of experimental and theoretical values is about 9 %.

Finally, let us examine flapper displacement characteristic. Figure 17 shows a measured example of current-displacement relationship. Other measured results, though they are not shown here, also revealed a good linearity between current and displacement. Referring to (33), (34) and (43), the gradient of the flapper displacement at the origin is

$$\frac{dY}{di} = \left(l_F - \frac{l_t}{2}\right)\frac{d\psi}{di} = \left(l_F - \frac{l_t}{2}\right)\frac{nU_0\eta}{K_EV_p}\cdot\frac{\alpha/r+\lambda}{(\alpha/r+\lambda)^2-(U_0/K_E)\eta^2}. \quad (46)$$

The theoretical and the experimental values of the gradient are compared in Fig.18. In this case, the theoretical values are about 7 percent less than the experimental values; standard deviation of the ratio of experimental and theoretical values is about 10 percent.

5 CONCLUSIONS

A theory to give a mathematical model of torque motor magnetic circuit is presented. Parameters expressing the linear model of the armature torque are also given. Experiments are carried out using a basic magnetic circuit model and a prototype torque motor. Nine permanent magnets of different materials with different geometric dimensions are used in the experiment. The basic magnetic circuit is made to evaluate characteristic values of magnetic circuits. For the prototype torque motor, the theory coincided well with experiment.
The following conclusions are concerned with not only a torque motor but also basic equations for magnetic circuits containing permanent magnets:

1. Basic equations for permanent magnet circuits should be modified to divide the magnetomotive force and the reluctance of the permanent magnet by a common factor λ.
2. The permanent magnet reluctance plays an important role on the characteristics of torque motors. Hence, a modeling ignoring the magnet reluctance is not appropriate.

Acknowledgements
Mrs. M. Kusakabe and H. Okazaki are thanked for their assistance in the experimental work. Materials used in the experiments were supplied from Ebara Research Co. Ltd.

REFERENCES

(1) Edited by J. H. Blackburn, G. Reethof and J. L. Shearer, Fluid Power Control, p.322, 1960.

(2) H. E. Merritt, Hydraulic Control Systems, John Wiley and Sons, Inc., p.183. 1967.

(3) C.-J. Liaw and F. T. Brown, Nonlinear dynamics of an Electrohydraulic Flapper Nozzle Valve, Trans. ASME, Journal of Dynamic Systems, Measurement and Control, Vol.112, pp.298-304, 1990.

(4) P. M. FizSimons and J. J. Parazzolo, Modeling of a One-Degree-of-Freedom Active Hydraulic Mount, Trans. ASME, Journal of Dynamic Systems, Measurement and Control, Vol.118, pp.286-294, 1996.

(5) S. J. Lin and A. Akers, A Dynamic Model of the Flapper Nozzle Component of an Electrohydraulic Servovalve, Trans. ASME, Journal of Dynamic Systems, Measurement and Control, Vol.111, pp.105-109, 1989.

(6) S. Chikazumi, Physics of Ferromagnetism, Vol.1, p.23, 1989, (in Japanese).

(7) J. H. Kraus, Electromagnetics, Fourth Ed., McGraw-Hill, p.319, 1992.

(8) J. F. Gieras and M. Wing, Permanent Magnet Motor Technology, Marcel Dekker, p.64, 1997.

(9) K. Ohkawa, Introduction to permanent magnet circuit, Sogo-Denshi Shuppan, p.110, 1994 (in Japanese).

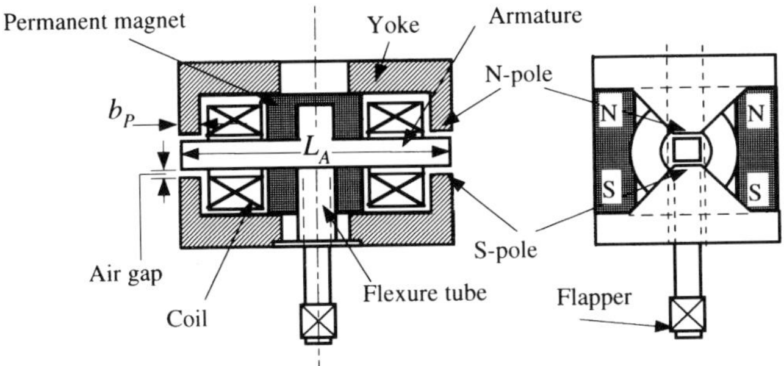

Fig.1 Torque motor

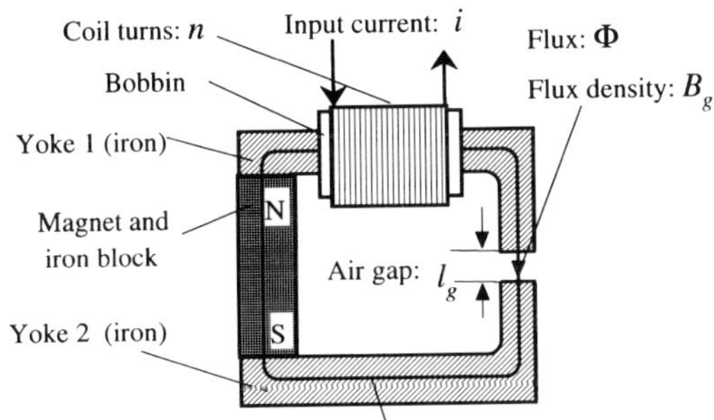

Fig. 2 Basic model of permanent magnetic circuit

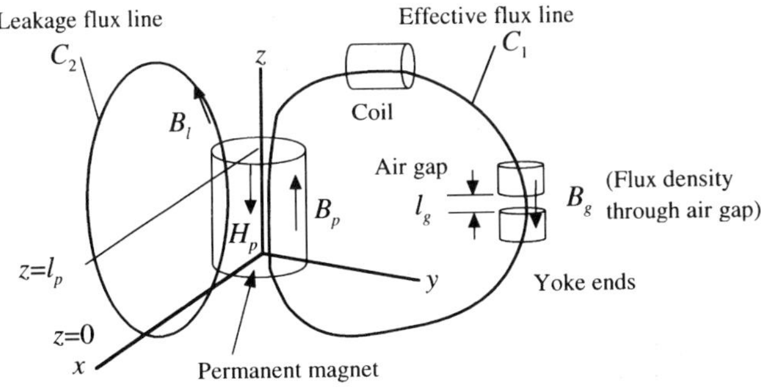

Fig. 3 Magnetic circuit and flux lines

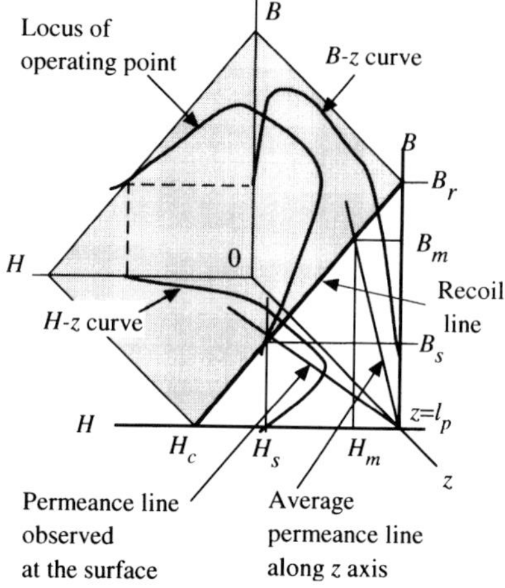

Fig.4 Operating point of permanent magnet

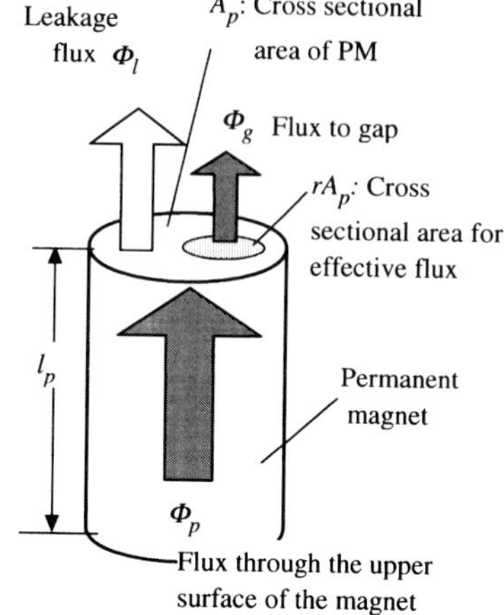

Fig. 5 Flux distribution

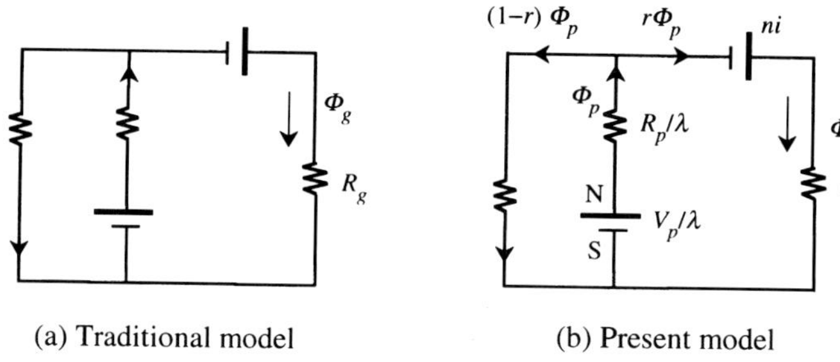

(a) Traditional model (b) Present model

Fig.6 Equivalent electric circuits for the basic magnetic circuit

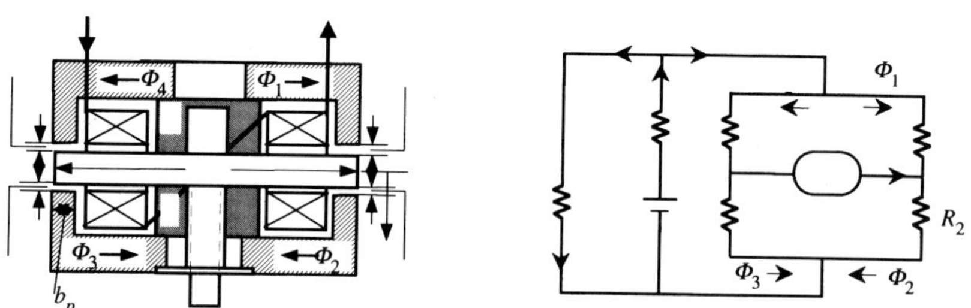

Fig.7 Magnetic structure of torque motor

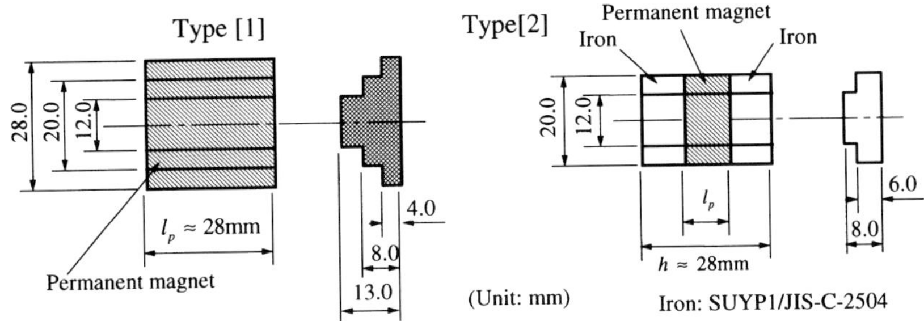

Fig.8 Testpieces used for Experiments

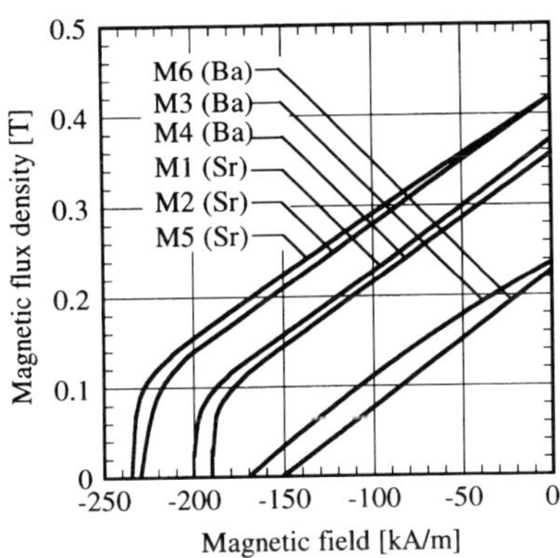

Fig.9 Demagnetization Curve

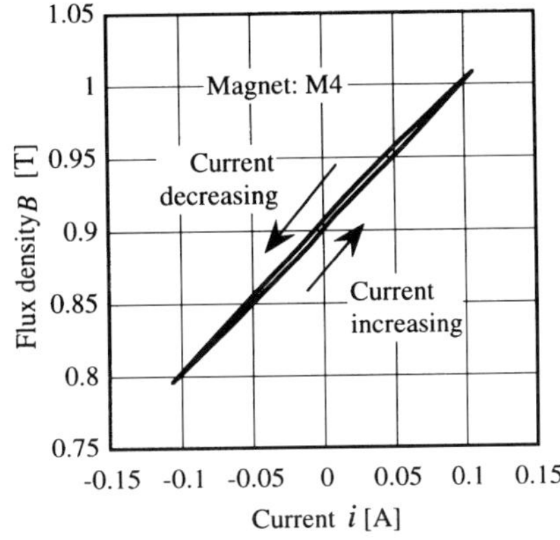

Fig.10 *B-i* characterisitc in air gap (hysteresis characterisitc)

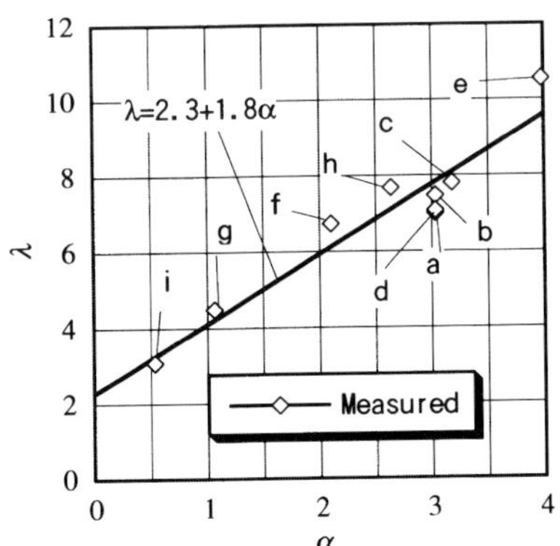

Fig.11 Length factor: λ

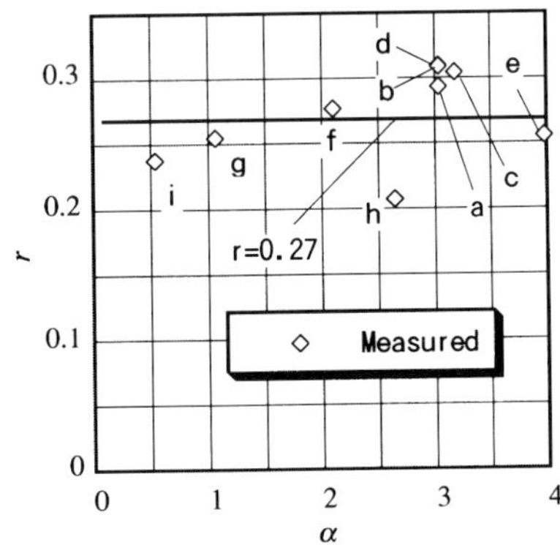

Fig.12 Area factor: *r*

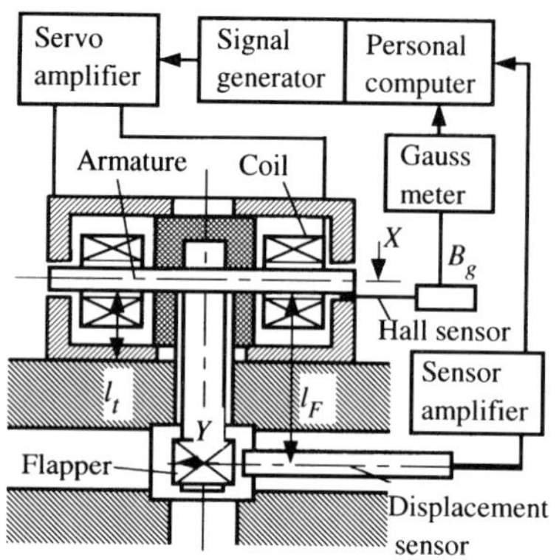

Fig.13 Experimental setup

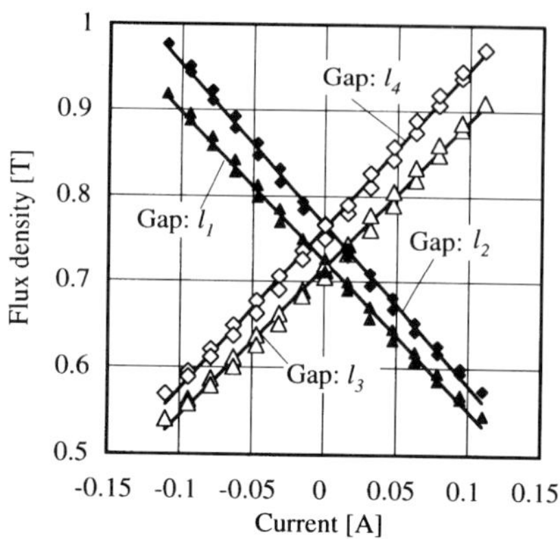

Fig.14 Flux density in torque motor air gaps (Code: Td)

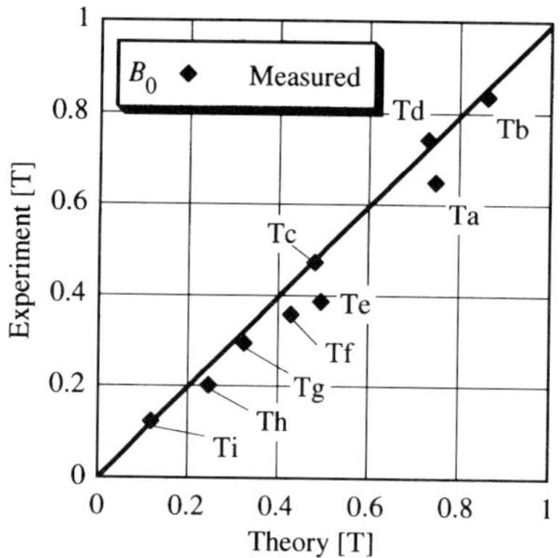

Fig.15 Flux density at the neutral

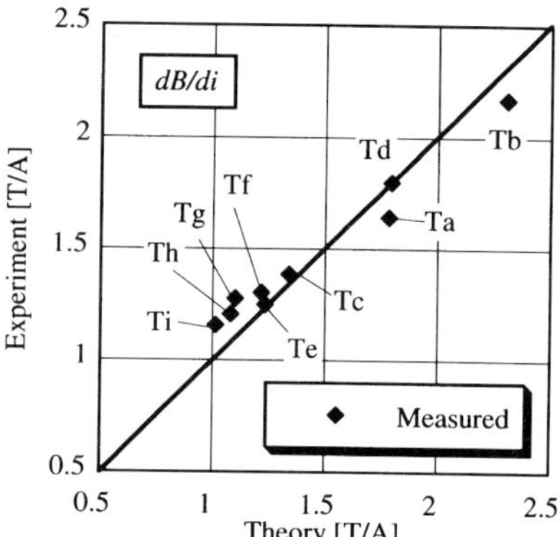

Fig.16 Gradient of flux desity

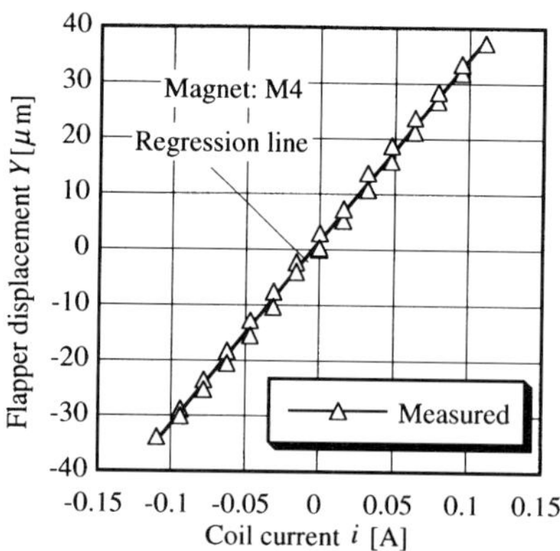

Fig.17 Flapper displacement

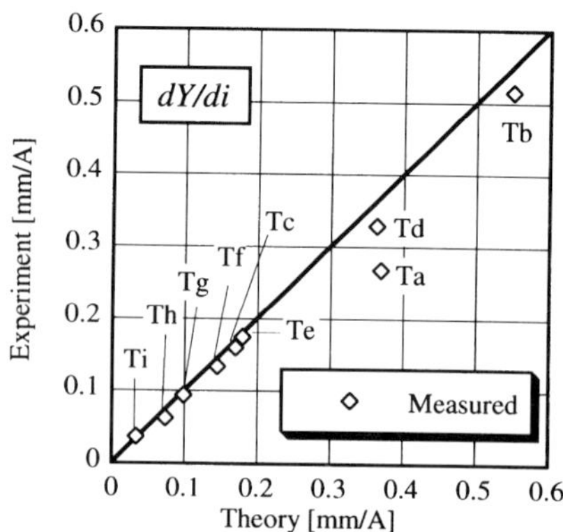

Fig.18 Flapper displacement gain

Improving the accuracy of the water hydraulic position servo by compensating servo-valve nonlinearities

E MÄKINEN and **T VIRVALO**
Institute of Hydraulic and Automation, Tampere University of Technology, Finland

ABSTRACT

A special concern in water hydraulic servo valves is leakage due to the low viscosity of water. To keep leakage low some manufacturers use a small overlap (dead band) of the main spool. Some manufactures allow a little higher leakage to linearize flow characteristics of valves. In this paper the position accuracy of a water hydraulic position servo is improved by compensating nonlinearities in the water hydraulic servo valve. The following nonlinearities are considered: 1) velocity gain is equalized in both directions; 2) the dead band of the servo valve is compensated; 3) the nonlinear mechanical flow characteristics of the servo valve is linearized; 4) and the dependancy of the volume flow on the square root of pressure difference is also linearized. The improvements achieved are shown by experimental tests.

KEYWORDS: water hydraulic, position servo, nonlinearities

NOMENCLATURE

A_1, A_2	Piston area [m^2]
$G_{open}(s)$	Open loop transfer function
K_{qa}	Velocity gain [m/sV]
p_1,p_2	Cylinder chamber pressure [Pa]
p_L	Load pressure [Pa]
p_s	Supply pressure [Pa]
p_T	Tank pressure [Pa]
Δp_N	Nominal pressure drop/control notch [Pa]
Q_1, Q_2	Cylinder chamber volume flow [m^3/s]
Q_N	Nominal flow rate of servo valve [m^3/s]
s	Laplace-operator [s]

uc	Controller output without compensations [V]
uc_{act+}	Actual controller output in extending piston movement [V]
uc_{act-}	Actual controller output in retracting piston movement [V]
x	Relative control signal [-]
β	Ratio of piston areas [-]
δ_n	Damping factor [-]
ω_n	Natural frequency [rad/s]

1. INTRODUCTION

Servo valves cause the greatest position errors in typical position servo applications in hydraulics and pneumatics, as well as in water hydraulics. There are different types of water hydraulic servo valves on the market. The well-known structure principle of the two-stage servo valve is also used in water hydraulic servo valves. Both force and electrical feedback of the main spool are in use. The flow gain of both valve types depends nonlinearly on load pressure. Hysteresis can be remarkable in force feedback servo valves while being quite small in servo valves with electrical feedback. Threshold and resolution behave in the same way as hysteresis in these valves as well. A special concern in water hydraulic servo valves is leakage due to the low viscosity of water. To keep leakage low some manufacturers use a small overlap of the main spool. Some manufactures allow a little higher leakage to linearize flow characteristics of valves. Quite often the mechanical characteristics of water hydraulic servo valves are linear at least in principle. The manufacturer's specifications of servo valves are often quite general and lack accuracy. Due to the manufacturing process there are also some uncertainties in the specifications. In practice this means that often the flow control notches of valves are not linear functions of the control signal, as it would appear in a small size characteristics curve. Typically zero overlaps are used in hydraulic servo valves. However with tight manufacturing tolerances low leakage is achieved. This results in very good pressure gain. However, in practice the effective control notches are seldom linear, especially in water hydraulic servo valves. In some cases the characteristics of the control notches have been made nonlinear on purpose. Because the viscosity of water is rather low some compromises have to be made between steady state performance and leakage in water hydraulic servo valves. Former studies (1), (2), (3), (4) have shown that, especially in position servo applications, the steady state characteristics of servo valves have a rather large influence on certain performance properties of position servos, for instance positioning accuracy, settling time and damping.

The linear model of a water hydraulic servo system is used in simple design processes (4), (5). the third order open loop model is used in a position servo design, Eq.(1).

$$G_{open}(s) = \frac{K_{qa}\omega_n^2}{s\left(s^2 + 2\delta_n\omega_n s + \omega_n^2\right)} \tag{1}$$

According to experiences in hydraulic servo design the velocity gain K_{qa} can be assumed to be constant in this kind of design process (3), (4), (5). The velocity gain depends on the characteristics of a servo valve. Obviously this is also a fact with some water hydraulic valves, but there are remarkable differences among water hydraulic servo valves (1). The other parameters, natural frequency ω_n and damping factor δ_n, can be evaluated in the same way as in oil hydraulics (2).

There are different ways to improve the performance of basic position servo systems. Most often sophisticated control algorithms like sliding mode, state controller, and observer, as well as different kinds of nonlinear controllers are used (6), especially in oil hydraulics. Experience with and knowledge of water hydraulic servo drives is still quite limited. In order to understand the characteristics and properties of water hydraulic cylinder servo drives the Proportional-controller (P-controller) is used in this study. A solution to the performance problems of the position servo system found in (1) has been attempted by compensating the worst nonlinearities of the servo valve.

The basic principles of compensating the nonlinearities of water hydraulic servo valves is first presented and discussed. Then some experimental results are presented with different compensation methods applied to the P-controlled water hydraulic position servo. Very basic and traditional tuning is used in the example position servo system. The position loop gain is one third of the critical open loop gain. This is a normal way to guarantee overshoot-free step response in oil hydraulics and as has been shown in (2). the same principle can be used in water hydraulics, too.

An example of nonlinear orifice characteristics of a virtual servo valve is shown in Figure 1.

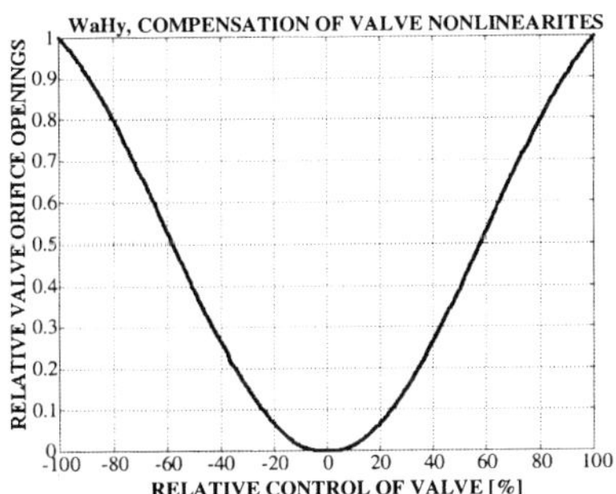

Figure 1 Relative flow control notches as a function of relative control signal of a virtual servo valve

In this case the following nonlinearities are included as an example, the dead band is ±4% and the valve flow orifices are nonlinear.

The actual nonlinearity of the control notches can be found through measurements and can be approximated according to the information given by valve manufacturers.

2. COMPENSATION OF DEAD BAND OF VALVES

The dead band of control valves affects accuracy, especially in position servo applications. Nonlinear control notches of control valves effect flow gain and dynamic response both in position and velocity servo applications.

The following factors have to be considered in compensating the mechanical dead band (overlap) of control valves. Care should be taken that the maximum flow rate of a valve doesn't change in dead band compensation. It is also good to notice that at least a little dead band ("electric dead band") should be left to avoid oscillations with small control signal noise.

The dead band compensation principle is shown in Figure 2. In this example case the dead band is ±10% (the effect can be seen more clearly). Large dead bands are typically found in proportional direction control valves.

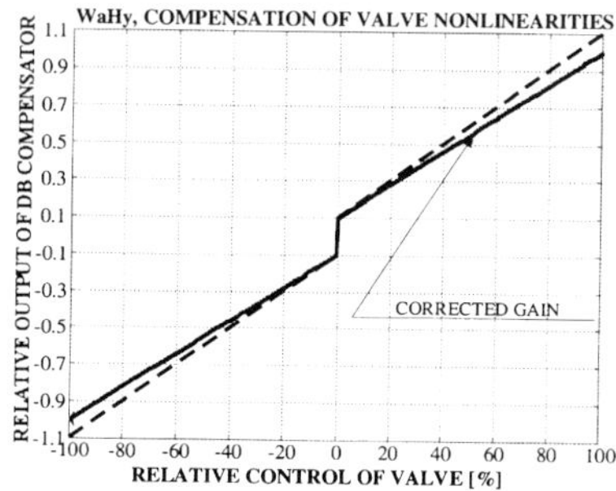

Figure 2 Principle of compensating mechanical dead band

When the valve control signal crosses zero the compensation constant is added. The constant is positive in the positive direction and negative in the negative direction. The actual control signal has to be modified small amount in order to get maximum flow rate with the maximum control signal. As can be seen in Figure 2 (solid line), the gain has to be changed as a function of the compensation amount.

When the zero control signal is applied to the valve, a small dead band should be arranged to avoid possible oscillations. An electrically realized dead band is the simplest way to arrange tuneable dead band, Figure 3.

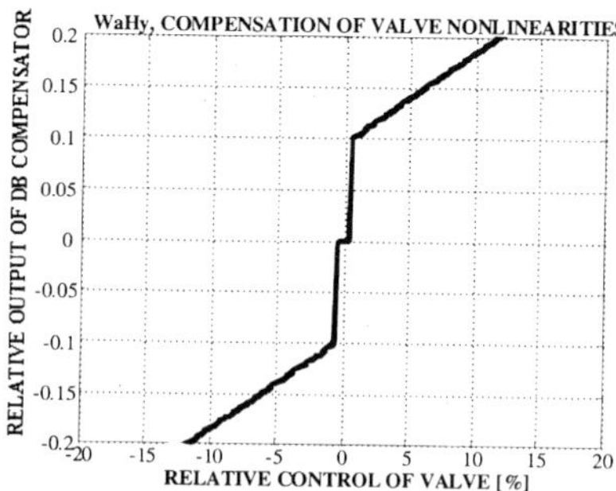

Figure 3 Electrical dead band 0.5% and compensation of mechanical dead band

The use of an electrical dead band makes the control more stable and also allows a little over compensation of a mechanical dead band. It is good to remember also that in the compensated case the valve spool has to move over the mechanical dead band. This means a time delay compared to the case where the spool is originally linear.

3. COMPENSATION OF NONLINEAR CONTROL NOTCHES OF VALVES

The linear flow gain is desirable in most fluid power servo applications. The volume flow through the analog fluid power valve depends on the control notches of the valve and pressure difference in the control notches. The same compensation method as in the compensation of the mechanical dead band can be used to compensate the nonlinear control notches.

The principle for modifying the control signal of an analog valve in order to linearize the nonlinearity of valve control notches is depicted in Figure 4.

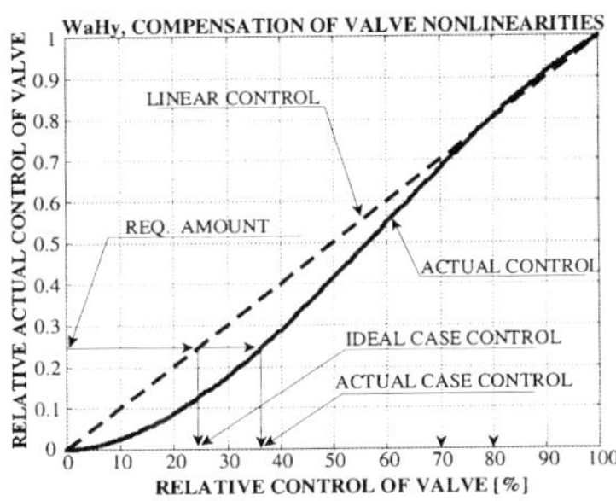

Figure 4 Principle to achieve the actual control signal to linearize flow orifice nonlinearity

The actual control signal can become higher order function of the original linear control signal or their relationship can be based on a two dimensional table. In this example case the nonlinearities of the valve control notches are linearized with a two dimensional, 21 cell table. The actual valve spool movements are shown in Figure 5 for the uncompensated and compensated cases.

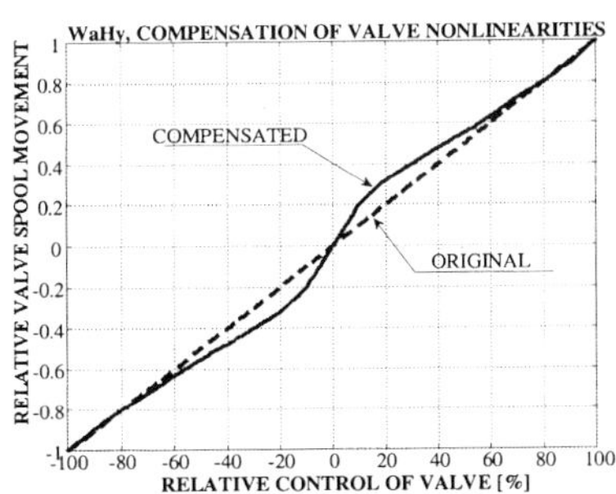

Figure 5 Valve spool movements in uncompensated and compensated cases

In practice the difference between the valve spool movements in these two cases is so small that it has no influence on the dynamics of the valve. This is the case with servo valves, but it has to be considered more carefully in the case of proportional valves.

4. COMPENSATION OF LOAD PRESSURE

Volume flow in fluid power systems depend on pressure differences in the control notches of valves. This means that the velocity of actuators and also the velocity gain of control valves depend on pressure differences in the service ports of control valves.

When the relative valve control signal x > 0 the following equations can be written Eq.(2) and Eq.(3),

$$Q_1 = \frac{xQ_N\sqrt{p_s - p_1}}{\sqrt{\Delta p_N}} \tag{2}$$

$$Q_2 = \frac{xQ_N\sqrt{p_2 - p_T}}{\sqrt{\Delta p_N}} \tag{3}$$

where Q_N is the nominal volume flow rate of the valve and Δp_N is the nominal pressure difference in the control notch. The following relationship is used for the ratio of piston areas, Eq.(4).

$$\beta = \frac{A_2}{A_1} \tag{4}$$

Let's define the load pressure p_L with Eq.(5).

$$p_L = \frac{A_1 p_1 - A_2 p_2}{A_1} = p_1 - \beta\, p_2 \tag{5}$$

With Eq.(2), Eq.(3), Eq.(4) and Eq.(5) the following relationships can be achieved assuming for simplicity $p_T = 0$, Eq.(6) and Eq.(7).

$$p_1 = \frac{p_L + \beta^3 p_s}{1 + \beta^3} \tag{6}$$

$$Q_1 = \frac{xQ_N}{\sqrt{\Delta p_N}}\sqrt{\frac{p_s - p_L}{1 + \beta^3}} \tag{7}$$

Respectively when the relative valve control signal x < 0 the following relationship can be achieved Eq.(8) and Eq.(9).

$$p_1 = \frac{p_L + \beta\, p_s}{1 + \beta^3} \tag{8}$$

$$Q_1 = \frac{xQ_N}{\sqrt{\Delta p_N}}\sqrt{\frac{p_L - \beta\, p_s}{1 + \beta^3}} \tag{9}$$

To get a linear relationship between the velocity of the piston and the controller output, the following modifications have to be made. The actual controller output is achieved according to Eq.(10) when the controller output >0 (extending piston movement).

$$uc_{act+} = uc\sqrt{\frac{\beta^3+1}{1-\frac{p_L}{p_s}}} \tag{10}$$

$$uc_{act-} = uc\sqrt{\frac{\beta^3+1}{\beta+\frac{p_L}{p_s}}} \tag{11}$$

5. RESULTS

The compensation methods are applied to the position servo system with the following specifications:

- Cylinder 63/36-500 mm
- Inertia load 2850 kg
- Supply pressure 8 MPa
- Servo valve
  - Nominal flow rate $8.3 \cong 10^{-4}$ m^3/s
  - Nominal pressure drop/notch 3.5 MPa
  - Hysteresis 0.1%
  - -90° bandwidth 80 Hz
- Resolution of position transducer 527000 pulses/m

The simple P-controller is used in order to better understand the behaviour of a water hydraulic position servo. The tuning of the controller is done according to the principle that no overshoot is allowed in short and long stroke step responses.

The performance of the uncompensated position servo is compared with the following three methods to compensate the nonlinearities of the servo valve:

- Control notch compensation
  - Linearization of the nonlinear control notches of the servo valve
- Compensation of the dead band of the servo valve
- Control notch and load pressure compensation
  - Linearization of the nonlinear control notches of the servo valve and compensation the load pressure

First each compensation method is tested using 10 mm (the servo valve doesn't saturate) and 100 mm (the servo valve saturates) piston strokes without any external load. Then the control notch compensation and the control notch and load pressure compensation methods are tested in two special cases: short 1mm piston strokes in both directions and piston strokes, when the spring force is applied.

5.1. Compensating the control notch nonlinearities of the servo valve

First only nonlinear control notches of the valve are compensated. The piston position error responses with the uncompensated and compensated valve are shown in Figure 6, when both outward (lower curves) and inward (upper curves) piston strokes are 10 mm.

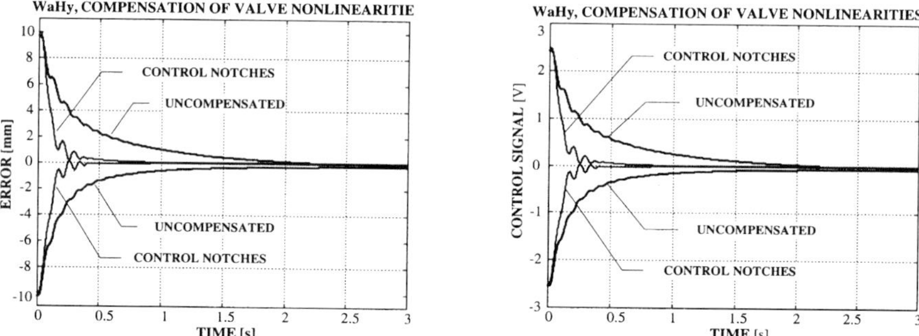

Figure 6 (left) Piston position error responses with uncompensated and control notch compensated valve, (right) control signal of servo valve

The well-known compensation of the asymmetric cylinder is used in both cases. When the valve nonlinear control notches are compensated the position responses are significantly faster than in the uncompensated case. The behaviour is also much more symmetrical in the compensated case. The steady state position errors of both cases (after 3 s) are on the same level. A closer look shows that with the compensated valve the position error in both directions is within 0.1 mm in 0.5 s whereas without compensation the position error is about 1 mm in inward motion and about 2 mm in outward motion in 0.5 s. Even after the 3 s motion, position error with the uncompensated valve is greater than with the compensated valve. Even so when the compensation is used there are more oscillations in the response but there is no overshoot. The controller output shown in the right part of Figure 6 is also stable.

The velocity responses of the long piston strokes (the valve saturates) are presented in Figure 7.

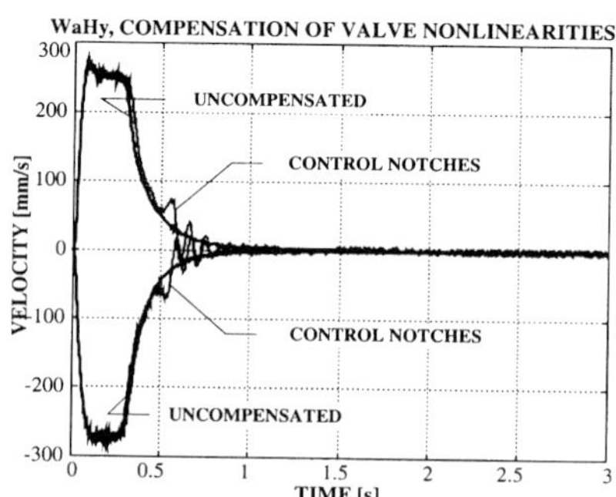

Figure 7 Velocity responses in long piston stroke cases

The maximum velocity is in practice the same in both directions showing that the compensation of the asymmetric cylinder works well. The velocity of both motion directions is almost the same in the case of compensated and uncompensated valves. The high inertia load causes some oscillations to the velocity response with the compensated valve, but still there is no overshoot in the position error responses. The final position is also reached significantly faster with the compensated valve in the case of the long piston strokes.

The other method to reduce response time is to use the nonlinear gain of the controller (1).

5.2. Compensating the dead band of the servo valve

The dead band of the valve can be compensated according to the principle shown in Figures 2 and 3. The dead band is very obvious and well specified in proportional valves. In water hydraulic servo valves there is typically no significant dead band, but due to the low viscosity of water small overlaps are used in control notches. The piston position error responses are shown in Figure 8, when different amounts of dead band compensation are used.

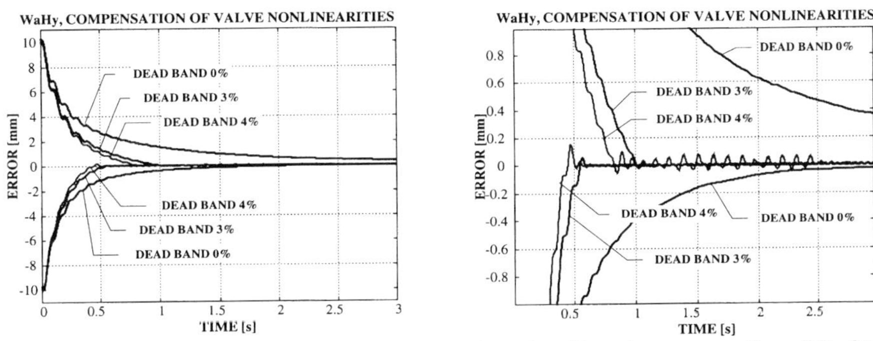

Figure 8 (left) Piston position error responses, when dead band compensations 0%, 3% and 4% are used, (right) zoomed piston error responses

The compensation values used are 0, 3 and 4 percent of the maximum control signal of the valve. The piston strokes are 10 mm in both directions. According to Figure 8 it can be said that in this case the dead band compensation improves the performance of the system in to some extent. The position response is remarkably faster with the dead band compensation, especially the settling time, which is much shorter. However there is obviously a risk of small amplitude oscillations as can be seen in the right part of Figure 8.

The steady state position accuracy is very good when the dead band compensation is used. The small position oscillations (< 0.1 mm) caused by too much dead band compensation can be seen in the right part of Figure 8, when the compensation value 4% is used. When the compensation value is 3% there are no oscillations. The other important factor that influences the behaviour of the dead band compensated responses can also be seen in Figure 8. The settling time of the compensated responses are quite different in different piston stroke directions. The main reason for the asymmetry in this case is that the symmetric compensation has been used and there has been a small zero drift of the valve spool.

To improve position accuracy and at the same time avoid oscillations the electrical dead band is quite often included in the compensation. The electrical dead band did not seem to have any significant influence on the performance of the system in the studied case. In general the dead band compensation doesn't seem to be a very effective way to compensate the zero stroke nonlinearities of servo valves.

5.3. Compensating mechanical nonlinearities and load pressure

The load pressure affects the pressure drops in the control notches of the valve and similarly the piston velocity. According to equations 9 and 10 the effect of the load pressure can be compensated. The method can be used if the following conditions are fulfilled:

- Supply pressure is constant, or it can be measured and fed back to the controller.
- The cylinder chamber pressure can be measured and fed back to the controller.

Typically it is necessary to use a low pass filtering in these kinds of pressure measurements.

For comparison purposes the position error responses are presented in Figure 9, when the control notch compensation and the control notch and load pressure compensation methods are used. The piston stroke is 10 mm in both directions. The output of the controller is also shown in Figure 9.

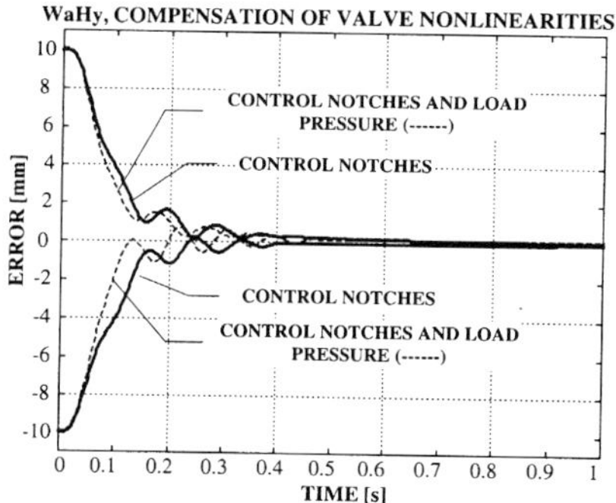

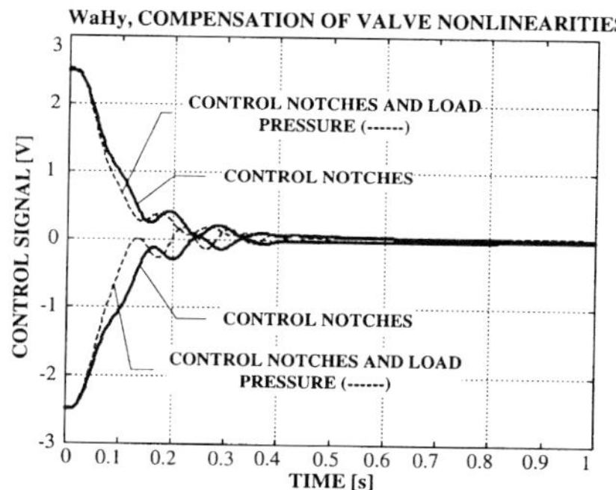

Figure 9 (left) Piston position error responses with control notch compensation and both control notch and load pressure compensation, (right) valve control signals

The response with the pressure compensation is a little bit faster than in the case of the compensating only the valve nonlinear control notches. The position accuracy is quite good, especially the repeatability, in the control notch and pressure compensation case, but there is a small position offset in these responses. The reason for this offset is still under the study. The output of the controller is quite stable in both cases.

5.4. Testing compensation in some special cases

Some of the serious drawbacks in water hydraulic servo systems are the nonlinearities of servo valves and the load dependable performance of systems. The nonlinearities of servo valves strongly influence the accuracy and resolution of water hydraulic position servo systems. The resolution of a servo system is very important in some applications like process control valves, assembly, and manipulators.

Hydraulics, as well as water hydraulics, are considered to be naturally good in heavily loaded systems. However, the load causes a position error also in water hydraulic position servos. The characteristics of servo valves effects the amount of position errors caused by loads.

Position accuracy under load is important in some applications like presses, mills, assembly applications, drilling booms, and material handling booms.

In this study the following tests were conducted in order to gain a better understanding of the basic performance of water hydraulic position servo systems and the influence of basic compensation methods on position accuracy:

- The position accuracy and the settling time of the short piston stroke
- The influence of spring force on the piston position accuracy

The P-controller was also used in these tests. The controller was tuned so that without the load there is no overshoot in the step response.

The short piston stroke (1 mm) responses are shown in Figure 10, when the uncompensated, control notch compensated and control notch and load pressure compensated servo valve have been used. In order to get a better view of piston position accuracy the zoomed position error responses are also presented in Figure 10.

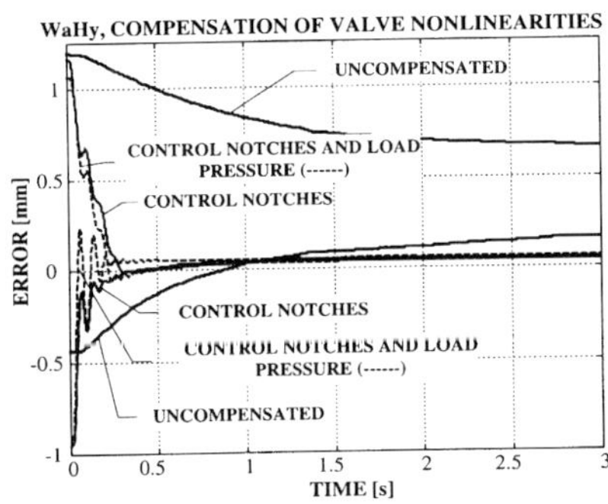

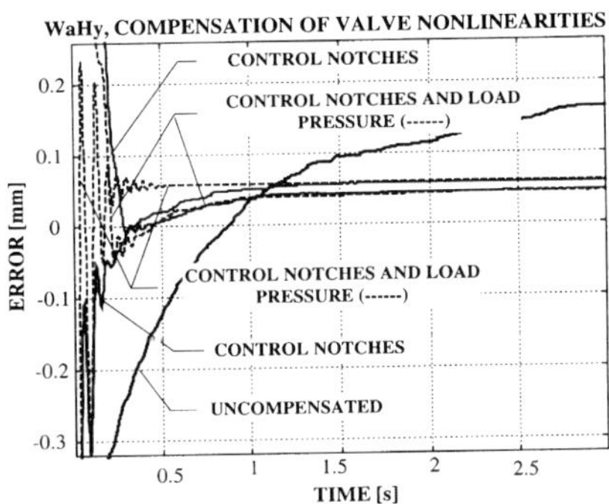

Figure 10 (left) Piston position error responses of short stroke with uncompensated, control notch compensated and control notch and load pressure compensated servo valve, (right) zoomed piston position error responses

The position error response with the uncompensated servo valve is poor. The response is very slow, position error is high, and hysteresis is quite large. The position error responses with the control notch compensated and the control notch and load pressure compensated servo valve are faster and more accurate. There are oscillations in the position error response when the control notch and load pressure compensated servo valve is used, Figure 10. Oscillations occur at the end of the response, but they damp quite quickly and the amplitude of the oscillations is only about 0.2 mm. The settling times inside of 0.1 mm are quite short.

The influence of the load force on the position accuracy was studied driving the piston and the inertia load against the spring. The position error responses are shown in Figure 11, when the spring force is about 10 kN (about 40% of the maximum force of the cylinder) and uncompensated and compensated servo valve were used. The zoomed position error response is shown in the right part of Figure 11.

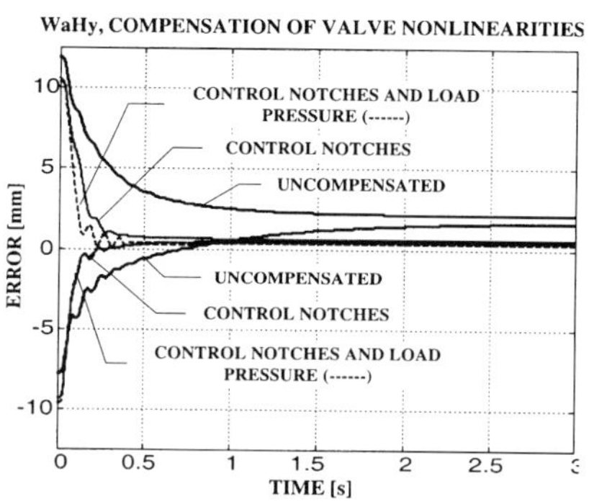

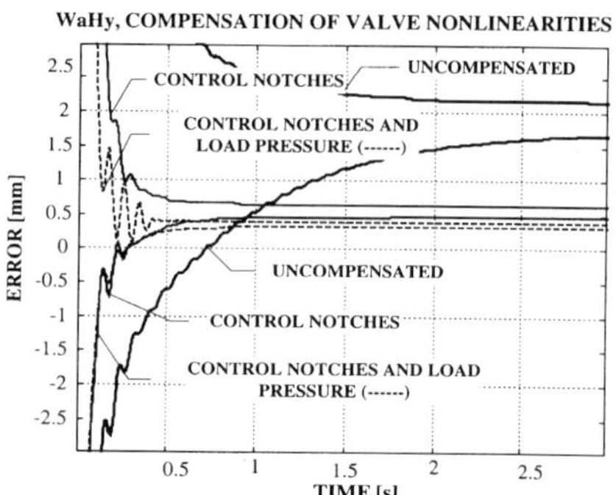

Figure 11 (left) Piston position error responses with uncompensated, control notch compensated and control notch and load pressure compensated servo valve, load is 10 kN, (right) zoomed position error responses

Again it can be seen that the responses of the uncompensated servo valve are slow and the position accuracies achieved are poor. The position error responses with the control notch compensated and with the control notch and load pressure compensated servo valve are better, especially with the load pressure compensated servo valve. The position error with uncompensated servo valve is about 2 mm and with the control notch compensated and load pressure compensated servo valve it is less than 0.5 mm.

6. SUMMARY

In water hydraulic position servo systems the servo valve is typically the worst nonlinearity. To be competitive with other techniques (oil hydraulic, pneumatic and electric drives) somehow the nonlinearities of water hydraulic servo valves have to be overcome. There are a few main methods to do this. The first one is to increase the position loop gain. The second one is to use a nonlinear controller. The third method, used also in this study, is to compensate the nonlinearities of the servo valve.

Three different compensation methods were used in this study. It was noticed that they all work and that they can be combined with each other at least to some extent. An obvious method is the compensation of the dead band. However, it was noticed that one should be very careful when applying this method, especially in servo valve applications. Proportional direction control valves have often a well-defined dead band or the overlaps of control notches, but in water hydraulic servo valve these are not so clear.

Sometimes, for different reasons, the control notch characteristics of water hydraulic servo valves are not linear. If nonlinearities are close to the zero position of the valve spool, they have quite a strong effect on the performance of position servo systems. The effects of the linearization of these nonlinearities of a servo valve were studied. The method was found to be quite effective and simple to realize.

The flow rate of servo valves depends on the pressure caused by a load. This influences the flow gain of the servo valve and hence the performance of the whole servo system. The possibilities to compensate the load pressure influences were studied by measuring the cylinder chamber pressures and calculating the compensation function. The method, when combined with linearization of nonlinearities of the valve control notches, was found effective, especially in short piston stroke cases and when an external load force was applied.

7. CONCLUSIONS

According to the studied water hydraulic position servo application and the compensation methods applied to the nonlinearities of the servo valve control notches the following conclusions can be made:

- Dead band compensation should be conducted very carefully and applied only to electrically feedback servo valves.
- Linearization of nonlinear control notches is an effective way to improve steady state and dynamic characteristics of position servo systems.
- Load pressure compensation is also an effective way to improve the performance of position servo systems, especially in the cases of external load forces.

8. REFERENCES

1. Mäkinen, E., Virvalo, T. 2000. The influence of characteristics of servo valves on the accuracy of a water hydraulic position servo . 6th Triennal International Symposium on Fluid Control Measurement and Visualization, April 13-17, 2000, Sherbrooke, Canada 6 p.
2. Mäkinen, E., Virvalo, T. & Vilenius, M. 1999. Comparison of water and oil hydraulic position servos. Yokota, s. (ed). Fluid power. Proceedings of the fourth JHPS International Symposium on Fluid Power, Tokyo, 15-17, November 1999, pp. 709-714
3. Mäkinen, E., Virvalo, T. & Vilenius, M. 1999. The effect of reference signal to the behavior of the water hydraulic position servo. Koskinen, KT, Vilenius, M. & Tikka, K. (eds.). Proceedings of the sixth Scandinavian International Conference on Fluid Power, may 26 - 28, 1999, Tampere, Finland vol: 1 pp. 261-270.
4. Virvalo, T., Mäkinen, E. & Vilenius, M. 1999. On the damping of water hydraulic cylinder drives. Yokota, S. (ed). Fluid power. Proceedings of the fourth JHPS International Symposium on Fluid Power, Tokyo, 15-17, November 1999, pp. 351-356
5. Virvalo, T. 1993. Modelling hydraulic position servo realized with commercial components. 3rd International Conference on Fluid Power Transmission and Control, Zhejiang University, Hangzhou China, 13-16 September 1993. 6 p.
6. Edge, K. 1996. The control of fluid power systems - responding to the challenges. Proc. Instn. Mech. Engn. Vol. 211 part 1, pp. 91-110

Authors' Index